Current Progress in Botany

Current Progress in Botany

Edited by Jerry Miller

SYRAWOOD
PUBLISHING HOUSE
New York

Published by Syrawood Publishing House,
750 Third Avenue, 9th Floor,
New York, NY 10017, USA
www.syrawoodpublishinghouse.com

Current Progress in Botany
Edited by Jerry Miller

International Standard Book Number: 978-1-68286-843-0 (Hardback)

Cataloging-in-Publication Data

Current progress in botany / edited by Jerry Miller.
p. cm.
Includes bibliographical references and index.
ISBN 978-1-68286-843-0
1. Botany. 2. Plants. I. Miller, Jerry.
QK45.2 .C87 2020
580--dc23

Printed in China.

TABLE OF CONTENTS

Preface VII

Chapter 1 **Nickel accumulation in leaves, floral organs and rewards varies by serpentine soil affinity** 1
George A. Meindl, Daniel J. Bain and Tia-Lynn Ashman

Chapter 2 **DNA analysis of soil extracts can be used to investigate fine root depth distribution of trees** 10
Sean L. Bithell, Lucy T. T. Tran-Nguyen, Mark N. Hearnden and Diana M. Hartley

Chapter 3 **Contrasting responses of root morphology and root-exuded organic acids to low phosphorus availability in three important food crops with divergent root traits** 23
Yan-Liang Wang, Marit Almvik, Nicholas Clarke, Susanne Eich-Greatorex, Anne Falk Øgaard, Tore Krogstad, Hans Lambers and Jihong Liu Clarke

Chapter 4 **Carbon dioxide stimulation of photosynthesis in *Liquidambar styraciflua* is not sustained during a 12-year field experiment** 34
Jeffrey M. Warren, Anna M. Jensen, Belinda E. Medlyn, Richard J. Norby and David T. Tissue

Chapter 5 **Climatic niche and flowering and fruiting phenology of an epiphytic plant** 47
Narayani Barve, Craig E. Martin and A. Townsend Peterson

Chapter 6 **Nitrogen addition and harvest frequency rather than initial plant species composition determine vertical structure and light interception** 57
Ute Petersen and Johannes Isselstein

Chapter 7 **Composition of fungal soil communities varies with plant abundance and geographic origin** 70
Vanessa Reininger, Laura B. Martinez-Garcia, Laura Sanderson and Pedro M. Antunes

Chapter 8 **Effects of elevated atmospheric CO_2 concentrations, clipping regimen and differential day/night atmospheric warming on tissue nitrogen concentrations of a perennial pasture grass** 84
Astrid Volder, Roger M. Gifford and John R. Evans

Chapter 9 **Developmental stages of cultivated strawberry flowers in relation to chilling sensitivity** 99
Maria Teresa Ariza, Carmen Soria and Elsa Martínez-Ferri

Chapter 10 **An empirical model that uses light attenuation and plant nitrogen status to predict within-canopy nitrogen distribution and upscale photosynthesis from leaf to whole canopy** 111
Gaëtan Louarn, Ela Frak, Serge Zaka, Jorge Prieto and Eric Lebon

Chapter 11 **Root hydraulic conductivity and adjustments in stomatal conductance: hydraulic strategy in response to salt stress in a halotolerant species** **127**
Victoria Vitali, Jorge Bellati, Gabriela Soto, Nicolás D. Ayub and Gabriela Amodeo

Chapter 12 **Collaboration between grass seedlings and rhizobacteria to scavenge organic nitrogen in soils** **142**
James F. White, Qiang Chen, Mónica S. Torres, Robert Mattera, Ivelisse Irizarry, Mariusz Tadych and Marshall Bergen

Chapter 13 **Supplementary calcium ameliorates ammonium toxicity by improving water status in agriculturally important species** **153**
Elvia Hernández-Gómez, Luis A. Valdez-Aguilar, Donita L. Cartmill, Andrew D. Cartmill and Irán Alia-Tajacal

Chapter 14 **Climate change reduces nectar secretion in two common Mediterranean plants** **161**
Krista Takkis, Thomas Tscheulin, Panagiotis Tsalkatis and Theodora Petanidou

Chapter 15 **Natural populations of *Arabidopsis thaliana* differ in seedling responses to high-temperature stress** **174**
Nana Zhang, Brian Belsterling, Jesse Raszewski and Stephen J. Tonsor

Chapter 16 **Giving drought the cold shoulder: a relationship between drought tolerance and fall dormancy in an agriculturally important crop** **187**
Keith G. Pembleton and Puthigae Sathish

Permissions

List of Contributors

Index

PREFACE

Plants are crucial for the sustainment of life on Earth. They provide the necessary oxygen and food, as well as medicine and other products. They are essential for binding and stabilizing soils, preventing soil erosion, and balancing carbon and water cycles. The branch of biology concerned with the study of plants including their structure, biochemical processes and properties is known as botany. It also encompasses plant physiology and pathology, plant classification and their interactions with the environment. The foundation of many applied sciences such as forestry, agriculture and horticulture is built on the principles and applications of botany. Modern botany is a multidisciplinary subject that integrates many different areas of science and technology. The study and investigation of plant growth and differentiation, structure, biochemistry, evolution, systematics, etc. are significant areas of research in this discipline. The field of botany has undergone rapid development over the past few decades. This book brings forth some of the most innovative concepts and elucidates the unexplored aspects of this area of study. Coherent flow of topics, student-friendly language and extensive use of examples make this book an invaluable source of knowledge.

After months of intensive research and writing, this book is the end result of all who devoted their time and efforts in the initiation and progress of this book. It will surely be a source of reference in enhancing the required knowledge of the new developments in the area. During the course of developing this book, certain measures such as accuracy, authenticity and research focused analytical studies were given preference in order to produce a comprehensive book in the area of study.

This book would not have been possible without the efforts of the authors and the publisher. I extend my sincere thanks to them. Secondly, I express my gratitude to my family and well-wishers. And most importantly, I thank my students for constantly expressing their willingness and curiosity in enhancing their knowledge in the field, which encourages me to take up further research projects for the advancement of the area.

Editor

1

Nickel accumulation in leaves, floral organs and rewards varies by serpentine soil affinity

George A. Meindl[1*], Daniel J. Bain[2] and Tia-Lynn Ashman[1]

[1] Department of Biological Sciences, University of Pittsburgh, Pittsburgh, PA 15260, USA

[2] Department of Geology and Planetary Science, University of Pittsburgh, Pittsburgh, PA 15260, USA

Associate Editor: Colin M. Orians

Abstract. Serpentine soils are edaphically stressful environments that host many endemic plant species. In particular, serpentine soils are high in several heavy metals (e.g. nickel, cobalt and chromium) and these high heavy metal concentrations are thought, in part, to lead to varying levels of plant adaptation and soil affinities (i.e. endemic vs. non-endemic plant species). It is unclear, however, whether serpentine endemics vs. non-endemics differ with respect to heavy metal uptake into either vegetative or reproductive organs. Here, we use nickel as a model to determine whether plant heavy metal uptake varies with the level of endemism in several non-hyperaccumulating species. Under controlled greenhouse conditions, we grew seven plant species from the Brassicaceae family that vary in their degrees of affinity to serpentine soil from low (indifferent) to medium (indicator) and high (endemic) in soil that was nickel supplemented or not. We quantified nickel concentrations in leaves, pistils, anthers, pollen and nectar. While nickel concentrations did not vary across organs or affinities when grown in control soils, under conditions of nickel supplementation endemic species had the lowest tissue concentrations of nickel, particularly when considering leaves and pistils, compared with indifferent/indicator species. Species indifferent to serpentines incorporated higher concentrations of nickel into reproductive organs relative to leaves, but this was not the case for indicator species and endemics where nickel concentration was similar in these organs. Our findings suggest that endemic species possess the ability to limit nickel uptake into above-ground tissues, particularly in reproductive organs where it may interfere with survival and reproduction. Indifferent species accumulated significantly more nickel into reproductive organs compared with leaves, which may limit their reproductive potential relative to endemic species when growing on serpentine soils. Additional work determining the fitness consequences of these differences will further our understanding of edaphic endemism.

Keywords: Brassicaceae; endemism; flowers; metal accumulation; nickel; serpentine soil.

Introduction

Edaphic factors, such as soil texture, depth and chemical composition, are a primary force in shaping the distributions of plant species (Silvertown 2004; Toledo *et al.* 2012; Dubuis *et al.* 2013). While many plant species can be found growing in a variety of habitats, some species become entirely restricted to a particular soil type (i.e. edaphic endemics; Macnair and Gardner 1998; Rajakaruna 2004). Serpentine soils, which can be found on every continent (Brooks 1987), provide one of the

* Corresponding author's e-mail address: gam35@pitt.edu

most remarkable examples of plant adaptation to atypical soils (O'Dell and Rajakaruna 2011) and geographic regions containing serpentine soil often harbour numerous endemic species (Brooks 1987; Safford *et al.* 2005; Anacker 2011). Serpentine-derived soils represent a nutritionally stressful growing environment for most plants because of a low Ca : Mg ratio, deficiency of essential nutrients (e.g. N, P, K) and high levels of potentially phytotoxic heavy metals (e.g. nickel [Ni], cobalt, chromium; Brooks 1987; Brady *et al.* 2005; Kazakou *et al.* 2008). While comparisons of plant tissue chemistry between endemic and non-endemic species can provide insight into the physiological features of edaphic endemics (Palacio *et al.* 2007), it is unclear whether soil affinity (i.e. endemic vs. non-endemic) affects plant tissue chemistry for serpentine plant species. For example, if endemic species are specifically adapted to the abiotic stresses of serpentine soil, then they might be better able to acquire limiting resources and/or exclude phytotoxic elements from the soil than non-endemic species. Considering that species span a gradient of affinity to serpentine soils, with some only occasionally found on serpentines (i.e. 'indifferent', $\leq$45 % occurrences on serpentines), some commonly found either on or off serpentines (i.e. 'indicator', $\sim$55–64 % occurrences on serpentines) and others entirely restricted to serpentines (i.e. 'endemic', $\geq$95 % occurrences on serpentines) (Safford *et al.* 2005), serpentine soils provide an ideal system to test whether soil affinity affects tissue chemistry.

The elevated heavy metal concentrations, in particular, in serpentine soils are thought to drive levels of plant adaptation (Lazarus *et al.* 2011), as non-adapted species lack physiological mechanisms to avoid metal toxicity (e.g. via metal exclusion or sequestration with chelating agents; Kazakou *et al.* 2008). However, of the few studies examining the differences in metal accumulation between endemic and non-endemic species, some have found decreased shoot metal accumulation in endemic species (Nagy and Proctor 1997; Burrell *et al.* 2012) while others have found no such difference (Fiedler 1985; Lee and Reeves 1989). Plant response to the heavy metal Ni found in serpentines is relatively well studied (Kazakou *et al.* 2008), and provides an excellent model to test whether endemic vs. non-endemic species vary in heavy metal accumulation. While some plants require Ni in trace quantities as an active component of the enzyme urease (Welch 1981), it is generally considered toxic to plants and is implicated in causing abnormal vegetative growth, necrosis and chlorosis of leaves and inhibiting photosynthesis (reviewed in Yusuf *et al.* 2011). Furthermore, Ni is known to negatively impact aspects of plant reproduction for non-hyperaccumulators, such as decreasing pollen germination (Tuna *et al.* 2002; Breygina *et al.* 2012) and seed production (Malan and Farrant 1998) when plants not adapted to elevated Ni conditions are grown in them. However, whether Ni is accumulated into reproductive organs, such as anthers and pistils, has only been studied in one serpentine plant species that is a known Ni hyperaccumulator (i.e. accumulates >1000 ppm Ni [*Streptanthus polygaloides*]; Meindl and Ashman 2014; Sánchez-Mata *et al.* 2014). While Ni hyperaccumulators are often specialized to serpentines (Reeves and Baker 2000), they represent an extreme minority, in both number of taxa and plant–soil interactions, of serpentine endemic plant species (Reeves 2006; Anacker 2011). Beyond these rare, yet relatively well-studied hyperaccumulators (Reeves 2006; Gall and Rajakaruna 2013), it is largely unknown whether the vast majority of serpentine species exhibit significant variation in Ni accumulation into above-ground organs. Therefore, studies are needed that focus on metal uptake for non-hyperaccumulating species to determine more general patterns of metal uptake or exclusion across serpentine plant species. Furthermore, it is unknown whether most serpentine plants accumulate Ni into pollen grains, despite evidence that plants growing in soils contaminated by metals via human activities can accumulate them into pollen (Moroń *et al.* 2012). Metals in pollen could reduce germination (citations above; Mohsenzadeh *et al.* 2011; Yousefi *et al.* 2011*a*) or pollinator attraction (Meindl and Ashman 2014), and Ni accumulation in nectar can affect pollinator foraging (Meindl and Ashman 2013, 2014). Thus, a first and necessary step towards understanding the reproductive consequences of growth on serpentine soil is documenting metal concentrations of reproductive organs and floral rewards of non-hyperaccumulating serpentine plants, as well as determining whether or not non-hyperaccumulating endemic species are better able to avoid potentially deleterious effects of metals by excluding them from reproductive organs than non-endemics. However, explicit experimental comparisons of metal accumulation across a range of species that vary in serpentine affinity, as well as across a range of vegetative and reproductive organs, are lacking.

To test these ideas, we grew seven species of plants from the Brassicaceae family that varied in serpentine soil affinity, but are not considered metal hyperaccumulators, in either control soils or soils supplemented with Ni to determine whether serpentine endemic and non-endemic plants differ with respect to Ni uptake (i.e. accumulate lower or similar concentrations of Ni into leaves, reproductive organs and rewards compared with non-endemic species, respectively). Using Ni as a model for plant response to serpentine heavy metals in general, we answered these questions: (i) Do serpentine endemic and non-endemic species differ in terms of Ni uptake into (a) leaves, (b) pistils, (c) anthers, and/or (d) nectar? (ii) Do serpentine endemics and non-endemics differ in the relative concentrations of Ni in vegetative vs. reproductive organs? (iii) Is Ni incorporated into pollen grains in any of these species?

Methods

Study system

Plants in the Brassicaceae family (mustard family) are well represented on California serpentine soils (Safford *et al.* 2005), including the seven species used here that differ in serpentine affinity from strictly endemic (i.e. $\geq$95 % occurrence on serpentine) to indifferent to serpentine soils (i.e. $\leq$45 % occurrence on serpentine soils): endemic: *Streptanthus morrisonii, S. breweri* var. *breweri*; indicator: *S. glandulosus* ssp. *glandulosus, S. tortuosus*; indifferent: *Hirschfeldia incana, Erysimum capitatum* var. *capitatum, Boechera breweri* (Table 1). To assign serpentine affinity scores to taxa, we follow the nomenclature used in Safford *et al.* (2005); however, in this work we follow the recent revised nomenclature for two taxa (i.e. *S. tortuosus* var. *suffrutescens* [now *S. tortuosus*] and *Arabis breweri* [now *Boechera breweri*]; Baldwin *et al.* 2012). All are spring flowering, insect pollinated herbaceous annuals or perennials that occur in North America, with four taxa being restricted to California (Table 1). Seeds from each taxon were bulk-collected from a single population per species in the summer of 2012.

Experimental design

Twenty plants per species (total $N = 140$) were grown at the University of Pittsburgh in the fall of 2012. Seeds were subjected to a 4 °C cold and dark treatment for 2 weeks prior to planting. Two weeks after germination, seedlings were transplanted to 27 cm^3 'rocket' pots (Deepots; Stuewe and Sons, Inc., Tangent, OR, USA) filled with standard potting soil (Fafard #4; Sun Gro Horticulture, Agawam, MA, USA) and six Nutricote® NPK 13-13-13 time-release fertilizer pellets (Arysta LifeScience Corporation, New York, NY, USA). One month after transplanting, all perennials (*S. morrisonii, E. capitatum* var. *capitatum, S. tortuosus, B. breweri*) received a 4 °C cold treatment for 1 month at 8D : 16N. Subsequently, these perennials and the annuals (*S. breweri* var. *breweri, H. incana, S. glandulosus* ssp. *glandulosus*) were grown under controlled conditions of 12D : 12N, between 70 and 80 °F, until flowering.

One month after potting (annuals), or 1 week after cold treatment (perennials), soil treatment solutions were applied to each plant weekly: either (i) Ni-supplemented (40 mL of 400 ppm Ni nitrate ($Ni(NO_3)_2$-$6H_2O$) solution) or (ii) control (40 mL of ammonium nitrate (NH_4NO_3) solution to compensate for 190 ppm nitrogen applied to plants in the Ni-supplemented soil treatment). This reflects a natural level of Ni, as serpentine soils contain bioavailable fractions of Ni ranging from 50 to 500 ppm (e.g. L'Huillier and Edighoffer 1996; Chardot *et al.* 2005). Soil treatments were conducted for 4–18 weeks, depending on time to flower. All plants were watered as needed.

Organ/reward collection and chemical analysis

Three organs (leaves, pistils, anthers) and two floral rewards (pollen, nectar) were collected from individual plants. A single fully developed leaf from the basal rosette was collected from each plant after four soil treatment applications. Pistils, anthers and nectar were collected from the first 5–15 flowers produced per plant. To collect nectar from several flowers per plant we folded a circular piece of filter paper (Whatman® Grade 1; GE Healthcare Bio-Sciences, Pittsburgh, PA, USA) in half and touched it to the floral nectaries in a circular pattern. Nectar volume

Table 1. Species descriptions and seed collection locations for all plant species studied. Plants were divided into three categories: serpentine endemic, serpentine indicator or serpentine indifferent. Serpentine affinity score is provided for all taxa, as defined by Safford *et al.* (2005)—species not discussed by Safford *et al.* (2005) are given a score of '<1'. Life history (annual or perennial) and distribution ranges are provided for all species (CA, California; NA, North America; AE, Afroeurasia).

Species	Plant category	Habitat affinity score	Life history	Range	Seed collection location
S. breweri var. *breweri*	Endemic	5.7	Annual	CA	N38°51′52.4″; W122°24′16.4″
S. morrisonii	Endemic	6.1	Perennial	CA	N38°48′45.3″; W122°22′54.9″
S. glandulosus ssp. *glandulosus*	Indicator	1.9	Annual	CA	N38°51′43.9″; W122°23′57.3″
S. tortuosus	Indicator	1.7	Perennial	Western NA	N39°59′18.4″; W121°17′19.8″
E. capitatum var. *capitatum*	Indifferent	<1	Perennial	NA	N41°16′32.5″; W122°41′54.4″
H. incana	Indifferent	<1	Annual	NA, AE	N38°51′30.0″; W122°24′35.2″
B. breweri	Indifferent	<1	Perennial	CA	N39°57′12.3″: W121°19′4.5″

was determined via the Baker's (1979) spot-staining method, as described in Kearns and Inouye (1993). The measured diameter (mm) of each nectar spot was compared with a standard table that relates spot diameter to nectar volume (μL). This technique is valid for nectars with sugar concentrations ranging from 10 to 50 % and nectar spot diameters ≤12 mm, which is true for many Brassicaceae members (e.g. Masierowska 2003; Nedić *et al.* 2013) including those in our study. Pistils and anthers were dissected from freshly opened whole flowers using forceps. While leaves were collected from every plant ($N = 10$ per species-soil treatment), some plants ($N = 5$) never flowered and thus 7–10 plants were sampled per species-soil treatment for floral organs and rewards. In addition, pollen was collected from two plants per species-soil treatment from an independent set of mature anthers.

Prior to chemical analysis, leaves and pistils were rinsed with diH_2O and dried at 60 °C for 48 h. Anther, pollen and nectar samples were allowed to air dry for 48 h in microcentrifuge tubes. Samples were weighed to the nearest 0.0001 g on a AE200 Mettler® analytical balance (Mettler-Toledo; LLC, Columbus, OH, USA) and microwave digested in 2–4 mL of trace metal grade HNO_3 and brought to a final volume of 12–14 mL with MilliQ (Millipore, Bedford, MA, USA) H_2O. Concentration of Ni is reported as ppm in organs and pollen (i.e. mg kg^{-1}) and nectar (i.e. μL L^{-1}) and was determined via inductively coupled plasma mass spectrometry (ICP-MS, NEXION 300X; PerkinElmer, Waltham, MA, USA) at the University of Pittsburgh (for details see Meindl and Ashman 2014).

Statistical analysis

All statistical analyses were conducted in SAS (2011; version 9.3; SAS Institute Inc., Cary, NC, USA). To evaluate the effect of soil treatment, serpentine habitat affinity and organ/reward type on plant Ni concentration, mixed-model ANCOVA was conducted (PROC MIXED). The model included the fixed effects of soil treatment (Ni supplement, control), serpentine habitat affinity (endemic, indicator or indifferent), organ/reward type (leaves, pistils, anthers or nectar) and their interactions and random factors of individual and species, where species was nested within serpentine habitat affinity (Table 1). The number of Ni applications to the soil was included as a covariate ('application number'). Denominator degrees of freedom for *F*-tests were determined using the Kenward–Roger approximation, which is preferred for small sample sizes and unbalanced data (Littell *et al.* 2002). For Ni-treated plants only, we used pre-planned contrasts to determine whether endemic species (i) incorporated less Ni than indicator/indifferent species in organs/rewards, and (ii) displayed lower concentrations of Ni in reproductive organs (anthers, pistils) relative to leaves than indicator/indifferent species using the CONTRAST option. We used a Student's *t*-test (PROC TTEST) to determine if pollen Ni concentration was higher in Ni-treated plants than in controls. For all analyses, Ni concentrations were natural-log transformed to improve normality of residuals. Back-transformed lsmeans (and 95 % confidence intervals) of Ni concentrations are presented for clarity.

Results

Soil Ni treatment was effective, as mean Ni concentrations in Ni-treated plants were 16 times higher than control plants across all organs and nectar (46.5 vs. 3.00 ppm), but the effect of habitat affinity on Ni in plant tissue was dependent on both soil treatment and organ/reward type (habitat affinity × soil treatment × organ/reward type: $P < 0.05$; Table 2; Fig. 1). Within the Ni soil treatment, endemic species had lower Ni concentrations in leaves and pistils compared with both indicator and indifferent species (leaves—endemic: 39.0 ppm; indicator and indifferent: 62.2 ppm; pistils—endemic: 55.3 ppm; indicator and indifferent: 99.5 ppm; Table 2). However, endemic species did not have significantly lower Ni concentrations compared with indicator/indifferent species in either anthers (endemic: 72.9 ppm; indicator/indifferent: 99.5 ppm; Table 2) or nectar (endemic: 13.1 ppm; indicator/indifferent: 12.8 ppm; Table 2). Furthermore, within the Ni soil treatment, indifferent species had Ni concentrations in anthers/pistils (126.5 ppm) that were twice as high as that in leaves (62.8 ppm; $P < 0.0001$, Table 2). Conversely, we did not detect a difference between endemic and indicator species with respect to concentrations of Ni in reproductive organs relative to vegetative organs (anthers/pistils vs. leaves: endemic 59.7 vs. 39.0 ppm; indicator: 72.2 vs. 66.7 ppm; Table 2).

Across all species, mean Ni concentrations in pollen were 10 times higher in Ni-treated plants than in controls (50.9 vs. 5.7 ppm; $t = -5.33$; df = 26; $P < 0.0001$). Nickel concentration of pollen from Ni-treated plants was highest for indicator species (59.1 ppm), followed by endemic (44.7 ppm) and indifferent (38.9 ppm) species.

Discussion

Serpentine endemic species incorporated Ni in lower concentrations in leaves and reproductive organs compared with indicator/indifferent species, suggesting that these species may be better adapted to the Ni-rich serpentine soil environment than the indicator/indifferent species studied. However, the magnitude of this difference depended on organ type, as endemics incorporated

Table 2. Results from mixed-model ANCOVA and pre-planned contrasts of Ni accumulation to leaves, pistils, anthers and nectar ('organ/reward type') of seven mustard species that vary in their affinity to serpentine soil ('habitat affinity') when grown in either Ni-supplemented or control soils ('soil treatment'). The number of soil treatment applications ('application number') was included as a covariate. Random effects of individual plant ('individual') and species (nested within habitat affinity; 'species (habitat affinity)' were also included in the model. Significance of fixed effects denoted as $^{*}P \leq 0.05$, $^{**}P \leq 0.01$ and $^{***}P \leq 0.0001$.

Source of variation	**df (num., den.)**	***F***
Habitat affinity	2, 1.86	10.42
Soil treatment	1, 119	1204.29***
Organ/reward type	3, 225	8.88***
Application number	1, 13.3	26.57**
Habitat affinity × soil treatment	2, 117	2.18
Habitat affinity × organ/reward type	6, 382	5.16***
Soil treatment × organ/reward type	3, 383	163.22***
Habitat affinity × soil treatment × organ/reward type	6, 383	2.24*
Random effects		*Z*
Individual		3.28**
Species (habitat affinity)		0.19
Pre-planned contrasts		
Endemic vs. non-endemic (indifferent/indicator)		
Leaves	1, 31.5	5.85*
Pistils	1, 32.6	8.24**
Anthers	1, 32.5	3.78
Nectar	1, 32.5	0.02
Vegetative (leaves) vs. reproductive (anthers/pistils)		
Endemic	1, 349	3.15
Indicator	1, 321	0.01
Indifferent	1, 384	20.42***

significantly less Ni into leaves and pistils, but not anthers, compared with indicator/indifferent species. Furthermore, while Ni exclusion is one possible mechanism to limit toxicity, effective sequestration in leaves is another (Yusuf *et al.* 2011). Species indifferent to serpentine had higher Ni concentrations in reproductive organs relative to leaves, whereas endemic and indicator species had similar Ni concentrations across all organs, again suggesting that plant species not regularly associated with serpentines do not possess mechanisms to limit uptake of Ni into reproductive organs.

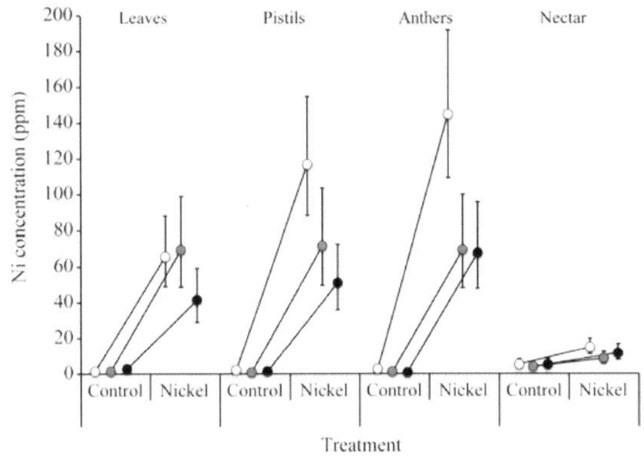

Figure 1. Nickel concentrations among serpentine indifferent (*H. incana*, *E. capitatum* var. *capitatum*, *B. breweri*), indicator (*S. tortuosus*, *S. glandulosus* ssp. *glandulosus*) and endemic (*S. morrisonii*, *S. breweri* var. *breweri*) plant species when grown in control vs. nickel-supplemented soils by organ/reward type (vegetative organ [leaves], two reproductive organs [pistils, anthers] and one floral reward [nectar]). Symbols represent back-transformed lsmeans (±95 % CI): white symbols, indifferent species; grey symbols, indicator species; black symbols, endemic species.

In the present study, serpentine endemic Brassicaceae exhibited the greatest degree of Ni exclusion, particularly in the leaves and pistils. Similarly, DeHart *et al.* (in review) found that while there was no difference in Ni concentrations between field-collected leaves, flowers or seeds of serpentine endemic and non-endemic species from the plant families Fabaceae, Phrymaceae and Ranunculaceae (possibly due to low levels of phytoavailable Ni in soils at their study sites), endemic species had significantly lower concentrations of the heavy metal cobalt across all organs than non-endemics (DeHart *et al.*, in review). Furthermore, studies of edaphic endemics in other soil environments also suggest that endemic plants may be specialized to their respective soil environment relative to non-endemics. By comparing leaf tissue chemistry of plant species that were either endemic or non-endemic to gypsum soils, which are high in sulfate ions and low in several macronutrients, Palacio *et al.* (2007) concluded that many gypsum soil endemic plants were more efficient at extracting limiting nutrients (e.g. N, P) from gypsum soils relative to non-endemic species, though range size also played a role in the level of specialization observed. Taken together, these findings suggest that many edaphic endemics may be physiologically better suited to their respective soil environments, e.g. by excluding heavy metals or acquiring limiting nutrients, than non-endemics. Our results contribute to ideas that support the specialist model of edaphic endemism (Meyer 1986; Palacio *et al.* 2007) rather than the refuge model, in which endemics are not specifically adapted to a particular soil type (Gankin and Major 1964), although studies of tissue chemistry would have to be coupled with measures of fitness to confirm this idea. It is important to recognize, however, that we intentionally excluded from our study serpentine soil endemics that are known to hyperaccumulate Ni, such as *S. polygaloides*. These species represent an exception to plant metal accumulation by serpentine soil endemics, rather than the rule (Reeves 2006; Anacker 2011), and thus predictions relating to heavy metal accumulation would clearly differ when considering metal hyperaccumulating taxa. However, because metal hyperaccumulation may impart chemical defences to plants (Rascio and Navari-Izzo 2011) and thus may impact plant fitness, documenting metal hyperaccumulation into reproductive organs and rewards in these species may provide valuable insight into the potential adaptive value of metal hyperaccumulation (Boyd and Martens 1992) and warrants additional study. In fact, recent experimental evidence suggests that two Ni hyperaccumulating taxa concentrate Ni in both vegetative and reproductive organs (*S. polygaloides* and *Noccaea fendleri*; Meindl *et al.*, in press). Furthermore, other chemical aspects of serpentine soils besides high Ni concentrations, such as low Ca : Mg ratios, may be equally important drivers of plant adaptation to the serpentine soil environment (Brady *et al.* 2005; O'Dell and Rajakaruna 2011; DeHart *et al.*, in review). Further comparisons of both macronutrient (e.g. Ca, Mg, N, K, P) and heavy metal (e.g. Ni, Co, Cr) concentrations between tissues of endemic and non-endemic species will provide a more comprehensive view of plant adaptation to serpentine soils.

Interestingly, Ni accumulation was, on average, higher in reproductive organs compared with leaves across all species in this study, corroborating similar findings of increased metal accumulation in flowers relative to leaves (Severne 1974; Gabbrielli *et al.* 1997). However, our work suggests that plants restricted to soils with elevated metals generally have lower metal concentrations in floral organs relative to plants not restricted to such soils. This pattern suggests a cost to floral metal accumulation, which could relate to decreased reproductive success via negative impact on pollen and ovule viability, as well as seed and fruit production (Maestri *et al.* 2010). Indeed, recent studies suggest that floral metal accumulation can decrease both pollen and ovule viability due to developmental abnormalities in anthers and ovaries (Yousefi *et al.* 2011*a*, *b*). While we show that Ni is incorporated into pistils, anthers, pollen and nectar, specifically, further work elucidating the effects of floral Ni accumulation on ovule and pollen viability and reproductive success in natural populations will provide necessary information towards understanding the adaptive value of both metal exclusion and metal accumulation. For example, pollen germination for some plant species known to accumulate high concentrations of the metalloid selenium is actually improved by increasing concentrations of selenium in pistils (Prins *et al.* 2011). However, the effects of floral metal accumulation on plant fitness for serpentine species are unknown, though experiments testing these effects are currently underway (G.A.M. and T.L.A., unpubl. res.).

Current data (DeHart *et al.* in review; this study) support the idea that serpentine endemics possess adaptations to elevated metal concentrations in serpentine soils (i.e. reduced uptake and translocation to leaves, reproductive organs and rewards) that non-endemics lack. These results suggest that non-endemic species may be at a fitness disadvantage compared with endemics when growing on serpentine soils. For example, in a series of experiments with *Mimulus guttatus* (Phrymaceae), Searcy and Mulcahy (1985) and Searcy and Macnair (1990) suggested that copper in the pistils of plants could act as a selective filter since seed production was reduced when pollen donors were not adapted to copper-rich soils. Floral metal accumulation may therefore produce a prezygotic isolating mechanism in non-endemic species compared with endemics by decreasing plant fitness when maternal and paternal plants are growing in different soil environments (i.e. serpentine and non-serpentine). In this way, floral metal accumulation may act as a reproductive barrier that favours reproduction between plants growing in similar soil environments, selecting against species that have serpentine and non-serpentine populations in close proximity to each other. Therefore, understanding metal accumulation into flowers and floral rewards is vital not only for identifying potential reproductive costs associated with plant growth on metal-rich soils, such as serpentine, but also for explaining patterns of species distributions, reproductive isolation and plant endemism.

Metal accumulation by plants can be influenced by many factors, including phylogeny (Broadley *et al.* 2001) and maternal effects (Macnair 2002). Therefore, it must be acknowledged that many of the species used in this study, including all of the species in the endemic and indicator categories, belong to the same genus within the Brassicaceae, *Streptanthus*. However, the main comparisons of this study involved endemics vs. both indifferent and indicator species, with the latter group including members of several genera spread across multiple Tribes (*Boechera*: Boechereae; *Erysimum*: Erysimeae; *Hirschfeldia*: Brassiceae; *Streptanthus*: Schizopetaleae; Al-Shehbaz 2010). Therefore, additional work comparing heavy metal accumulation across vegetative and reproductive organs of endemic and non-endemic plants from a variety of plant families (e.g. Asteraceae, Caryophyllaceae, Phrymaceae), and thus taking phylogenetic relationships into account regarding variation in metal accumulation, will contribute towards a more general understanding of edaphic endemism. Though not incorporated in the present study, the application of phylogenetically independent contrasts with paired endemic and non-endemic taxa (e.g. Cunningham *et al.* 1999) would be particularly informative when comparing Ni accumulation across levels of serpentine affinity. Furthermore, because seeds from each taxon were bulk-collected from a single population per species, intra-population level variation, if it exists (e.g. Macnair 2002), would be confounded with species. Our conclusions, however, are robust across affinity groups as each group includes two or more species. In addition, it is unlikely that maternal environmental effects influenced our findings as tissues were only collected from adult plants, and maternal affects generally manifest in earlier life stages (i.e. seeds and seedlings) and decrease with plant age (Roach and Wulff 1987; Lopez *et al.* 2003; Donohue 2009). However, phytoavailable Ni concentrations in serpentine soils are well known to vary, both within and across regions containing serpentine soil (Echevarria *et al.* 2006). This variation can lead to ecotypic variation within species, with some populations being adapted to high phytoavailable Ni concentrations, while others are not (O'Dell and Rajakaruna 2011). Thus, future studies incorporating multiple populations will allow for further resolution of genetic and maternal environment effects on plant Ni accumulation, and whether these vary by serpentine affinity.

Conclusions

Although the current study does not assess differences in plant fitness or competitive ability between endemic and non-endemic plants, our results highlight consistent differences in heavy metal uptake between endemic and non-endemic serpentine species. While edaphic features of serpentine soils are known to influence plant fitness for non-endemic plants, both directly (e.g. Swope and Stein 2012) and indirectly (e.g. Meindl *et al.* 2013), the specific effects of Ni accumulation on plant reproduction for non-hyperaccumulating serpentine species are not fully understood. Nickel tolerance has been identified as a key feature of serpentine soil tolerance (Brady *et al.* 2005; Alexander *et al.* 2007; O'Dell and Rajakaruna 2011) and is generally accomplished through root sequestration or exclusion, though not all serpentine plant species effectively exclude Ni from above-ground tissues (reviewed in Alexander *et al.* 2007; O'Dell and Rajakaruna 2011). Given the known toxicity of Ni for plant growth and reproduction (see citations in Introduction), our study suggests that endemic and non-endemic plants may differ in reproductive potential (e.g. differences in seed production or pollinator visitation) when grown in serpentine soils due to differential Ni uptake and translocation. While our findings suggest that endemic species possess the ability to limit Ni uptake into above-ground tissues, future work assessing the fitness consequences of growth on serpentine soils will provide valuable information towards understanding edaphic endemism. However, studies like ours are necessary prerequisites for determining whether serpentine endemic and non-endemic species differ in reproductive or competitive capabilities (e.g. Imbert *et al.* 2011) due to differences in physiological response to soils.

Sources of Funding

This research was funded by a Botany In Action Fellowship from the Phipps Conservatory and Botanical Gardens, an Ivy McManus Diversity Fellowship (University of Pittsburgh) and an Andrew Mellon Predoctoral Fellowship (University of Pittsburgh) to G.A.M., NSF (EAR-IF 0948366) to D.J.B., and NSF (DEB 1020523, 1241006) to T.-L.A.

Contributions by the Authors

G.A.M. and T.-L.A. designed the experiment and wrote the paper, G.A.M. performed the experiment, and G.A.M. and D.J.B. performed all chemical analyses.

Acknowledgements

K. DeHart and C. Murray assisted with plant organ/reward analysis, and E. York and K. Schuller provided greenhouse assistance. M. Koski, M. Hanley, C. Orians, five anonymous reviewers and members of the Ashman lab provided useful comments on earlier versions of this manuscript.

Literature Cited

Alexander EB, Coleman RG, Keeler-Wolf T, Harrison SP. 2007. *Serpentine geoecology of western North America*. New York: Oxford University Press.

Al-Shehbaz IA. 2010. Brassicaceae. In: Flora of North America Editorial Committee, ed. *Flora of North America*. New York and Oxford: Oxford University Press, 224–746.

Anacker B. 2011. Phylogenetic patterns of endemism and diversity. In: Harrison SP, Rajakaruna N, eds. *Serpentine: the evolution and ecology of a model system*. Berkeley: University of California Press, 49–79.

Baker I. 1979. Methods for the determination of volumes and sugar concentrations from nectar spots on paper. *Phytochemical Bulletin* **12**:40–42.

Baldwin BG, Goldman DH, Keil DJ, Patterson R, Rosatti TJ, Wilken DH, eds. 2012. *The Jepson manual: vascular plants of California*. Berkeley: University of California Press.

Boyd RS, Martens S. 1992. The raison d'etre for metal hyperaccumulation by plants. In: Baker AJM, Proctor J, Reeves RD, eds. *The vegetation of ultramafic (serpentine) soils*. Andover: Intercept, 279–289.

Brady KU, Kruckeberg AR, Bradshaw HD Jr. 2005. Evolutionary ecology of plant adaptation to serpentine soils. *Annual Review of Ecology, Evolution, and Systematics* **36**:243–266.

Breygina M, Matveyeva N, Polevova S, Meychik N, Nikolaeva Y, Mamaeva A, Yermakov I. 2012. $Ni2+$ effects on *Nicotiana tabacum* L. pollen germination and pollen tube growth. *BioMetals* **25**:1221–1233.

Broadley MR, Willey NJ, Wilkins JC, Baker AJM, Mead A, White PJ. 2001. Phylogenetic variation in heavy metal accumulation in angiosperms. *New Phytologist* **152**:9–27.

Brooks RR. 1987. *Serpentine and its vegetation: a multidisciplinary approach*. Portland: Dioscorides Press.

Burrell AM, Hawkins AK, Pepper AE. 2012. Genetic analyses of nickel tolerance in a North American serpentine endemic plant, *Caulanthus amplexicaulis* var. *barbarae* (Brassicaceae). *American Journal of Botany* **99**:1875–1883.

Chardot V, Massoura ST, Echevarria G, Reeves RD, Morel J-L. 2005. Phytoextraction potential of the nickel hyperaccumulators *Leptoplax emarginata* and *Bornmuellera tymphaea*. *International Journal of Phytoremediation* **7**:323–335.

Cunningham SA, Summerhayes B, Westoby M. 1999. Evolutionary divergences in leaf structure and chemistry, comparing rainfall and soil nutrient gradients. *Ecology* **69**:569–588.

Donohue K. 2009. Completing the cycle: maternal effects as the missing link in plant life histories. *Philosophical Transactions of the Royal Society B* **364**:1059–1074.

Dubuis A, Giovanettina S, Pellissier L, Pottier J, Vittoz P, Guisan A. 2013. Improving the prediction of plant species distribution and community composition by adding edaphic to topo-climatic variables. *Journal of Vegetation Science* **24**:593–606.

Echevarria G, Massoura ST, Sterckeman T, Becquer T, Schwarz C, Morel JL. 2006. Assessment and control of the bioavailability of nickel in soils. *Environmental Toxicology and Chemistry* **25**: 643–651.

Fiedler PL. 1985. Heavy metal accumulation and the nature of edaphic endemism in the genus *Calochortus* (Liliaceae). *American Journal of Botany* **72**:1712–1718.

Gabbrielli R, Gremigni P, Morassi LB, Pandolfini T, Medeghini P. 1997. Some aspects of Ni tolerance in *Alyssum bertolonii* Desv.: strategies of metal distribution and accumulation. In: Jaffré T, Reeves RD, Becquer T, eds. *The ecology of ultramafic and metalliferous areas. Proceedings of the Second International Conference on Serpentine Ecology*. Nouméa: ORSTOM, 225–227.

Gall J, Rajakaruna N. 2013. The physiology, functional genomics, and applied ecology of heavy metal-tolerant Brassicaceae. In: Lang M, ed. *Brassicaceae*. Hauppauge: Nova, 121–148.

Gankin R, Major J. 1964. *Arctostaphylos myrtifolia*, its biology and relationship to the problem of endemism. *Ecology* **45**:792–808.

Imbert E, Youssef S, Carbonell D, Baumel A. 2011. Do endemic species always have a low competitive ability? A test for two Mediterranean plant species under controlled conditions. *Journal of Plant Ecology* **5**:305–312.

Kazakou E, Dimitrakopoulos PG, Baker AJM, Reeves RD, Troumbis AY. 2008. Hypotheses, mechanisms and trade-offs of tolerance and adaptation to serpentine soils: from species to ecosystem level. *Biological Reviews of the Cambridge Philosophical Society* **83**: 495–508.

Kearns CA, Inouye DW. 1993. *Techniques for pollination biologists*. Niwot: University Press of Colorado.

L'Huillier L, Edighoffer S. 1996. Extractability of nickel and its concentration in cultivated plants in Ni rich ultramafic soils of New Caledonia. *Plant and Soil* **186**:255–264.

Lazarus BE, Richards JH, Claassen VP, O'Dell RE, Ferrell MA. 2011. Species specific plant-soil interactions influence plant distribution on serpentine soils. *Plant and Soil* **342**:327–344.

Lee WG, Reeves RD. 1989. Growth and chemical composition of *Celmisia spedenii*, an ultramafic endemic, and *Celmisia markii* on ultramafic soil and garden loam. *New Zealand Journal of Botany* **27**:595–598.

Littell RC, Stroup WW, Freund RJ. 2002. *SAS for linear models*. Cary: SAS Institute.

Lopez GA, Potts BM, Vaillancourt RE, Apiolaza LA. 2003. Maternal and carryover effects on early growth of *Eucalyptus globulus*. *Canadian Journal of Forest Research* **33**:2108–2115.

Macnair MR. 2002. Within and between population genetic variation for zinc accumulation in *Arabidopsis halleri*. *New Phytologist* **155**: 59–66.

Macnair MR, Gardner M. 1998. The evolution of edaphic endemics. In: Howard DJ, Berlocher SH, eds. *Endless forms: species and speciation*. New York: Oxford University Press, 157–171.

Maestri E, Marmiroli M, Visioli G, Marmiroli N. 2010. Metal tolerance and hyperaccumulation: costs and trade-offs between traits and environment. *Environmental and Experimental Botany* **68**: 1–13.

Malan HL, Farrant JM. 1998. Effects of the metal pollutants cadmium and nickel on soybean seed development. *Seed Science Research* **8**:445–453.

Masierowska ML. 2003. Floral nectaries and nectar production in brown mustard (*Brassica juncea*) and white mustard (*Sinapis alba*) (Brassicaceae). *Plant Systematics and Evolution* **238**:97–107.

Meindl GA, Ashman T-L. 2013. The effects of aluminum and nickel in nectar on the foraging behavior of bumblebees. *Environmental Pollution* **177**:78–81.

Meindl GA, Ashman T-L. 2014. Nickel accumulation by *Streptanthus polygaloides* (Brassicaceae) reduces floral visitation rate. *Journal of Chemical Ecology* **40**:128–135.

Meindl GA, Bain DJ, Ashman T-L. 2013. Edaphic factors and plant–insect interactions: effects of soil chemistry on pollination and florivory. *Oecologia* **173**:1355–1366.

Meindl GA, Bain DJ, Ashman T-L. Variation in nickel accumulation in leaves, reproductive organs and floral rewards in two hyper-accumulating Brassicaceae species. *Plant and Soil*. In Press. DOI:10.1007/s11104-014-2184-8.

Meyer SE. 1986. The ecology of gypsophile endemism in the eastern Mojave Desert. *Ecology* **67**:1303–1313.

Mohsenzadeh F, Chehregani A, Yousefi N. 2011. Effect of the heavy metals on developmental stages of ovule, pollen, and root proteins in *Reseda lutea* L. (Resedaceae). *Biological Trace Element Research* **143**:1777–1788.

Moroń D, Grześ IM, Skórka P, Szentgyörgyi H, Laskowski R, Potts SG, Woyciechowski M. 2012. Abundance and diversity of wild bees along gradients of heavy metal pollution. *Journal of Applied Ecology* **49**:118–125.

Nagy L, Proctor J. 1997. Soil Mg and Ni as causal factors of plant occurrence and distribution at the Meikle Kilrannoch ultramafic site in Scotland. *New Phytologist* **135**:561–566.

Nedić N, Mačukanović-Jocić M, Rančić D, Rørslett B, Šoštarić I, Stevanović ZD, Mladenović M. 2013. Melliferous potential of *Brassica napus* L. subsp. *napus* (Cruciferae). *Arthropod-Plant Interactions* **7**:323–333.

O'Dell RE, Rajakaruna N. 2011. Intraspecific variation, adaptation, and evolution. In: Harrison S, Rajakaruna N, eds. *Serpentine: evolution and ecology in a model system*. Berkeley: University of California Press, 97–137.

Palacio S, Escudero A, Montserrat-Martí G, Maestro M, Milla R, Albert MJ. 2007. Plants living on gypsum: beyond the specialist model. *Annals of Botany* **99**:333–343.

Prins CN, Hantzis LJ, Quinn CF, Pilon-Smits EAH. 2011. Effects of selenium accumulation on reproductive functions in *Brassica juncea* and *Stanleya pinnata*. *Journal of Experimental Botany* **62**:5633–5640.

Rajakaruna N. 2004. The edaphic factor in the origin of plant species. *International Geology Review* **46**:471–478.

Rascio N, Navari-Izzo F. 2011. Heavy metal hyperaccumulating plants: how and why do they do it? And what makes them so interesting? *Plant Science* **180**:169–181.

Reeves RD. 2006. Hyperaccumulation of trace elements by plants. In: Morel JL, Echevarria G, Goncharova N, eds. *Phytoremediation of metal-contaminated soils. NATO science series: IV: earth and environmental sciences*, Vol. 68. New York: Springer, 1–25.

Reeves RD, Baker AJM. 2000. Metal-accumulating plants. In: Raskin I, Ensley BD, eds. *Phytoremediation of toxic metals: using plants to clean up the environment*. New York: John Wiley and Sons, 193–229.

Roach DA, Wulff RD. 1987. Maternal effects in plants. *Annual Review of Ecology and Systematics* **18**:209–223.

Safford HD, Viers JH, Harrison SP. 2005. Serpentine endemism in the California flora: a database of serpentine affinity. *Madroño* **52**: 222–257.

Sánchez-Mata D, de la Fuente V, Rufo L, Rodriguez N, Amilis R. 2014. Localization of nickel in tissues of *Streptanthus polygaloides* Gray (Cruciferae) and endemic nickel hyperaccumulators from California. *Biological Trace Element Research* **157**:75–83.

SAS Insitute Inc. 2011. *SAS/STAT 9.3 user's guide*. Cary, NC: SAS Institute Inc.

Searcy KB, Macnair MR. 1990. Differential seed production in *Mimulus guttatus* in response to increasing concentrations of copper in the pistil by pollen from copper tolerant and sensitive sources. *Evolution* **44**:1424–1435.

Searcy KB, Mulcahy DL. 1985. Pollen selection and the gametophytic expression of metal tolerance in *Silene dioica* (Caryophyllaceae) and *Mimulus guttatus* (Scrophulariaceae). *American Journal of Botany* **72**:1700–1706.

Severne BC. 1974. Nickel accumulation by *Hybanthus floribundus*. *Nature* **248**:807–808.

Silvertown J. 2004. Plant coexistence and the niche. *Trends in Ecology & Evolution* **19**:605–611.

Swope SM, Stein IR. 2012. Soil type mediates indirect interactions between *Centaurea solstitialis* and its biocontrol agents. *Biological Invasions* **14**:1697–1710.

Toledo M, Peña-Claros M, Bongers F, Alarcón A, Balcázar J, Chuviña J, Leaño C, Licona JC, Poorter L. 2012. Distribution patterns of tropical woody species in response to climatic and edaphic gradients. *Journal of Ecology* **100**:253–263.

Tuna AL, Burun B, Yokas I, Coban E. 2002. The effects of heavy metals on pollen germination and pollen tube length in the tobacco plant. *Turkish Journal of Biology* **26**:109–113.

Welch RM. 1981. The biological significance of nickel. *Journal of Plant Nutrition* **3**:345–356.

Yousefi N, Chehregani A, Malayeri B, Lorestani B, Cheraghi M. 2011*a*. Investigating the effect of heavy metals on developmental stages of anther and pollen in *Chenopodium botrys* L. (Chenopodiaceae). *Biological Trace Element Research* **140**: 368–376.

Yousefi N, Chehregani A, Malayeri B, Lorestani B, Cheraghi M. 2011*b*. Effect of heavy metals on the developmental stages of ovule and seed proteins in *Chenopodium botrys* L. Chenopodiaceae). *Biological Trace Element Research* **144**:1142–1149.

Yusuf M, Fariduddin Q, Hayat S, Ahmad A. 2011. Nickel: an overview of uptake, essentiality and toxicity in plants. *Bulletin of Environmental Contamination and Toxicology* **86**:1–17.

2

DNA analysis of soil extracts can be used to investigate fine root depth distribution of trees

Sean L. Bithell[1,3]*, Lucy T. T. Tran-Nguyen[1], Mark N. Hearnden[1] and Diana M. Hartley[2]

[1] Plant Industries, Northern Territory Department of Primary Industry and Fisheries, GPO Box 3000, Darwin, NT 0801, Australia

[2] CSIRO Ecosystem Sciences, GPO Box 1700, Canberra, ACT 2601, Australia

[3] Present address: New South Wales Department of Primary Industries, Tamworth Agricultural Institute, 4 Marsden Park Rd, Tamworth, NSW 2340, Australia

Associate Editor: Astrid Volder

Abstract. Understanding the root distribution of trees by soil coring is time-consuming as it requires the separation of roots from soil and classification of roots into particular size classes. This labour-intensive process can limit sample throughput and therefore sampling intensity. We investigated the use of quantitative polymerase chain reaction (qPCR) on soil DNA extractions to determine live fine root DNA density (RDD, mg DNA m^{-2}) for mango (*Mangifera indica*) trees. The specificity of the qPCR was tested against DNA extracted from 10 mango cultivars and 14 weed species. All mango cultivars and no weeds were detected. Mango DNA was successfully quantified from control soil spiked with mango roots and weed species. The DNA yield of mango root sections stored in moist soil at 23–28 °C declined after 15 days to low concentrations as roots decayed, indicating that dead root materials in moist soil would not cause false-positive results. To separate large roots from samples, a root separation method for field samples was used to target the root fragments remaining in sieved (minimum 2 mm aperture) soil for RDD comparisons. Using this method we compared the seasonal RDD values of fine roots for five mango rootstock cultivars in a field trial. The mean cultivar DNA yields by depth from root fragments in the sieved soil samples had the strongest relationship (adjusted multiple $R^2 = 0.9307$, $P < 0.001$) with the dry matter (g m^{-2}) of fine (diameter $<$0.64 mm) roots removed from the soil by sieving. This method provides a species-specific and rapid means of comparing the distribution and concentration of live fine roots of trees in orchards using soil samples up to 500 g.

Keywords: DNA persistence; fine-roots; root DNA concentration; root DNA density (RDD).

Introduction

Developing information on fine root and associated biomass distribution of trees is important to a number of fundamental research areas, including within season and annual growth cycles (Hendricks *et al.* 2006; Milchunas 2009); root longevity and carbon turnover (Eissenstat and Yanai 1997; Valverde-Barrantes *et al.* 2007; Goebel *et al.* 2011); and water or nutrient uptake (Green and Clothier 1999; Coehlo and Borges 2004; Morgan *et al.* 2007). Understanding of how different species or cultivars effect fine root growth (Hendrick and Pregitzer 1996), root longevity and turnover (Eissenstat and Yanai 1997; Valverde-Barrantes *et al.* 2007) and water or nutrient uptake (Morgan *et al.* 2007; da Silva *et al.* 2011) is also required.

* Corresponding author's e-mail address: sean.bithell@dpi.nsw.gov.au

To improve the productivity of horticultural tree crops, information on the root traits of rootstock cultivars associated with desirable breeding objectives such as vigour control, fruit size and yield efficiency is required (Gregory *et al.* 2013). For fruit tree crops, knowledge of the distribution and density of fine roots is also important to optimize the efficiency of irrigation and fertigation methods (Coehlo and Borges 2004; Morgan *et al.* 2007). As such there is a need for new methods that allow the rapid analysis of large numbers of field samples. In addition, it would be useful if techniques to study fine roots enable live roots to be distinguished from dead roots. Especially, as the proportion of live to dead roots and the rate of root decomposition can vary among species, as found for tropical tree species (Valverde-Barrantes *et al.* 2007). Therefore, methods that characterize the live fine root component of tropical trees could be more meaningful than methods for the total fine root component. Such techniques will also be useful to assess the impacts of tree root diseases (Graham *et al.* 2013).

Our study species was mango. In production regions such as the semi-arid environment of the Northern Territory (NT) of Australia, mango fruit production occurs in the dry season and irrigation is dependant on finite sources of groundwater for irrigation and fertigation (Crane *et al.* 1997; Pigram 2006).

In live root studies of other tree species, roots were removed from soil samples by wet sieving, then separated based on colour, texture or chemical staining (Persson 1990; Hendrick and Pregitzer 1996; Schroth 2003). The use of root colour for mango is difficult as new growth is white but becomes suberized and dark-brown to black in colour (Sharma and Sharma 1988). The window to assess washed roots before they start to decay is only 48 h for refrigerated samples (Valverde-Barrantes *et al.* 2007) or several weeks for samples stored at 1–2 °C (Persson 1990). Roots can be stored longer for microscopic assessments when stained and stored at 5 °C in 17 % acetic acid (Livesley *et al.* 1999). Using such methods, a single core can take 4–8 h to assess (Persson 1990). This time requirement is a significant constraint for large-scale field studies (Mackie-Dawson and Atkinson 1991; Schroth and Kolbe 1994; Schroth 2003).

Another method such as ground-penetrating radar is not suitable for studying fine roots, distinguishing roots of mixed species or separating live from dead roots (Nadezhdina and Cermak 2003; Butnor *et al.* 2008). An X-ray method detects fine roots, but includes combined measurement of live and dead roots (Pierret *et al.* 2005). In-growth core techniques record fine root production, such as a study of mango where core cavities were filled with commercial potting mix (Blaikie *et al.* 2004). However, artefacts in comparison to non-disturbed samples are found in some study systems with this method (Ostonen *et al.* 2005; Milchunas 2009). Minirhizotrons are widely used with visual assessments used to determine live roots from dead roots (Eissenstat and Yanai 1997; Johnson *et al.* 2001). Although the insertion of fluorescence genes has provided a means of detecting the roots of a target species (Faget *et al.* 2010), that method cannot be used to assess mature trees planted prior to this insertion-reliant technique being available.

A soil DNA extraction-root qPCR method to quantify root DNA density (RDD) provides several advantages for root studies. The high throughput soil extraction-qPCR technique, capable of processing large (300–500 g) and numerous (minimum 160 samples day^{-1}) soil samples (Ophel-Keller *et al.* 2008), is critical for field trial sample processing. DNA methods enable the targeting of specific species in mixed species samples and qPCR methods provide the ability to detect small amounts (25 mg of dry root (kg soil)$^{-1}$) of root material (Mommer *et al.* 2008; Riley *et al.* 2010; Haling *et al.* 2011). Traditionally wet washing accompanies root sorting (Persson 1990), but if the qPCR method uses soil DNA extractions there is no need to separate all root material from soil. This is advantageous as root condition degradation has been implicated by reduced DNA yields for the washing processes (Haling *et al.* 2012).

The RDD method provides a different way of studying root responses and is not equivalent to root dry weight or root length. As the DNA content of different aged roots, sizes or root morphologies can differ within grass and tree species (Fisk *et al.* 2010; Haling *et al.* 2011), RDD values will be specific to the root type, size and age of the material used. Soil DNA-based RDD methods require development to study the particular root classes of trees in field trials. In field studies of grasses and wheat, the DNA yields of undisturbed soil cores have been used to study the RDD of the entire root system (Haling *et al.* 2012; Huang *et al.* 2013). However, such methods cannot be used for trees where the objective is to compare specific root size classes as the DNA yields of large roots cannot be distinguished from those of smaller roots [e.g. mango roots >20 mm in diameter occur (Coehlo *et al.* 2001)]. Large structural tree roots also have a more irregular distribution compared with fine roots (Schroth 2003; Levillain *et al.* 2011). Removing the larger root classes from the soil samples is thus necessary to study the differences in the fine root systems of trees.

Prior work had shown that root DNA degrades in moist soil (Riley *et al.* 2010), but the effects of root storage in dried soil were unknown. It was also shown that different sieve sizes separate different root diameter size classes from soil samples (Livesley *et al.* 1999) and qPCR analyses of soil DNA extractions could be used to quantify root DNA in soil samples (Riley *et al.* 2010; Haling *et al.* 2011, 2012). Many studies employing molecular methods of mango cultivars have been undertaken to determine the relationships between cultivars and wild relatives (see Dillon *et al.* 2013 and references within); however, there are no universal primers or probes available to detect a wide range of mango cultivars. A universal probe/primer for mango was identified as challenging due to polymorphism among cultivars (Yonemori *et al.* 2002). The research questions investigated were: (i) can a probe/primer for detecting a wide range of mango cultivars be developed; (ii) what are the effects of dry storage of soil on root DNA and (iii) which root diameter size classes are represented by soil qPCR analyses of sieved soil samples containing root fragments. The results of these studies were used in unison to investigate mango fine root RDD using qPCR analyses of soil DNA extractions from sieved dried soil samples. Research focused on Kensington Pride (KP), which is the most commonly used rootstock cultivar in Australia.

Methods

We developed a field-capable mango RDD method through a series of experiments. Following the development of a mango-specific probe, its specificity using qPCR was evaluated on DNA extracts of mango and weed tissue (Exp. 1), to compare the root DNA concentration of mango rootstock cultivars (Exp. 2). Other experiments were performed to evaluate the qPCR method on soil samples, including effects of soil type on root DNA yield (Exp. 3); root tissue concentration in soil and DNA yield relationship (Exp. 4); and root tissue degradation conditions on DNA yield (Exp. 5). The field capability for fine RDD quantification was trialled using sieved soil samples containing root fragments where we investigated the fine RDD of five mango rootstock cultivars in a field trial (Exp. 6).

Soil and plant samples

Soil collections. Soil samples were collected from Darwin and Katherine, NT, Australia, at sites where mango had never been grown (control soils) and from under mango trees (mango soils) (Table 1). Samples were collected using hand augers (Dormer Soil Samplers, SA5010C, hole diameter 58 mm) and surface debris was removed before coring. Augers were thoroughly cleaned between the sampling of different trees and control areas. All soils were dried in forced air ovens and then stored in double-layered plastic bags in an air-conditioned facility.

Darwin soils were collected from Berrimah Agricultural Research Centre (BARC) [Lateritic Red Earth, Deep Red Petroferric Kandosol (Hill *et al.* 2011)]. Darwin control samples were collected from an area that had been used for vegetable trials and pasture for ~15 years before a forage sorghum trial from December 2009 to April 2010. Two separate sets of samples from this site were used. The first set was prepared from samples collected in March 2010 (D Field site Control) (Table 1). Composite samples from duplicate cores for each 30 cm depth

Table 1. Listing of four soil sources from two localities, Berrimah Agricultural Research Centre, Darwin (D) and Fox Road[1], Katherine (K) used in this study with land use history, plant species present at sampling and sample depth details. Label names including 'Control' indicate that mangoes had never been grown at those sites, label names including 'Mango' were sites where mangoes were growing and sampling occurred under mango trees. Experimental numbers (Exp. no.) also indicate which soil source was used in experiments.

Label and Exp. no.	Land use	Species present at sampling	Sample depths (cm)
D Field site Control, Exp. 3	Vegetables and pasture for >15 years	*Sorghum bicolor, Alysicarpus vaginealis, Macroptilium gracile, Chamaechrista rotundifolia, Stylosanthes humilis, Crotalaria goreensis, Oldenlandia corymbosa, Corchorus trilocularis, Boerhavia dominii, Sida acuta, Cyperus rotundus, Cynodon dactylon, Urochloa mosambicennsis*	0–180 in steps of 30
K Field site Control, Exps. 2, 4, 5	Forage cropping	*Centrosema pascuorum, Cenchrus echinatus, Brachiaria pubigera, Urochloa mosambicensis, Boerhavia dominii*	0–60
K Field site Mango 1, Exp. 3	Mango	Mango, KP rootstock	0–180 in steps of 30
K Field site Mango 2, Exp. 6	Mango	Mango, rootstock cultivars: Water tank (NT14), KP (NT16), Kurukan (NT21), MYP (NT50), Vellaikulamban (NT51). *Cajans geminatus, Tridax procumbens, Bracheria decumbens, Vigna lanceolata* var. *filiformus, Urochloa mosambicensis*	0–60 cm in steps of 15

[1]Mango trees at this site had paclobutrazol applied annually as a collar drench in late January or February at 20 mL (a.i. 250 g L^{-1}) per tree.

zone down to 1.8 m were prepared from soil dried at 65 °C for 48 h. Analyses of 0–15 cm samples from this site had soil total carbon of 1.91 % and pH (water method) of 5.48. The percentage of clay in the top 20 cm ranged from 29 to 34 %, while coarse sand ranged from 30 to 32 % and fine sand was 29 % (Hill *et al.* 2011).

Katherine soils were collected from Fox Road [Deep Red Magnesic Kandosol, Australian classification system (Hill *et al.* 2011)]. Control samples (K Field site Control) were collected in November 2010 from a field that had never been used for growing mango. Sampling in this field followed a transect at four 20-m intervals where four auger samples were collected from within a 2.5 m^2 area. After the control soils were dried as previously described, soils were sieved (2 mm aperture) and rocky material removed, plant material collected in the sieve was returned to the sieved soil.

Mango soil samples were collected in September 2010, about 50 m from the control field in a mango rootstock trial planting. The first mango soil samples (K Field site Mango 1) were collected from neighbouring guard row trees (14-year-old KP scion on KP rootstock) surrounding the trial. These samples were collected and prepared as described for D Field site Control.

Plant samples for sequencing, qPCR development and specificity testing. Mango root tissue was sampled from 6- to 9-month-old potted seedlings held in the NT, Department of Primary Industry and Fisheries (DPIF), Plant Industries Mango rootstock collection (Table 2).

Mango samples: Roots (~2–6 mm diameter) were sampled and any damaged or discoloured material was omitted. All samples were surface sterilized with a KIMWIPES® tissue pre-wetted with 70 % ethanol. Root samples were then cut in fine cross-sectional slices stored wrapped in KIMWIPES® tissue over calcium chloride granules in plastic vials sealed with parafilm and stored at 4 °C until required for use. All root diameter measurements in this study were made with a digital calliper (Lufkin model 103192).

Weed root samples: The seed of weed species (Table 2) were sourced and sown into heat sterilized coarse river sand in 1 L pots. These were propagated on a raised table with no pot to soil contact in a covered nursery. Seedlings were harvested after 5–6 weeks and root samples were prepared and stored as described for the mango root samples.

DNA extractions. DNA extractions from root tissues were conducted using the DNeasy Plant Mini Kit (QIAGEN, Australia) according to the manufacturer's instructions. Crushed glass was used as an abrasive to assist with grinding the mango material. Preliminary extractions used both KP leaf and root materials for comparison.

Mango PCR and primer design. In total 13 mango samples (10 cultivars, including 2 accessions of cv. NDM) were sequenced (Table 2). Root samples used for sequencing had the epidermis removed. The extracted DNA samples were first subjected to PCR amplification targeting the ITS region (ITS1—5.8S ribosomal gene—ITS2), using the universal ITS 5 and ITS 4 primers (White *et al.* 1990). PCRs were conducted in a 50 μL mixture containing 1× ImmoMix Red with 3 mM $MgCl_2$ (Bioline, Australia) and 0.2 μM of each primer and 5–10 ng of DNA template. The PCR conditions used were: initial enzyme activation of 95 °C for 10 min followed by 35 cycles of denaturing at 95 °C for 1 min, annealing at 56 °C for 1 min, extension at 72 °C for 1 min and final extension of 72 °C for 10 min. The PCR products were separated on a 1 % agarose/ethidium bromide gel using 1× TAE buffer and visualized under UV light.

The PCR products were purified using the QIAquick PCR purification kit (QIAGEN) according to the manufacturer's instructions, with final elution in 30 μL sterile distilled water, pH 7.3. DNA concentration was estimated using agarose gel electrophoresis with the low mass ladder (Life Technologies, Australia) as a reference. The products were sequenced using the Big Dye Terminator Mix (Bioscience North Australia, Darwin, NT). Nucleotide sequences were analysed and edited using Geneious Pro (Drummond *et al.* 2011). Multiple sequence alignment was conducted and included GenBank accession numbers for *Mangifera indica*, cv. Irwin AB071669 and cv. Nam Doc Mai (NDM) AB071672 (Yonemori *et al.* 2002). Primer3 program (part of Geneious package) was used to determine candidates for qPCR primers and probes.

Regions within the ITS region were analysed and conserved regions of the internal transcribed region were used to determine the mango-specific probe and primer sequences for the qPCR test. The probe–primer combination was required to cover a range of mango rootstock cultivars. Sequences were subjected to BlastN searches against the non-redundant nucleotide database (http://blast.ncbi.nlm.nih.gov/Blast.cgi?PROGRAM=blastn&BLAST_PROGRAMS=megaBlast&PAGE_TYPE=BlastSearch&SHOW_DEFAULTS=on&LINK_LOC=blasthome) to identify species of highest similarity. The nucleotide sequence of cv. NDM (Hamilton) was compared with cv. NDM, GenBank Accession Number AB071672 (Yonemori *et al.* 2002). This sample was used as a reference for comparison with previously un-sequenced mango rootstocks.

Mango qPCR specificity test (Exp. 1). This experiment evaluated the specificity of the probe–primer combination. Following preliminary studies on the development of probes and primers for qPCR, we developed a mango-specific

Table 2. Mango qPCR specificity test using DNA of 12 mango (*Mangifera indica*) cultivars and 14 weed species (Exp. 1). Details of plant species, cultivar (NT accession code), source, sample code and tissue type are shown. DNA at a concentration of 100 pg μL^{-1} was used and the results were expressed as detected (+) and undetected (−) at cycle 40.

Species	Cultivar	Source	Sample code	Tissue	Detection
Mangifera indica	KP (NT16)	DPIF[1] Darwin, NT	3[S2], 8[S], 9b, 21a[S]	Root	+
M. indica	NDM (NT36)	K. Rayner, Katherine, NT	23a[S]	Root	+
M. indica	NDM	D. Hamilton, Howard Springs, NT	11[S]	Root	+
M. indica	NDM	Viet Ma, Lambells Lagoon, NT	SAR 5[3]	Root	+
M. indica	NDM	Viet Ma, Lambells Lagoon, NT	SAR 1[3]	Stem	+
M. indica	MYP (NT50)	DPIF Darwin, NT	16a[S]	Root	+
M. indica	B (NT63)	DPIF Darwin, NT	17a[S]	Root	+
M. indica	Chok Anan	DPIF Darwin, NT	18a[S]	Root	+
M. indica	Pancho (NT10)[4]	DPIF Darwin, NT	19a[S]	Root	[2]
M. indica	Chandrakaran (NT55)	DPIF Darwin, NT	20[S]	Root	+
M. indica	Vellaikulamban (NT51)	DPIF Darwin, NT	22a[S]	Root	+
M. indica	13/1 (NT52)	DPIF Darwin, NT	24a[S]	Root	+
M. indica	Brodie (NT9)	DPIF Darwin, NT	25a[S]	Root	+
Dactyloctenium aegyptium		Darwin, NT	35	Root	−
Digitaria spp.		Darwin, NT	36	Root	−
Chloris inflata		Darwin, NT	37	Root	−
Pennisetum typhoides	Katherine Pearl	Darwin, NT	30	Root	−
Sorghum bicolor	Jumbo	Landmark, Queensland	32	Root	−
Melinis repens		Darwin, NT	38	Root	−
Chamaecrista rotundifolia		Darwin, NT	39	Root	−
Boerhavia dominii		Darwin, NT	40	Root	−
Cleome viscose		Darwin, NT	41	Root	−
Sida acuta		Darwin, NT	42	Root	−
Crotalaria goreenis		Darwin, NT	43	Root	−
Ludwigia parviflora		Darwin, NT	44	Root	−
Ludwigia species		Darwin, NT	45	Root	−
Eleusine indica		Darwin, NT	46	Root	−
Mitrocarpus hirtus		Darwin, NT	47	Root	−

All mango cultivars in the table are polyembryonic.
[S]Samples sequenced.
[1]Department of Primary Industry and Fisheries (DPIF).
[2]Samples sequenced only.
[3]DNA of young root and stem collected from seed germinated under aseptic conditions.
[4]NT accession of cv. Red Haromanis.

probe–primer F1 (5′-TCGAGTCTTTGAACGCAAGTTG-3′) and R3 (5′-CCACGCCGAAAGATCGTT-3′) primers, and probe (6FAM-5′-ACACCCAGGCAGACGT-3′). The specificity of this probe–primer was trialled using qPCR testing on DNA from the roots of a wide range of mango cultivars and from roots of common tropical weed species (Table 2). DNA from roots of cv. NDM (Viet Ma) seed, sprouted under laboratory conditions in sterile media, was used to prepare a reference standard and to establish a standard curve for the qPCR. The mango qPCR was performed on a 7900HT real-time PCR system using TaqMan® MGB (Minor Groove Binder) probe (Applied Biosystems, CA, USA) and QuantiTect Probe PCR Master Mix (QIAGEN GmbH, Hilden, Germany). PCRs were conducted in a 10 μL mixture containing 1× QuantiTect Probe Master Mix, 0.4 μM of each primer, 0.2 μM of probe and 0.4 ng of DNA template. The PCR conditions used were: step 1, 50 °C for 2 min; step 2, 95 °C for 15 min; step 3, 95 °C for 15 s; step 4, 60 °C for 1 min; steps 3–4 were repeated for 45 cycles.

Quantification of mango root DNA (Exp. 2). This experiment compared the root DNA concentrations of mango rootstock cultivars. DNA was extracted from root tissue (minimum diameter range 1.1–2.8 mm, maximum diameter range 1.7–4.2 mm before epidermis was removed) from five cultivars (Watertank, KP, Kurakan, MYP and Vellaikulamban, Table 2). For all cultivars except Watertank, samples were prepared separately from each of the three 5- to 6-month-old potted seedlings. Kurakan, KP, MYP and Vellaikulamban seedlings were propagated from seeds collected from the same mother trees used as seed sources for the rootstock field trial. The original mother tree of NT14 had died; therefore, roots for NT14 were collected from a single tree which was planted as a nucellar seedling from the original mother tree. Root sample preparation followed the methods described for mango samples used in the sequencing and specificity sections. DNA was extracted using the DNeasy Plant Mini Kit (QIAGEN) from ~500 mg samples of root tissue, to provide three root extracts per cultivar. Neat DNA extracts were diluted to 200 pg μL^{-1} and used for qPCR. The cycle threshold (Ct) value, the number of cycles for the fluorescent signal to cross the threshold, was determined for all samples. The Ct values were then converted to mg mango DNA per mg dry root.

Quantification of mango DNA in soil. Four separate batches of dried soil samples (300–500 g) from five experiments (Exps. 3–7) were shipped from Darwin to Adelaide for DNA testing by a commercial service, Root Disease Testing Service (RDTS, South Australian Research and Development Institute (SARDI), Adelaide, Australia) (Ophel-Keller *et al.* 2008). Each batch of samples included three batch check soil samples (300 g K Field site Control) without amendment and another three with 100 mg kg^{-1} of KP (NT16) root added. These batch checks were prepared in Darwin and shipped to RDTS with each batch.

To determine the presence of PCR inhibitors in the soil extracts, a known amount of an exogenous organism (internal control) was included in each sample by SARDI before extraction. Lower amplification of this internal control indicates that the extraction has a poor DNA recovery or contains PCR inhibitors (Ophel-Keller *et al.* 2008). Work by Haling *et al.* (2011) showed that DNA extraction efficiency, measured by the internal control, was similar across soil types and rarely affected by soil inhibitors.

All DNA values from soil qPCR analysis were provided as pg DNA per g soil. For field samples the total pg value of the entire sample was calculated from the total sample weight of soil and converted to m^2 values as described for soil carbon by McKenzie *et al.* (2000) and expressed as mg DNA m^{-2} for each sample depth.

Soil type effects (Exp. 3). This experiment evaluated the effects of soil type on DNA recovery and presence of PCR inhibitors in soil extracts. Samples of two soil types commonly used for mango production in the NT (D Field site Control and K Field site Mango 1) were used. The soil samples (300 g) were spiked with known amount of the internal control before DNA extraction as described previously.

Mango root concentration: DNA yield and cultivar detection in control soil (Exp. 4). This experiment evaluated the qPCR quantification of root tissue in soil samples. Ten each of 1-month-old first- and second-order KP (NT16) potted seedlings were placed in an automated mist house on a raised metal grid for 8–12 weeks. Roots grew freely through the bottom of the pots and grid apertures, and also remained moist due to misting. Sliced root tissue (minimum diameter range 1.1–1.8 mm, maximum diameter range 3.2–4.4 mm) was dried at 40 °C for 48 h before being used to spike control (K Field site Control) soil samples. Soil samples (300 g per sample) were spiked with different concentrations of the dried aerially produced KP root slices (0, 25, 50, 100, 300, 800, 1600 mg kg^{-1} of soil) with four replicates of each concentration.

Root tissue from mango cultivars or weed species (Table 4) was sampled from potted plants, prepared and dried as described previously. This material was used to spike (300 mg root per kg soil) duplicate control soil samples (K Field site Control, 300 g per sample) for each of the mango cultivars or weed species.

Effects of root tissue degradation on DNA yield (Exp. 5). This experiment evaluated the effects of root tissue degradation and soil conditions on DNA yield. Moderate (sectioned) and severe (sliced) root tissue disturbance treatments were compared for tissue stored in moist and dry soil over time. Aerially produced roots (as described in Exp. 4) from three potted 1-year-old NDM (Hamilton) seedlings were harvested and divided into three diameter classes: 1.5–2, 1–1.5 and 0.8–0.9 mm, and each class was divided into two equal wet weight groups. Sub-samples from each of the three size classes were then dried at 40 °C for 48 h. The average dry matter (DM) content across the size classes was 5.5 %. Half of each group was cut into 10 mm sections and the other half was finely sliced. A factorial experiment included three treatments: (i) soil moisture: dry soil (0.46 % gravimetric moisture content) or watered soil [to 75 % of field capacity (FC), 9.3 %] every 2–3 days; (ii) root fragment size: 10 mm root sections or finely cut slices of root; and (iii) time of sampling: Time 0, sampled at experimental setup; Time 1, after 2 weeks; and Time 2, after 4 weeks. A randomized complete block design with four replicates totalling 48 experimental units was used. Sectioned or sliced roots (220 mg wet weight, equivalent to DM concentration of 129 mg $(\text{kg soil})^{-1}$) were added to plastic bags containing 300 g of K Field site Control soil and mixed thoroughly. Distilled water was then added to the FC treatments to bring the sample to 75 % of FC. FC treatment bags were not sealed, dry treatments were sealed. Units for Time 1 and Time 2 were placed in an incubator (14 h 23 °C, 10 h 28 °C). Units from Time 0 were transferred to paper bags following setup and dried at 40 °C for 48 h before sealing in plastic bags for DNA analysis. FC treatments were removed and weighed every 2–3 days (average 48 h moisture loss 0.14 %, range 0.06–0.39 %) and distilled water added to a 75 % FC weight. Units for Time 1 and Time 2 were removed at 15 and 30 days, respectively, dried in paper bags, prepared for DNA analysis as described for Time 0.

Field study

Rootstock cultivar field trial (Exp. 6). This experiment evaluated the mango DNA yields of sieved root-removed soil samples of five rootstock cultivars in the field trial, for three sample depths and two time points. Mango soil samples were collected from under selected polyembryonic rootstock cultivar treatments in 11–12 November 2010 and 22–23 February 2011 from a mango rootstock trial in Katherine (K Field site Mango 2) (Table 1). The trial (five replicates of a balanced lattice design) was planted in 1996, and trees were spaced at 9 m between rows and 5 m within rows. Orchard management practices were intensive and semi-mechanized (Crane *et al.* 1997; Bally 2011). By 2010, the trees in the trial had filled their allocated space and were pruned post-harvest each year to maintain separate trees rather than a hedgerow system.

Soil-root samples were collected from five replicates of rootstock cultivars from five cultivar treatments (KP, MYP, Kurukan, Watertank and Vellaikulamban). Six cores were collected 0.5 m from the trunk of each tree using a sampling template providing equally spaced sampling positions. For the February sample, the template was rotated to position cores 0.3 m from the November 2010 positions. The six samples for each depth (0–15, 15–30 and 30–45 cm) were bulked providing a sample volume of 2378 cm^3 per 15 cm depth per tree. Weed species up to a distance of 1.5 m from the trunk of each tree were recorded at sampling. The soil samples were dried (40 °C for 48 h) within 2.5 h of collection. Before sieving-root sorting and mango DNA analyses, the average soil moisture values from three selected samples were 0.47 and 0.48 %, respectively.

After drying, each sample was successively sieved through 8, 4 and 2 mm sieves (Endecott, London, UK). The sieving took a maximum of 15 min per sample. Roots were removed at each stage of sieving, and the root-removed sieved soil sorted for root fragments 2 mm or greater in diameter or length and these were removed. For the November samples, roots were then shaken in a 0.23 mm sieve to remove adhering sand and soil then sorted by eye on a white background. Sorting was aided by comparison with manufactured cylinders 0.64, 1.88 and 7.5 mm diameter. Roots were placed into four size categories of <0.64, 0.64–1.88, >1.88–7.5 and >7.5 mm. Root size sorting took ca. 1–2 h per sample. These root samples were dried (65 °C for 48 h) and weighed.

Following sieving and root removal, each soil sample was then sub-sampled for DNA analysis. To provide a representative proportion, a sub-sample representing 15 % of the weight of total sample was sub-sampled in a 70 mm deep tray using a level surface sub-sampling method described by Schroth (2003). The qPCR analysis of these soil samples for mango DNA was carried out as described in Exp. 2.

Root and root fragment removed sieved soil samples from one sampled tree per block ($n = 5$) were pooled for analyses from each of the 0–15, 15–30, 30–45 cm zones and an additional depth zone of 45–60 cm. Total carbon decreased from 0.52 to 0.08 % with increasing depth, the percentage of clay in samples increased with depth from 5.0 to 17.6 %, coarse sands decreased with depth from 79.1 to 67.4 %, fine sands increased with depth from 10.9 to 14.0 %.

Storage time effects on the analyses of field samples were investigated using all KP soil samples from the November collection ($n = 15$), these had qPCR analyses made on two separate sub-samples, analysed at 6 weeks interval. Samples were stored after sieving and root removal had been carried out in double-layered plastic bags in an air-conditioned facility.

Analyses

Analysis of variance (ANOVA) was used to compare cultivar effects on root Ct and DNA concentrations (Exp. 2). The relationships between (i) root DM soil concentration and mango DNA yields for KP was examined with linear regression for origin constrained to zero (Exp. 4) and

(ii) soil pg DNA values and root DM values from the rootstock trial was examined with linear regression (non-constrained) (Exp. 6). ANOVA was used for analysis of the following factors; tissue treatment, soil moisture, assessment time and their interactions in the DNA persistence experiment (Exp. 5). For the rootstock study (Exp. 6), weed densities were examined for their effect on soil DNA values by comparing trees with and without weeds. A generalized linear model with a Gaussian link function was used, with individual models for each depth, separate repeated-measure models were used for the November and February data. Comparison of soil DNA values for storage time effects in the rootstock study was made using a repeated-measures ANOVA for time effects and the factor sample depth, block was included. Comparison of soil DNA values at post-wet season and post-harvest periods for cultivars the rootstock study was made using a repeated-measures ANOVA model for the factors such as time and sample depth, including block. Tukey's HSD method was used for post-hoc pairwise comparisons where required. Analyses were carried out with S-PLUS (Insightful Corporation); for more detailed information on analyses **[see Supporting Information]**.

Results

Cultivar sequencing

A multiple sequence alignment of the 113 bp qPCR target region showed that there was sequence variation between the mango cultivars studied. In total, there were six positions with variability. One variation was located on the Mango F1 primer, and a second in the Mango R3 primer and the remainder was found in the conserved areas. There was no sequence variability found in the probe region **[see Supporting Information]**.

Plant tissue-only experiments

Mango qPCR specificity test (Exp. 1). The mango qPCR was specific and sensitive. It detected all the mango DNA tested (average Ct of 23.8 for 100 pg μL^{-1} mango DNA) but not the weed species (Table 2).

Quantification of mango root DNA (Exp. 2). The Ct values differed between cultivars, whereby KP and Vellaikulamban were significantly lower than Kurakan and MYP. The Ct of Watertank was also lower than that of MYP (Table 3). The mango root DNA concentration differed significantly; the cv. KP had higher concentrations than the four other cultivars.

Table 3. Determination of Ct values (with standard deviations, std. dev.) and root DNA concentrations (ng mango DNA mg^{-1} dry root) for root samples of five mango cultivars (Exp. 2), df = 10 for all comparisons, standard error of the mean (SEM) and Tukey's HSD (HSD) values also presented.

Cultivar	Ct (std. dev.)	Root DNA of mango (ng mango DNA per mg dry root)
Watertank (NT14)	22.89 (0.28)	0.873
KP (NT16)	22.22 (0.26)	3.965
Kurukan (NT21)	23.50 (0.19)	0.858
MYP (NT50)	24.14 (0.35)	0.576
Vellaikulamban (NT51)	22.31 (0.34)	1.199
SEM	0.168	0.4646
HSD	0.783	2.0767

Experiments with soil

For the soil DNA analysis, all negative batch check samples included had values below the detection limit (5 pg (g soil)$^{-1}$) with a range of 0–2.9 pg (g soil)$^{-1}$. For the positive batch check samples, the average value was 798.7 pg (g soil)$^{-1}$ (range 624.1–1029.6).

Soil type effects (Exp. 3). The concentrations of the internal control DNA extracted from soil samples from the two sites and various depths were similar [D Field site Control, average 758 641, range 179 424 (fg well^{-1}) and K Field site Mango 1, average 681 072, range 88 757 (fg well^{-1})].

Mango root concentration: DNA yield and cultivar detection in control soil (Exp. 4). Control soil spiked with KP root concentrations from 25 to 1600 mg kg^{-1} had a significant linear relationship with mango DNA yields (Fig. 1). The soil qPCR method also detected samples of other polyembryonic cultivars (MYP, B, Chok Anan, Pancho, Chandrakaran, Vellaikulamban, 13/1 and Brodie) or source of NDM (NT36), and a monoembryonic cultivar (Irwin) (Table 4). Picogram values among cultivars were variable, the minimum value of which was 367 pg DNA (g soil)$^{-1}$ across all cultivars.

Effects of root tissue degradation on DNA yield (Exp. 5). There was a significant three-way interaction ($P = 0.014$) between the tissue disturbance treatment, soil moisture and assessment time (Fig. 2). The pg g^{-1} values for the sliced-75 % FC treatment were significantly lower at Day 0 in comparison to the three other treatments. The sectioned-75 % FC also had significantly less DNA than the sectioned-dry treatment. At Day 15 the sectioned-75 % FC had similar pg values to the sliced-75 % FC treatment, with large differences between the soil moisture treatments. The trend for large soil moisture effects continued to Day 30 when both the pg values for

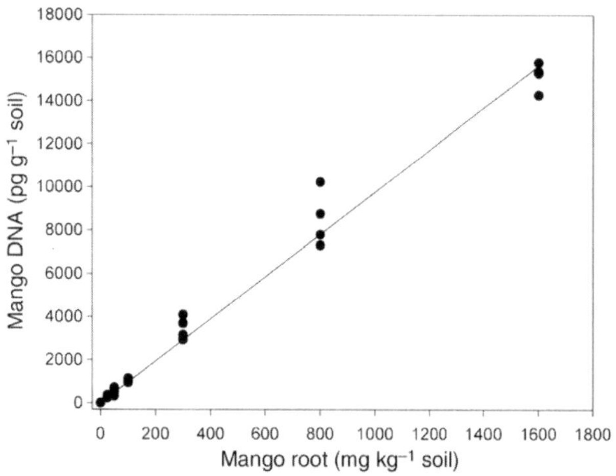

Figure 1. Root dry matter concentrations (0, 25, 50, 100, 300, 800 and 1600 mg kg^{-1}) of KP root tissue in control soil plotted against pg DNA (g soil)$^{-1}$ values (Exp. 4), with a fitted linear regression, $y = 9.7666$ (x), $P < 0.001$, adjusted multiple $R^2 = 0.9909$, standard error of coefficient 0.1769, df = 27 (Exp. 4).

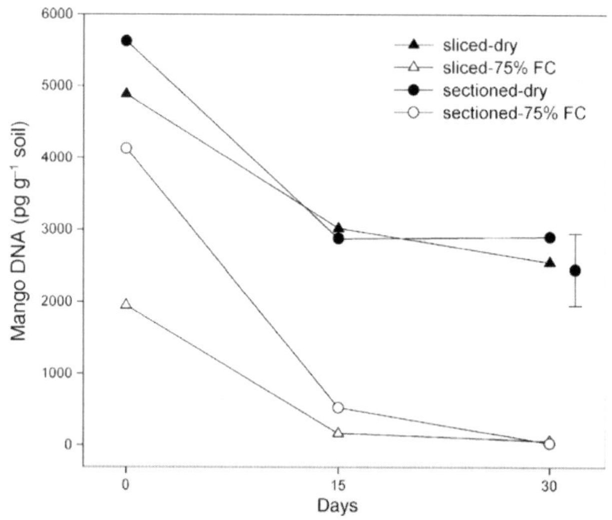

Figure 2. Results of a mango root DNA (pg DNA g^{-1} soil) persistence experiment (Exp. 5) over 30 days for NDM (Hamilton) root tissue subjected to four treatments: sliced root in dry soil; sliced root in 75 % FC soil; 10 mm root sectioned in dry soil; 10 mm root sectioned in 75 % FC soil, df = 33, standard error of the mean = 208.1, Tukey's HSD = 998.8.

Table 4. Quantification of mango root DNA (pg g^{-1} soil) in soil spiked with root material of each of 11 mango cultivars and six weed species (Exp. 4), mango cultivars (NT accession code). Source details of root material except for cv. Irwin is provided in Table 2.

Species	Cultivars	Root DNA (pg g^{-1} soil)
Mangifera indica	KP (NT16)	1270.9
M. indica	NDM (NT36)	2646.7
M. indica	MYP (NT50)	367.5
M. indica	B (NT63)	3244.3
M. indica	Chok Anan	1221.5
M. indica	Pancho (NT10)	516.1
M. indica	Chandrakaran (NT55)	977.4
M. indica	Vellaikulamban (NT51)	1606.3
M. indica	13/1 (NT52)	1318.8
M. indica	Brodie (NT9)	844.0
M. indica	Irwin[1]	1589.3
Digitaria spp.	–	0.0
Chloris inflata	–	0.0
Pennisetum typhoides	Katherine Pearl	0.0
Sorghum bicolor	Jumbo	0.0
Melinis repens	–	0.0
Crotalaria goreenis	–	0.0

[1]The only monoembryonic cultivar included in the study.

the 75 % FC treatments were approaching zero with final values of 60.3 pg for sliced-75 % FC and 30.2 pg for sectioned-75 % FC. In contrast, the average value of the two dry treatments was >2700 pg.

Rootstock cultivar field trial (Exp. 6). Weed species were present in the rootstock field trial. Twelve and nine of the 25 trees had weed seedlings present in November and February, respectively. Weed densities were low with ranges of 0.14–1.84 and 0.14–2.12 weeds m^{-2}, respectively. In November, five weed species were present, and in February, a sixth species (*Urochloa mosambicensis*) was also present (Table 1, K Field site Mango 2). There were no significant differences ($P = 0.353–0.910$) in soil mango DNA concentrations for November or February samples at any of the sample depths between trees with and without weeds present **[see Supporting Information].**

For the comparison of pg DNA values from sieved soil samples collected in November and the weight of roots removed from these samples, the mean mango DNA (dependant variable) values for each cultivar by sample depth had the strongest relationship (adjusted multiple $R^2 = 0.9307$, $P < 0.001$, df = 13) with the DM of Class 1 roots (Fig. 3). The relationship between the larger three root classes and total root DM was poor (multiple R^2 range 0.0022–0.2463, $P > 0.05$ in all cases).

Soil samples collected in November for rootstock KP (NT16, $n = 15$) had a separate analysis of duplicate sub-samples made 6 weeks after the first analyses. A repeated-measures analysis for mean mango DNA values gave no significant time effect ($P = 0.940$) or time by depth interaction effect ($P = 0.460$) between the two dates.

For the separate analysis of soil DNA concentrations from November and February samples for each cultivar, cv.

Watertank had a significant date by depth interaction ($P = 0.035$, tHSD = 0.2559). For cv. Watertank the November sample (0.717 mg DNA m^{-2}) had higher concentrations than the February sample (0.395 mg DNA m^{-2}) for the 0–15 cm depth (Fig. 4). For all cultivars depth effects were significant ($P < 0.05$), including significantly higher DNA values in the 0–15 than the 15–30 cm zone, except for cv. KP which only had a higher DNA concentration in the 0–15 than the 30–45 cm zone.

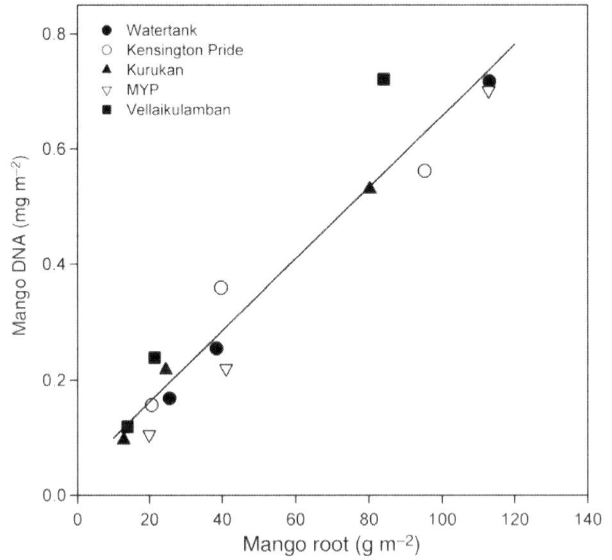

Figure 3. Regression results of average Class 1 (diameter <0.64 mm) mango root dry matter (DM g m^{-2}) values for samples of five rootstock cultivars (see legend) sampled in November from three sample depths (0–15, 15–30 and 30–45 cm) in the rootstock trial (Exp. 6) between average mango DNA concentrations in soil (mg m^{-2}) of sieved soil after root removal with fitted regression, $y = 0.0062x + 0.0376$.

Discussion

We developed a RDD method that detected the root DNA of all mango rootstock cultivars tested. We showed that the method could quantify depth and site differences in RDD values for fine roots for the predominate Australian rootstock cultivar, KP (Crane *et al.* 1997). Further field testing showed that the method also had potential for rootstock cultivar comparisons.

The root DNA persistence experiment (Exp. 5) demonstrated that the contribution of DNA from dead roots after 15–30 days in moist soil was minimal (Fig. 2). Riley *et al.* (2010) also found that the DNA concentrations of lyophilized *Lolium perenne* and *Trifolium subterraneum* roots in moist soil declined to low concentrations over a 7-day period. In our study, moist soil conditions also assisted initial DNA degradation, as at Time 0, a greater degradation of the finely sliced root material in moist soil than finely sliced root material in dry soil occurred. The DNA degradation results for a moist soil environment indicated that the method was suitable for sampling irrigated trees or for sampling during the wet (monsoonal) season.

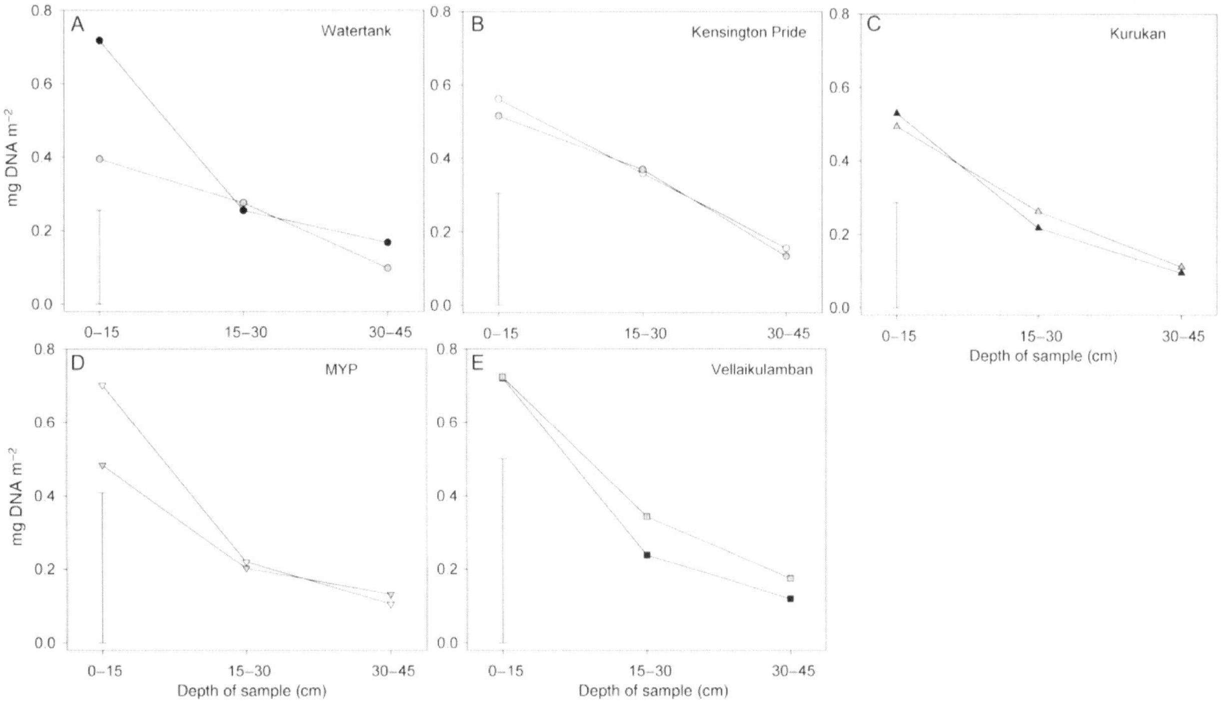

Figure 4. Mango DNA (mg DNA m^{-2}) of sieved soil for the five cultivars (plots A–E) for three sample depths sampled in November (post-harvest) and the following February (post-wet season) in the rootstock trial (Exp. 6), df = 12 for each comparisons, HSD values shown by vertical bars.

To provide a focus on the DNA of fine roots and effects of an uneven distribution of larger roots on DNA yields, the soil samples were dried and progressively passed through smaller sieves (min 2 mm aperture). The soil DNA analysis was carried out on the sieved soil that had passed through all sieves. The mango DNA concentration in the sieved soil was strongly related but not equal to the DM of the finest root class removed by the smallest root class in sieves (Fig. 3). Although there were substantial DM values for other root classes (data not presented), the poor relationship of the soil DNA yield with the DM of larger root classes confirmed that the DNA present in the root material was not associated with larger root classes.

To address tree root distribution heterogeneity issues, our method required analyses of large soil samples. High levels of sub-sampling and replication are often required in root studies due to uneven root distribution (Mackie-Dawson and Atkinson 1991; Schroth 2003), but such requirements may act as a sampling intensity limitation. Root size sorting (1–2 h per sample) was required to establish which size class of root the fragments remaining in sieved soil were associated with (Fig. 3); however, this more time-consuming root size sorting would not be required in further experiments under the same conditions.

The method was also used to compare the seasonal RDD of mango rootstock cultivars. One cultivar, Watertank, had greater DNA concentrations at the post-harvest sample than the post-wet season sample (Fig. 4). This indicated that cv. Watertank could, for example, have a greater mass of live fine roots during the fruit production period. A trial that included the five same rootstock cultivars grafted to KP (Fig. 3) found that cv. Watertank had higher cumulative fruit yields than the four other rootstock cultivars (Smith *et al.* 2008). In India, non-grafted mango trees had individual preflowering and post-harvest fine (<1.5 mm) root densities that were positively associated with individual tree fruit yield across three biennial bearing mango cultivars (Rao and Mukherjee 1988). Our RDD findings for cv. Watertank grafted to KP [an erratic non-alternate bearing scion (de Souza *et al.* 2004)] showed rootstock cultivar differences in seasonal root traits, rather than associations with the on–off bearing patterns of individual trees for non-grafted alternate bearing cultivars (Rao and Mukherjee 1988). Further work would be required to establish which root parameter (root age, root mass etc.) was responsible for the RDD finding and how this may relate to productivity. Importantly, the RDD method provided a rapid means of identifying cultivar root traits associated with productivity from which specific hypothesis could be investigated in future work.

Differences were detected in mango root tissue DNA concentrations between five rootstock cultivars. Although we targeted a conserved region to design the qPCR test, a multiple sequence alignment of the target region of 10 cultivars showed six positions with sequence variability **[see Supporting Information]**. As already stated, the DNA contents from different age roots, sizes or root morphologies can differ within species (Fisk *et al.* 2010; Haling *et al.* 2011), therefore some differences in root morphology or age among mango cultivars could have contributed to DNA concentration differences. Ct values among cultivars can vary through variable sequences or single-nucleotide polymorphisms within a sequence (Wu *et al.* 2010). Intraspecific polymorphism and sites of nucleotide additivity in the ITS region do occur among mango cultivars (Yonemori *et al.* 2002). To ensure these sequence differences and potential differences in the copy number of the reference genes do not artificially create cultivar differences. A reference single copy qPCR assay will need to be developed as outlined in Huang *et al.* (2013) for cultivar to cultivar comparisons. Because of this potential influence, we adopted a conservative approach of only comparing between timepoints for each cultivar in our field study of rootstock cultivars rather than among cultivars.

Some studies have investigated the root distribution of individual mango trees (Avilan and Meneses 1979; Coehlo *et al.* 2001; de Almeida *et al.* 2009). These studies have focused on the entire root system making interpretations about the size range of fine roots of mango difficult to interpret. However, results from one study indicate that roots <0.5 mm in diameter occur at low incidence (Coehlo *et al.* 2001). Our method was developed to focus on the fine root fragments of mango, for species with finer diameter roots the method may have to be adapted using smaller aperture sieves. Our method was also developed on high sand content soils. For plant species occurring on soil types that have high clay contents and bind heavily when dried (e.g. Vertosols), the sieving of soil to separate roots may not be easily possible.

Conclusions

We have developed a method to study the differences in distribution and density of fine root systems of the cv. KP using soil samples from mango orchards. The advantages of this method include: a sensitive and species-specific qPCR assay that is not affected by dead roots in moist soil; allows rapid pre-processing by using sieved soil samples (300–500 g) to remove the larger roots that would have increased between sample variation; and is suitable for a high throughput. The method also shows potential for detecting differences in cultivar RDD values.

Accession Numbers

Nucleotide sequences were submitted to GenBank and were assigned the following accession numbers. Kensington Pride NT16 (8) KJ833763; Nam Doc Mai NT36 (23a) KJ833765; MYP NT50 (16a) KJ833764; B NT63 (17a) KJ833759; Chok Anan (18a) KJ833762; Pancho NT10 (19a) KJ833766; Chandrakaran NT55 (20) KJ833761; Vellaikulamban NT51 (22a) KJ833767; 13/1 NT52 (24a) KJ833758 and Brodie NT9 (25a) KJ833760.

Sources of Funding

Our work was funded by the Northern Territory Research & Innovation Fund, Northern Territory Government, Australia.

Contributions by the Authors

S.L.B. and L.T.T.T-N. initiated and obtained funding for the conceptual qPCR mango root detection method and the resulting experimental path. They carried out all preliminary plant tissue-only DNA and initial probe–primer development work, worked on all experiments and co-wrote this manuscript. S.L.B. conducted the field studies, collected all material (soil and roots) for the experiments and conducted preliminary statistical analysis. L.T.T.T-N. developed an initial range of probes and primers. D.M.H. developed the final version of the probe and primer. M.N.H. contributed to the design of all experiments and carried out all statistical analyses.

Acknowledgements

We thank, Mark Hoult and Nick Hartley of Plant Industries, DPIF who provided substantial assistance with this study, and Herdina, Alan McKay and Nathan Bott of SARDI who contributed to the development of successful probe and primers. Herdina and Alan McKay also provided valuable advice on experimental design and execution. We also thank Viet Ma of Lambells Lagoon, Darwin, for kindly providing fruit for developing the standards.

Supporting Information

The following Supporting Information is available in the online version of this article –

Figure S1. Provides a multiple sequence alignment for root tissue samples from ten mango cultivars depicting the targeted ITS region (ITS1-5.8S rRNA-ITS2) used in the qPCR test.

File S2. Provides a statistical summary of experiments including ANOVA and regression tables.

Literature Cited

Avilan L, Meneses L. 1979. Effect of the physical characteristics of the soil on the root distribution of mango trees (*Mangifera indica* L.). *Turrialba* **29**:117–122.

Bally ISE. 2011. Advances in research and development of mango industry. *Revista Brasileira de Fruticultura* **33**:57–63.

Blaikie SJ, Kulkarni VJ, Muller WJ. 2004. Effects of morphactin and paclobutrazol flowering treatments on shoot and root phenology in mango cv. Kensington Pride. *Scientia Horticulturae* **101**:51–68.

Butnor JR, Johnsen KH, McInnis D, Stover DB, Roth BE, Day FP. 2008. Using ground-penetrating radar to estimate tree root mass—comparing results from two Florida surveys. In: Allred BJ, Daniels JJ, Reza Ehsani ME, eds. *Handbook of agricultural geophysics*. Boca Raton, FL: CRC Press, 375–382.

Coehlo EF, Borges AL. 2004. Irrigation and fertirrigation of mango. *Acta Horticulturae* **645**:119–128.

Coehlo EF, Oliveira FDC, Araujo ECE, Vasconcelos LFL, Lima DM. 2001. Root distribution of mango under trickle irrigation in a coast sand soil. *Revista Brasileira de Fruticultura* **23**:250–256.

Crane JH, Bally I, Mosqueda-Vaquez RV, Tomer E. 1997. Crop production. In: Litz RE, ed. *The mango*. Wallingford: CAB International, 203–256.

da Silva EV, Bouillet JP, de Moraes Gonçalves JL, Junior CHA, Trivelin PCO, Hinsinger P, Jourdan C, Nouvellon Y, Stape JL, Laclau J-P. 2011. Functional specialization of Eucalyptus fine roots: contrasting potential uptake rates for nitrogen, potassium and calcium tracers at varying soil depths. *Functional Ecology* **25**: 996–1006.

de Almeida EV, Fernandes FM, Boliani AC. 2009. Root system distribution of Coquinho rootstock on top of Haden mango cultivar. *Revista Brasileira de Fruticultura* **31**:798–804.

de Souza MP, de Queiroz MA, de Possidio EL, Pereira FA, Nunes RFD. 2004. Study of flowering and alternate bearing of mango varieties in the Sao Francisco valley. *Acta Horticulturae* **645**: 353–358.

Dillon NL, Bally ISE, Wright CL, Hucks L, Innes DJ, Dietzgen RG. 2013. Genetic diversity of the Australian National Mango Genebank. *Scientia Horticulturae* **150**:213–226.

Drummond AJ, Ashton B, Buxton S, Cheung M, Cooper A, Duran C, Field M, Heled J, Kearse M, Markowitz S, Moir R, Stones-Havas S, Sturrock S, Thierer T, Wilson A. 2011. Geneious v5.4. http://www.geneious.com/.

Eissenstat DM, Yanai RD. 1997. The ecology of root lifespan. In: Begon M, Fitter AH, eds. *Advances in ecological research*, Vol. 27. London: Academic Press Ltd-Elsevier Science Ltd, 1–60.

Faget M, Liedgens M, Stamp P, Flutsch P, Herrera JM. 2010. A minirhizotron imaging system to identify roots expressing the green fluorescent protein. *Computers and Electronics in Agriculture* **74**:163–167.

Fisk MC, Yanai RD, Fierer N. 2010. A molecular approach to quantify root community composition in a northern hardwood forest—testing effects of root species, relative abundance, and diameter. *Canadian Journal of Forest Research* **40**:836–841.

Goebel M, Hobbie SE, Bulaj B, Zadworny M, Archibald DD, Oleksyn J, Reich PB, Eissenstat DM. 2011. Decomposition of the finest root branching orders: linking belowground dynamics to fine-root function and structure. *Ecological Monographs* **81**:89–102.

Graham JH, Johnson EG, Gottwald TR, Irey MS. 2013. Presymptomatic fibrous root decline in citrus trees caused by Huanglongbing and potential interaction with *Phytophthora* spp. *Plant Disease* **97**: 1195–1199.

Green S, Clothier B. 1999. The root zone dynamics of water uptake by a mature apple tree. *Plant and Soil* **206**:61–77.

Gregory PJ, Atkinson CJ, Bengough AG, Else MA, Fernández-Fernández F, Harrison RJ, Schmidt S. 2013. Contributions of roots and rootstocks to sustainable, intensified crop production. *Journal of Experimental Botany* **64**:1209–1222.

Haling E, Simpson RJ, McKay AC, Hartley D, Lambers H, Ophel-Keller K, Wiebkin S, Herdina H, Riley IT, Richardson AE. 2011. Direct measurement of roots in soil for single and mixed species using a quantitative DNA-based method. *Plant and Soil* **348**:123–137.

Haling RE, Simpson RJ, Culvenor RA, Lambers H, Richardson AE. 2012. Field application of a DNA-based assay to the measurement of roots of perennial grasses. *Plant and Soil* **358**:176–192.

Hendrick RL, Pregitzer KS. 1996. Temporal and depth-related patterns of fine root dynamics in northern hardwood forests. *Journal of Ecology* **84**:167–176.

Hendricks JJ, Hendrick RL, Wilson CA, Mitchell RJ, Pecot SD, Guo DL. 2006. Assessing the patterns and controls of fine root dynamics: an empirical test and methodological review. *Journal of Ecology* **94**:40–57.

Hill J, Edmeades BF, Owens G, Hignett C. 2011. *Plant industry soils of the Northern Territory 'A Northern Territory companion to the healthy soils for sustainable vegetable farms'*. Palmerston: NRETAS, NT Government.

Huang CY, Kuchel H, Edwards J, Hall S, Parent B, Eckermann P, Herdina H, Hartley DM, Langridge P, McKay AC. 2013. A DNA-based method for studying root responses to drought in field-grown wheat genotypes. *Scientific Reports* **3**:3194.

Johnson MG, Tingey DT, Phillips DL, Storm MJ. 2001. Advancing fine root research with minirhizotrons. *Environmental and Experimental Botany* **45**:263–289.

Levillain J, M'Bou AT, Deleporte P, Saint-Andre L, Jourdan C. 2011. Is the simple auger coring method reliable for below-ground standing biomass estimation in Eucalyptus forest plantations? *Annals of Botany* **108**:221–230.

Livesley SJ, Stacey CL, Gregory PJ, Buresh RJ. 1999. Sieve size effects on root length and biomass measurements of maize (*Zea mays*) and *Grevillea robusta*. *Plant and Soil* **207**:183–193.

Mackie-Dawson LA, Atkinson D. 1991. Methodology for the study of roots in field experiments and the interpretation of results. In: Atkinson D, ed. *Plant root growth: an ecological perspective*. Oxford: Blackwell Scientific Publications, 25–47.

McKenzie N, Ryan P, Fogarty P, Wood J. 2000. Sampling, measurement and analytical protocols for carbon estimation in soil, litter and coarse woody debris. National Carbon Accounting System Technical Report No. 14. Canberra: Australian Greenhouse Office.

Milchunas DG. 2009. Estimating root production: comparison of 11 methods in shortgrass steppe and review of biases. *Ecosystems* **12**:1381–1402.

Mommer L, Wagemaker CAM, De Kroon H, Ouborg NJ. 2008. Unravelling below-ground plant distributions: a real-time polymerase chain reaction method for quantifying species proportions in mixed root samples. *Molecular Ecology Resources* **8**:947–953.

Morgan KT, Obreza TA, Scholberg JMS. 2007. Orange tree fibrous root length distribution in space and time. *Journal of the American Society for Horticultural Science* **132**:262–269.

Nadezhdina N, Cermak J. 2003. Instrumental methods for studies of structure and function of root systems of large trees. *Journal of Experimental Botany* **54**:1511–1521.

Ophel-Keller K, McKay A, Hartley D, Herdina H, Curran J. 2008. Development of a routine DNA-based testing service for soilborne diseases in Australia. *Australasian Plant Pathology* **37**:243–253.

Ostonen I, Lohmus K, Pajuste K. 2005. Fine root biomass, production and its proportion of NPP in a fertile middle-aged Norway spruce forest: comparison of soil core and ingrowth core methods. *Forest Ecology and Management* **212**:264–277.

Persson H. 1990. Methods of studying root dynamics in relation to nutrient cycling. In: Harrison AF, Ineson P, Heal OW, eds. *Nutrient cycling in terrestrial ecosystems: field methods, application and interpretation*. New York: Elsevier Applied Science, 198–217.

Pierret A, Moran CJ, Doussan C. 2005. Conventional detection methodology is limiting our ability to understand the roles and functions of fine roots. *New Phytologist* **166**:967–980.

Pigram JJ. 2006. *Australia's water resources: from use to management*. Collingwood, Victoria: CSIRO Publishing.

Rao DP, Mukherjee SK. 1988. Feeder root distribution of different varieties of mango. *Acta Horticulturae* **231**:325–334.

Riley IT, Wiebkin S, Hartley D, McKay A. 2010. Quantification of roots and seeds in soil with real time PCR. *Plant and Soil* **331**: 151–163.

Schroth G. 2003. Root systems. In: Schroth G, Sinclair FL, eds. *Trees, crops and soil fertility concepts and research methods*. Wallingford: CABI, 235–257.

Schroth G, Kolbe D. 1994. A method of processing soil core samples for root studies by subsampling. *Biology and Fertility of Soils* **18**: 60–62.

Sharma M, Sharma KC. 1988. Organization of root apical meristems in some angiosperms. *Botanical Bulletin of Academia Sinica* **29**: 223–229.

Smith MW, Bright JD, Hoult MD, Renfree RA, Maddern T, Coombes N. 2008. Field evaluation of 64 rootstocks for growth and yield of 'Kensington Pride' mango. *Hortscience* **43**:1720–1725.

Valverde-Barrantes OJ, Raich JW, Russell AE. 2007. Fine-root mass, growth and nitrogen content for six tropical tree species. *Plant and Soil* **290**:357–370.

White TJ, Bruns T, Lee SB, Taylor JW. 1990. Amplification and direct sequencing of fungal ribosomal DNA genes for phylogenetics. In: Innis MA, Gelfand DH, Sninsky JJ, White TJ, eds. *PCR protocols: a guide to methods and applications*. New York: Academic Press, 315–322.

Wu G, Zhang L, Wu YH, Cao YL, Lu CM. 2010. Comparison of five endogenous reference genes for specific PCR detection and quantification of *Brassica napus*. *Journal of Agricultural and Food Chemistry* **58**:2812–2817.

Yonemori K, Honsho C, Kanzaki S, Eiadthong W, Sugiura A. 2002. Phylogenetic relationships of *Mangifera* species revealed by ITS sequences of nuclear ribosomal DNA and a possibility of their hybrid origin. *Plant Systematics and Evolution* **231**:59–75.

3

Contrasting responses of root morphology and root-exuded organic acids to low phosphorus availability in three important food crops with divergent root traits

Yan-Liang Wang[1,2], Marit Almvik[1], Nicholas Clarke[1], Susanne Eich-Greatorex[2], Anne Falk Øgaard[1], Tore Krogstad[2], Hans Lambers[3]* and Jihong Liu Clarke[1]*

[1] Norwegian Institute of Bioeconomy Research (NIBIO), PO Box 115, N-1431 Ås, Norway
[2] Department of Environmental Sciences, Norwegian University of Life Sciences, PO Box 5003, N-1432 Ås, Norway
[3] School of Plant Biology and Institute of Agriculture, University of Western Australia, Crawley (Perth), WA 6009, Australia

Associate Editor: Mikio Nakazono

Abstract. Phosphorus (P) is an important element for crop productivity and is widely applied in fertilizers. Most P fertilizers applied to land are sorbed onto soil particles, so research on improving plant uptake of less easily available P is important. In the current study, we investigated the responses in root morphology and root-exuded organic acids (OAs) to low available P (1 μM P) and sufficient P (50 μM P) in barley, canola and micropropagated seedlings of potato—three important food crops with divergent root traits, using a hydroponic plant growth system. We hypothesized that the dicots canola and tuber-producing potato and the monocot barley would respond differently under various P availabilities. WinRHIZO and liquid chromatography triple quadrupole mass spectrometry results suggested that under low P availability, canola developed longer roots and exhibited the fastest root exudation rate for citric acid. Barley showed a reduction in root length and root surface area and an increase in root-exuded malic acid under low-P conditions. Potato exuded relatively small amounts of OAs under low P, while there was a marked increase in root tips. Based on the results, we conclude that different crops show divergent morphological and physiological responses to low P availability, having evolved specific traits of root morphology and root exudation that enhance their P-uptake capacity under low-P conditions. These results could underpin future efforts to improve P uptake of the three crops that are of importance for future sustainable crop production.

Keywords: Crops; hydroponic culture; phosphorus; root exudates; root morphology.

Introduction

Phosphorus (P) deficiency is one of the major limitations for crop productivity globally. In modern agriculture, mineral P fertilizers are extensively used, and large economic benefits have been generated. However, P fertilizers have also been added in amounts that far exceed the amount of P removed at harvest: up to ~90 % of P applied as fertilizer may be strongly sorbed to the soil to become less available to plants, or eroded and lost in run-off (Gerke *et al.* 1994; Smith and Schindler 2009; Andersson

* Corresponding author's e-mail address: hans.lambers@uwa.edu.au (H.L.); jihong.liu-clarke@nibio.no (J.L.C.)

et al. 2013). As a consequence, high P input has produced many problems, such as eutrophication and hypoxia of waterways (Withers *et al.* 2001) as well as eutrophication of oligotrophic natural terrestrial systems (Lambers *et al.* 2013). Hence, it is important to understand P-acquisition mechanisms and genetically improve crop plants to be able to acquire P more efficiently, which would enable reduced P-fertilizer application.

Higher plants have developed various responses to low P availability, including modified gene expression, morphological responses especially in root architecture (e.g. reduced primary root length, more lateral roots and root hairs) and physiological modification of the rhizosphere by root exudation and pH changes, as well as metabolic responses (Raghothama 1999; Lynch 2011). To improve P-acquisition efficiency, understanding changes in root structure and activities under various P availabilities is necessary. During the past decades, three main types of changes have been revealed: (i) changes in root morphology, such as root length, root hairs, root distribution and root diameter (Williamson *et al.* 2001; López-Bucio *et al.* 2002; Lambers *et al.* 2006; Desnos 2008); (ii) modification of root physiology, important for release of water and protons into soil and exudation of nucleases, acid phosphatases and carboxylic acids to change soil properties and mobilize organic and inorganic P sources (Read *et al.* 2003; Lambers *et al.* 2006, 2008; Prieto *et al.* 2012) and (iii) colonization by arbuscular mycorrhizal fungi, since P uptake by the mycorrhizal hyphae is the dominant pathway for P acquisition when plant roots are colonized by arbuscular mycorrhizal fungi (Smith *et al.* 2003). Finally, these factors might not operate alone but interact with each other to improve P uptake.

Although significant progress has been made in understanding plant processes associated with soil P mobilization and acquisition, each plant species is different and likely responds to P supply differently. Furthermore, there are a number of issues that are not well understood, such as the complex coordination of root morphology, and physiological and biochemical responses under variable P availability (Shen *et al.* 2011). Root exudation represents a significant carbon cost to plants, and root exudates are influenced by plant age, species and genotype, root structure and environmental factors including both biotic and abiotic stressors (Badri and Vivanco 2009). Phosphorus deficiency increases the release of organic acids (OAs) by roots of certain plants (López-Bucio *et al.* 2000), which could play a key role in mobilizing P from mineral surfaces and from oxides and hydroxides of Al and Fe, as well as Ca-phosphates (Neumann and Römheld 2001; Ryan *et al.* 2001; Neumann *et al.* 2014).

In this study, we aimed to investigate contrasting responses of root morphology and root exudation with regards to acquiring P. Therefore, we selected three economically important crops with different root systems as plant materials: barley, canola and potato. Barley, a typical monocot, has a wide-ranging network of fibrous roots, whereas the dicots (canola and potato) tend to have a long tap-root with thick lateral roots. Barley and potato are capable of forming an arbuscular mycorrhizal symbiosis to acquire soil P, while the non-mycorrhizal canola cannot do this (Brundrett 2009; Lambers and Teste 2013). We compared the root structure and root-exuded OAs of these three crop plants under different P supply using hydroponic culture. We hypothesized that under low P, (1) canola has a more developed root system with longer roots, greater root surface area and more lateral roots, or increased root exudates to compensate for its lack of mycorrhizas; (2) barley and potato exude less OAs than canola and thus save carbon for releasing other compounds to promote rhizosphere microbial activity and (3) compared with canola and barley, potato releases the least OAs and stores most of its belowground carbon resources in tubers as starch. The information generated in this study will be of value for a more sustainable production of canola, barley and potato.

Methods

Plant material and experimental design

Canola (*Brassica napus* cv. MARIE), barley (*Hordeum vulgare* cv. HEDER) and micropropagated seedlings of potato (*Solanum tuberosum* cv. PIMPERNEL) were selected for this study. Hydroponic culture experiments were conducted in the greenhouse of the Norwegian University of Life Sciences (NMBU). Germinated canola and barley seeds were first grown in full-strength nutrient solution containing 1.5 mM KCl, 2 mM $Ca(NO_3)_2$, 1.0 mM $MgSO_4$, 1 μM H_3BO_3, 1 μM $MnSO_4$, 1 μM $ZnSO_4$, 0.5 μM $CuSO_4$, 0.37 μM Na_2MoO_4 and 50 μM Fe-EDTA. Phosphorus was added as KH_2PO_4 in a concentration of 1 or 50 μM, designated as low P availability (P1) and P sufficient (P50), respectively (concentrations used by Cheng *et al.* 2014). Five-day-old uniform seedlings without seed residues were then carefully transferred to 1-L plastic pots (one plant per pot) filled with nutrient solution containing P1 or P50, with pH adjusted to 5.8 ± 0.2. Solutions were replaced every third day; at the same time, pots were rearranged randomly. For potato, ~10 cm heights of tissue culture-derived seedlings were selected for hydroponic culture, and no root tuber was produced in this 4-week experiment. Plants were grown at 25 °C/16 °C day/night temperature with a 16-h photoperiod at a light intensity of 200 ± 20 $\mu mol\ m^{-2}\ s^{-1}$ and 50–75 % relative humidity. Three independent replications were carried out.

Root exudate collection

Root exudates were collected as described by Khorassani *et al.* (2011). Briefly, after plants had been grown hydroponically under P1 or P50 for 2 and 4 weeks, whole root systems of intact plants were carefully washed with deionized water to remove the nutrient solution. Then the whole root system was placed into ultrapure Milli-Q water (Millipore, Billerica, MA, USA) in a container to collect root exudates; the volume varied from 20 to 50 mL depending on the size of the root system. The roots were kept in the water for 2 h (between 10:00 and 13:00) under the same controlled-climate conditions as described for plant growth. Neumann and Römheld (2001) reported that roots are not harmed by Milli-Q water and no significant degradation of the exudates occurs in a short time period of 2 h. As Valentinuzzi *et al.* (2015) reported, water is the most effective and suitable trap solution to collect exudates like OAs, especially in a short time period such as 2 h. Micropur (0.01 g L^{-1}, Katadyn Products, Kemptthal, Switzerland) was then added to the solution to inhibit the activity of microorganisms (Cheng *et al.* 2014). The collected root exudates were immediately frozen at −20 °C. Before analysis with liquid chromatography triple quadrupole mass spectrometry (LC–MS/MS), collected root exudate samples were filtered with Phenex regenerated cellulose syringe filters (pore size: 0.45 μm, filter diameter: 15 mm) (Phenomenex, Torrance, CA, USA), and 0.2 μg deuterium-labelled succinic acid was added to each sample to be used as an internal standard (IS).

Determination of OAs released from roots

The OA analysis was performed using a Waters Alliance 2695 LC system coupled to a Quattro Ultima Pt triple quadrupole mass spectrometer (Micromass, Manchester, UK), equipped with an electrospray ionization (ESI) source. The auto-sampler in the LC system cooled the samples to 5 °C. The injection volume was 50 μL and sample constituents were separated using an Acquity XSelect HSS T3 (50 × 3 mm, 2.5 μm) analytical column (Waters, Milford, MA, USA) at a column oven temperature of 40 °C and a sample run time of 6 min. The trifunctional alkyl C18-bonded phase used for the HSS T3 sorbent is compatible with the 100 % aqueous mobile phase; so, aqueous formic acid (0.08 M, pH 2.4) was used as the sole mobile phase solution at a flow rate of 0.3 mL min^{-1}. Electrospray ionization conditions were as follows: capillary voltage of 3.0 kV, source temperature of 120 °C, desolvation temperature of 350 °C, cone gas flow of 80 L h^{-1} and desolvation gas flow of 620 L h^{-1}. The cone voltage, collision energy and selection of product ions were optimized for each OA for maximum intensities. Two characteristic fragmentations of the OA precursor ion ($[M-H]^-$) were monitored. The most abundant product ion was used for quantification, and the second product ion was used for verification of the OA (Table 1). The optimal quantifier ions in our MS system were identical to the quantifier ions used by Erro *et al.* (2009). For malonic acid, which has a small molecular mass, only one product ion was used. Pure OA standards (citric acid (>99 %), L-(−) malic acid (>99.5 %), succinic acid (>99.5 %), malonic acid (99 %), L-(+)-tartaric acid (>99.5 %) and deuterium-labelled succinic acid-2,2,3,3-d_4 (98 %)) were purchased from Sigma-Aldrich (St Louis, MO, USA). The OAs were mixed in Milli-Q water in the range 0.05–2.0 μg mL^{-1}, containing a fixed amount (μg mL^{-1}) of the IS deuteriated succinic acid. Internal standard calibration was performed using weighted (1/*x*) quadratic regression analysis of the peak area ratios (analyte/IS) versus the concentration ratios. Limits of quantification were 0.015–0.04 μg mL^{-1}, corresponding to signal-to-noise ratios of 15.

Plant harvest and measurements of root morphology

Plants were harvested 28 days after transfer to hydroponics, and roots and shoots were sampled separately for subsequent analysis. The total number of green leaves and senesced leaves was recorded. Root length and root surface area were measured using WinRHIZO (EPSON 1680, WinRHIZO Pro2003b, Regent Instruments Inc., Quebec, Canada). Shoot and root dry weight (DW) were measured separately after being oven-dried for 48 h at 65 °C.

Determination of P

Total P concentrations in dry root and shoot tissues were determined by inductively coupled plasma atomic emission spectroscopy (AtomComp 1100, Thermo Jarrell-Ash, MA, USA) according to Ogner *et al.* (1999) after digestion in a mixture of 65 % (v/v) HNO_3/72 % (v/v) $HClO_4$ (5 : 1, v/v) at 220 °C in a microwave oven.

Statistical analyses

Data were statistically analysed by R software (version 3.1.3; software and commanders were downloaded from the NMBU library). Two-way ANOVAs were used to study main effects of P level, species and their interaction on all the parameters involved in this study. For multiple comparisons to determine which of the six P/species combinations were significantly different from each other, *post hoc* pair-wise Tukey honest significant difference tests were used after ANOVA. For all analyses, the significance α level of 0.05 was used.

Results

Plant growth at different P availability

After 4 weeks of P1 treatment, a 65 % lower total dry biomass was found in barley and >87 % lower biomass in

Table 1. Multiple reaction monitoring (MRM) conditions for the LC-ESI-MS/MS analysis of OAs.

OA	Retention time (min)	MRM transitions; precursor > product ions	Cone voltage (V)	Collision energy (V)
Tartaric acid	1.15	149 > 87 + 73	35	12
Malic acid	1.30	133 > 115 + 71	35	12
Malonic acid	1.45	103 > 59	35	6
Citric acid	1.91	191 > 111 + 87	35	18
Succinic acid	2.43	117 > 73 + 99	35	12
IS succinic acid-d_4	2.39	121 > 77 + 102	35	12

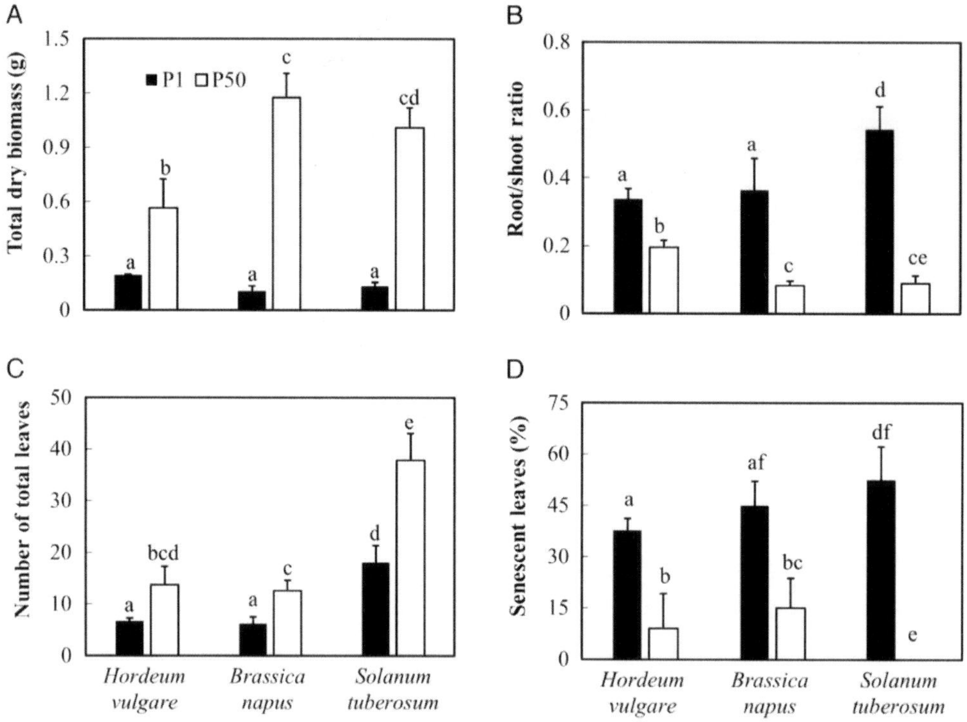

Figure 1. (A) Total dry biomass, (B) the root : shoot DW ratio, (C) total leaf number and (D) percentage senesced leaves of *H. vulgare*, *B. napus* and *S. tuberosum* grown in nutrient solution at two levels of P supply (P1 = 1 μM P, P50 = 50 μM P). Plants were harvested after 28 days growth. Values are means ± SE of $n = 9$. Different letters indicate significant differences ($P < 0.05$) as shown by Tukey tests.

canola and potato, compared with plants grown under P50; no significant differences were found between these three crops at a low P level, while canola and potato showed 102 and 68 % more biomass than barley at P50, respectively (Fig. 1A; Table 2). The root : shoot DW ratio increased considerably in plants grown under P1, with >70 % increase in barley, and >300 and 400 % increase in canola and potato, respectively; at P50, barley showed a significantly greater (~100 %) root/shoot ratio than canola and potato (Fig. 1B; Table 2). Consistent with the results for biomass, the number of total leaves was decreased by ~42–53 % at P1 compared with that at P50 for all three crops; potato (with an average of 38 leaves at P50 and 18 leaves at P1) had more leaves than canola (12 leaves at P50 and 7 leaves at P1) and barley (13 leaves at P50 and 7 leaves at P1) under both P50 and P1 conditions (Fig. 1C; Table 2). Low P availability also led to earlier leaf senescence in all three species, especially in potato (>53 % senescent leaves), with many shed leaves after 3 weeks, while there was almost no leaf senescence in potato at P50 (Fig. 1D; Table 2).

Root morphology

Changes in root morphology varied significantly among the studied crop species (Figs 2 and 3; Table 2). Under P1, compared with P50, barley showed a marked reduction in total root length (44 %) and root surface area (58 %), whereas canola showed a significant increase in total

Table 2. *F* and *P* values of a two-way ANOVA on the effects of P level (P1 versus P50), species (barley, canola and potato) and their interactions on the parameters determined in this study. P, P level; S, species; df, degrees of freedom; Error, error df values; D15, day 15; D28, day 28.

Parameters	Factors	df	*F*	*P*
Total DW	P	1	1048.1	<0.001
	S	2	46.7	<0.001
	P × S	2	83.8	<0.001
	Error	48		
Root/shoot ratio	P	1	397.2	<0.001
	S	2	11.0	<0.001
	P × S	2	27.4	<0.001
	Error	48		
Number of total leaves	P	1	245.4	<0.001
	S	2	248.2	<0.001
	P × S	2	36.9	<0.001
	Error	48		
Percentage of senescent leaves	P	1	546.2	<0.001
	S	2	6.1	0.004
	P × S	2	20.8	<0.001
	Error	48		
Total root length	P	1	2.2	0.154
	S	2	1.9	0.173
	P × S	2	34.9	<0.001
	Error	24		
Root surface area	P	1	32.0	<0.001
	S	2	3.6	0.045
	P × S	2	36.4	<0.001
	Error	24		
Root tip number	P	1	0.01	0.91
	S	2	11.8	<0.001
	P × S	2	22.9	<0.001
	Error	24		
D15 Citric acid	P	1	7.7	0.009
	S	2	6.3	0.017
	P × S	2	13.8	0.001
	Error	32		
D15 Malic acid	P	1	45.6	<0.001
	S	2	3.6	0.036
	P × S	2	1.7	0.199
	Error	48		
D15 Succinic acid	P	1	2.5	0.123
	S	2	9.6	<0.001
	P × S	2	8.1	0.001
	Error	48		
D28 Citric acid	P	1	47.3	<0.001
	S	2	28.4	<0.001
	P × S	2	23.7	<0.001
	Error	32		
D28 Malic acid	P	1	22.4	<0.001
	S	2	9.5	<0.001
	P × S	2	8.9	0.001
	Error	48		
D28 Succinic acid	P	1	7.7	0.008
	S	2	11.9	<0.001
	P × S	2	15.2	<0.001
	Error	48		
Shoot P concentration	P	1	1419.2	<0.001
	S	2	92.3	<0.001
	P × S	2	91.4	<0.001
	Error	30		
Shoot P content	P	1	676.1	<0.001
	S	2	1.1	0.340
	P × S	2	0.43	0.654
	Error	30		
Root P concentration	P	1	264.8	<0.001
	S	2	19.4	<0.001
	P × S	2	12.8	<0.001
	Error	30		
Root P content	P	1	127.5	<0.001
	S	2	8.4	0.001
	P × S	2	6.0	0.006
	Error	30		

root length (37 %), but a decrease in number of root tips (45 %). Total root length in potato and root surface area in canola did not change significantly. For potato, the number of root tips almost doubled under P deficiency. At P1, canola and potato showed 58 and 49 % greater total root length than barley, respectively; canola had 18 % greater root surface area than potato, and potato had 57 % greater root surface area than barley, but the average number of root tips decreased in the order potato (1892) > barley (883) > canola (627). On the other hand, barley showed 36 and 55 % greater root length than potato and canola, respectively, and ~55 % greater root surface area than canola and potato at P50. These three crops had almost the same number of root tips at P50.

Root exudate analysis

Three OAs—citric, malic and succinic acids—were detected in the root exudates of both P1 and P50 plants (Fig. 4; Table 2). All these three OAs were found in canola and barley exudates, while citric acid was not detected in potato root exudates (Fig. 4A and D). Under P1 treatment, the greatest amounts of total OAs (citric + malic + succinic acids) were found in canola root exudates (54 nmol $plant^{-1}$ h^{-1}), while potato plants released the

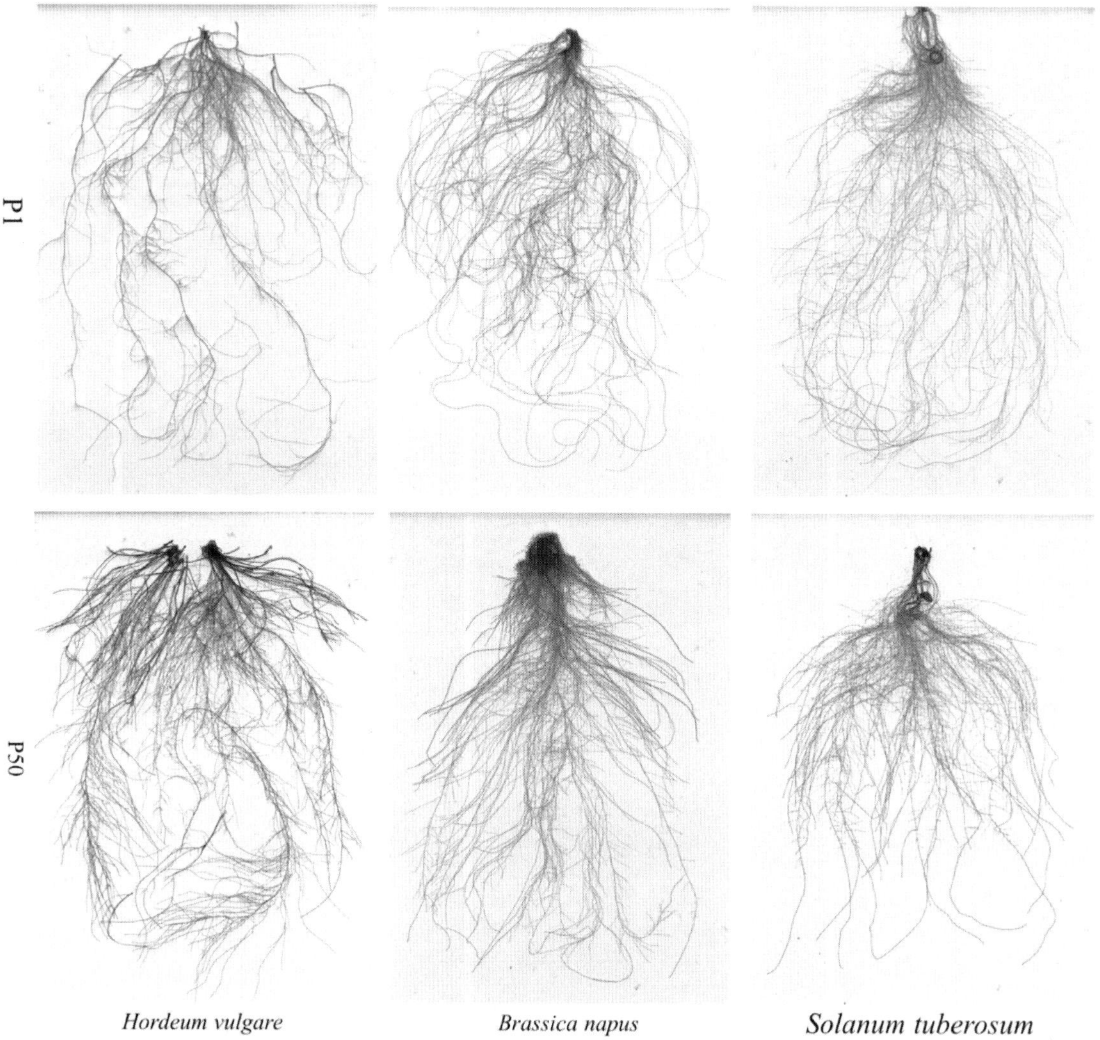

Figure 2. Root structure of *H. vulgare*, *B. napus* and *S. tuberosum* grown in nutrient solution at two levels of P supply (P1 = 1 μM P, P50 = 50 μM P). Plants were harvested after 28 days growth. Photos were taken by a WinRHIZO scanner and one representative photo of each crop was selected.

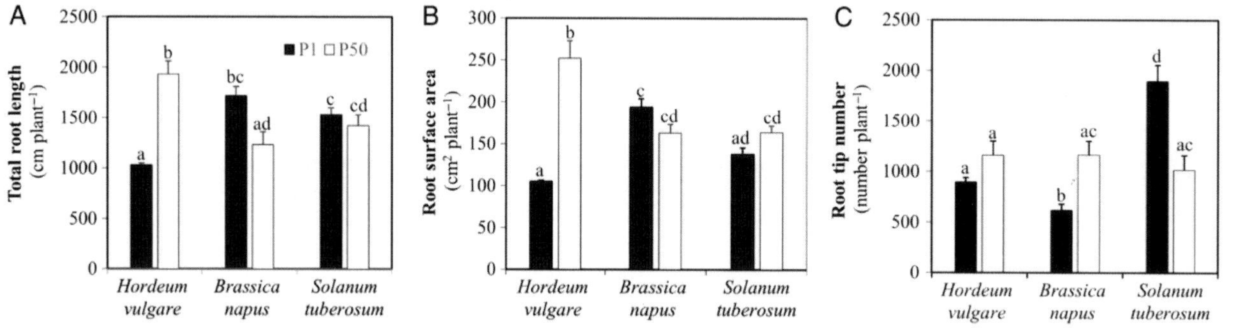

Figure 3. (A) Total root length, (B) root surface area and (C) root tip number of *H. vulgare*, *B. napus* and *S. tuberosum* grown in nutrient solution at two levels of P supply (P1 = 1 μM P, P50 = 50 μM P). Plants were harvested after 28 days growth. Values are means ± SE of $n = 6$. Different letters indicate significant differences ($P < 0.05$) as shown by Tukey tests.

lowest amount of OAs (11 nmol plant^{-1} h^{-1}). When root exudates were collected on the 15th day (just after solutions were replaced), P-deficient canola plants exuded over 400 % more citric acid and 1300 % more malic acid than the plants with sufficient P supply (Fig. 4A and B). However, succinic acid exudation at P1 in canola was only ~25 % of that at P50 (Fig. 4C). The exudation of all three OAs increased in the P1 treatment of barley, but

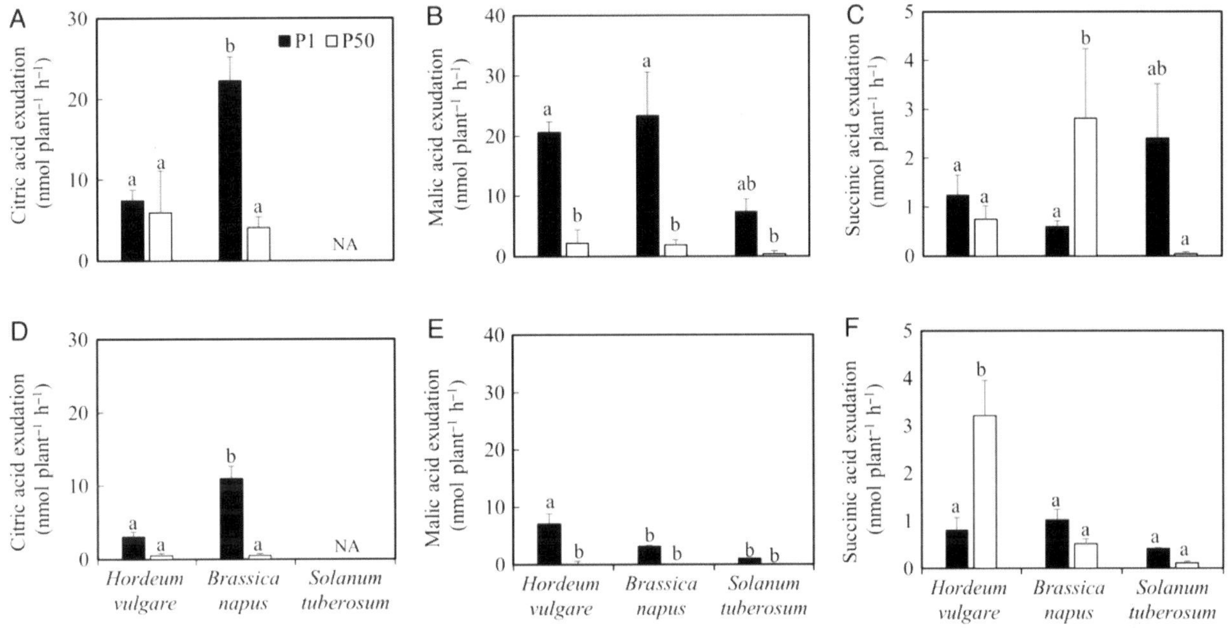

Figure 4. (A and D) Citric acid exudation, (B and E) malic acid exudation and (C and F) succinic acid exudation of *H. vulgare*, *B. napus* and *S. tuberosum* grown in nutrient solution at two levels of P supply (P1 = 1 μM P, P50 = 50 μM P). Root exudates were collected after 15 days growth (A–C) and 28 days growth (D–F). Values are means ± SE of $n = 8$–9. Different letters indicate significant differences ($P < 0.05$) as shown by Tukey tests. NA, not available.

for citric and succinic acids, this increase was not significant (Fig. 4A–C). Canola roots exuded over three times more citric acid than barley roots under P deficiency (Fig. 4A). For potato, the amounts of released malic and succinic acid were in general only 10–50 % of those for barley or canola (except for succinic acid at P1), and no significant differences between P1 and P50 were found (Fig. 4B and C). When root exudates were collected on the 28th day (just before harvesting), the three studied OAs released by roots in all studied crops decreased dramatically, with ~30–70 % reduction compared with those collected on the 15th day (Fig. 4D–F), and canola exuded the greatest amounts of citric acid, while barley exuded the greatest amounts of malic acid under P1 conditions; at P50, barley released significantly more succinic acid than canola and potato did (~3.2 nmol plant^{-1} h^{-1} in barley versus 0.6 nmol plant^{-1} h^{-1} in canola and 0.1 nmol plant^{-1} h^{-1} in potato).

Phosphorus status in shoots and roots

Unsurprisingly, the P concentrations and P contents in both shoot and root tissues were significantly higher (between 3 and 81 times) for plants supplied with P50 compared with those of plants supplied with P1 (Fig. 5; Table 2). Under P50 supply, the P concentrations and P contents in shoots decreased in the order barley > potato > canola (Fig. 5A and B), while in roots, the order was potato > barley > canola (Fig. 5C and D). The shoot and root P concentrations varied from 4 to 8 mg g^{-1} DW, comparing well with a P concentration in agricultural crops that generally varies from 1 to 5 mg P g^{-1} DW (Anonymous 1999). This shows that 50 μM was a P concentration sufficient for plant growth. Under P1 conditions, potato had the highest P concentrations and P contents in shoots, although no statistically significant difference was found in shoot P content. Canola had shoot P concentrations equal to those of barley, but it had a P content that was only 45 % as high due to having only 55 % as much shoot biomass (Figs 1A and 5A and B). In root tissues, potato and canola had almost the same P concentrations, which were 121 % higher ($P < 0.05$, Student's *t*-test) than in barley (Fig. 5C), and potato roots showed the greatest P content (0.06 mg versus 0.03 mg P in barley and canola), but these differences were not significant (Fig. 5D).

Discussion

Phosphorus availability and uptake directly affect crop productivity. Understanding the relationships between root architecture, exudates and P availability was thus the main objective of this study. To understand the complex adaptive responses of root morphology and root exudates to low P availability, we carried out hydroponic culture experiments on canola, barley and potato, three crop species with different root systems. Although a hydroponic culture system does not reflect natural growth conditions for plants compared with a soil experiment, it is widely used (Gahoonia *et al.* 2000; Dechassa and Schenk

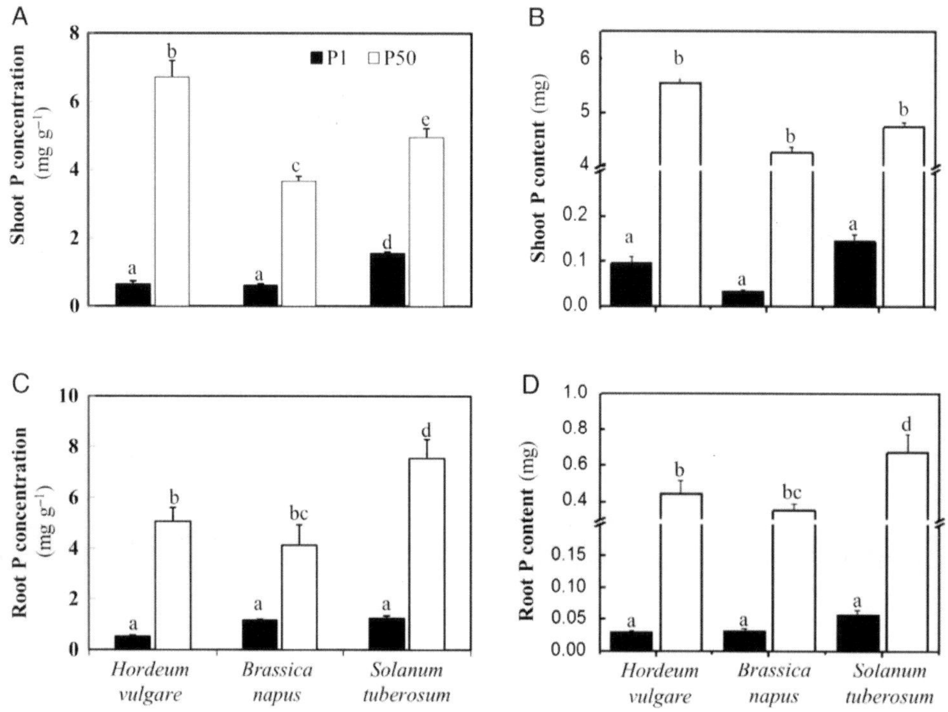

Figure 5. (A) Phosphorus concentration in shoot tissue, (B) P content in shoot tissue, (C) P concentration in root tissue and (D) P content in root tissue of *H. vulgare*, *B. napus* and *S. tuberosum* grown in nutrient solution at two levels of P supply (P1 = 1 μM P, P50 = 50 μM P). Plants were harvested after 28 days growth. Values are means ± SE of $n = 6-7$. Different letters indicate significant differences ($P < 0.05$) as shown by Tukey tests.

2004; Ligaba *et al.* 2004; Wang *et al.* 2013; Cheng *et al.* 2014) and useful for studying root architecture, including tiny root tips, and for collecting root exudates for biochemical analysis of OAs. Furthermore, the effects of plant–plant interactions were minimized in our system, since we grew a single plant per pot. Contrasting responses of root morphology and root-exuded OAs to low P availability in these three important food crops were revealed. These results could underpin future efforts to improve P uptake of the three crops.

The role of root morphology in improving P uptake

Root architecture plays an important role in P uptake. Previous studies have shown that at reduced P availability, most species allocate more biomass to roots and allocate root biomass in shallow soil horizons, as well as increase root length and develop more and longer root hairs and lateral roots; some even produce cluster roots, thereby promoting P uptake (Nielsen *et al.* 2001; Lambers *et al.* 2006, 2015; Brown *et al.* 2013). In this study, our results showed a pronounced increase in root/shoot biomass ratio for the three studied species (Fig. 1B) under low P supply, which corresponds well with previous reports (Hermans *et al.* 2006; Hammond and White 2008). However, we found some differences in root traits like root length, lateral root numbers and root surface area among different crops (Fig. 3): under low P availability, barley showed a 44 % reduction in root length, 58 % smaller root surface area and 21 % lower root tip number; canola showed a 37 % increase in root length and a 45 % reduction in root tips and potato showed a doubling of the number of root tips, compared with under P50 conditions. However, Steingrobe *et al.* (2001) observed enhanced root-length production at a low P supply in barley in a field experiment, which differs from our results. This could be due to the two different methodologies used (i.e. we used a hydroponic culture, whereas Steingrobe *et al.* performed field experiments involving a more complicated environment around the roots). Moreover, the effects of micropropagated plantlets on root systems are not clear so far.

Phosphorus uptake by plants is dependent on the surface area and length of the root system and on lateral roots to explore a large soil volume (Richardson *et al.* 2009; Balemi and Negisho 2012; Lambers *et al.* 2015). In our study, as we used KH_2PO_4 as P source, root morphology plays a dominant role in P acquisition. At P50, barley had greater shoot P concentrations and P contents than canola and potato (Fig. 5A and B), probably due to its significantly longer total root length and greater root surface area, enabling greater P uptake (Fig. 3A and B), because these three crops had the same root tip number (Fig. 3C). Another explanation is its smaller biomass and higher root/shoot ratio (Fig. 1A and B), resulting in higher shoot

P concentrations compared with those of canola and potato. On the other hand, potato showed a significantly greater root P concentration and P content than barley and canola did at P50 (Fig. 5C and D), suggesting that potato roots may have a greater P uptake capacity than barley and canola. Further studies are needed to reveal the mechanisms for this. In addition, under low P supply, canola roots showed greater root length and larger root surface area than barley and potato, while potato showed a remarkable increase in root tips compared with that at P50, and hence had greater root P concentrations and contents than canola and barley (Fig. 3). This also suggests that root tips had a major effect on improving P uptake, as found by Fitter *et al.* (2002). However, no significant differences were found in root P concentrations and P contents under P1, which could be due to the very limited P supply. Our results for potato root architecture agree with those of McArthur and Knowles (1993*a*), who reported that root growth in potato is less influenced by P deficiency than either leaf or stem growth. Furthermore, they found a greater root colonization level of arbuscular mycorrhizal fungi for P-stressed potato plants, and P uptake by roots was enhanced by different kinds of arbuscular mycorrhizal fungi at all levels of P supply (McArthur and Knowles 1993*b*). Brundrett (2002) reviewed the major influence that root morphology characteristics such as cortex cell properties and root features like root length/biomass ratio, root branching and root hairs have on mycorrhiza formation. Highly mycorrhiza-dependent plants tend to have coarser root systems and would not be as responsive to changes in nutrient availability. Hence, these root traits of potato, especially abundant root tips, might favour mycorrhizal colonization.

The possible role of root exudation in low-P responses

Root exudation of OAs is considered as an important mechanism to mobilize P sources and alleviate P starvation, because OAs can replace organic and inorganic P that is bound to soil particles (Lambers *et al.* 2006). The fact that those plants that release more OAs could take up more P from growth media and soil has also been confirmed using genetically modified plants (Lü *et al.* 2012; Wang *et al.* 2013). However, root exudates are influenced by many factors (Dechassa and Schenk 2004). We used the most suitable and effective trap solution—water—to collect the exudates (Valentinuzzi *et al.* 2015), but we could not completely avoid some microbial breakdown, so the root-exuded OAs were possibly somewhat underestimated in our data (Kuijken *et al.* 2015).

Great differences in OA exudates were found among species in our system (Fig. 4). For canola, the absolute rate of citric acid exudation corresponded well with that reported by Hoffland *et al.* (1989). Furthermore, an increase in citric and malic acids and a decrease in succinic acid were found for canola under P starvation, which compares well with the results of Hoffland *et al.* (1989), except that malic acid was the dominant OA. On the other hand, no OAs were induced by P deficiency from canola roots, as reported by Ligaba *et al.* (2004) using the same system.

For barley, citric, acetic and fumaric acids were detected in root exudates of two cultivars under P deficiency by Gahoonia *et al.* (2000) using a hydroponic culture, but they only showed a difference between two cultivars; no information about the difference between P supply and P deficiency was given. Our results showed no significant difference in citric acid exudation in barley between P-sufficient and P-deficient plants, but there was an increase in malic acid exudation in the P deficiency treatment (Fig. 4A, B and E), thus showing that there are differences in the types and quantities of OAs that are released from different crop species and varieties in response to P starvation.

For potato, few references about OA exudates can be found in the literature, but the increase in succinic acid exudation in response to P deficiency was in accordance with the results of Dechassa and Schenk (2004), while the absolute value in our results was lower ($\sim$0.002 versus 0.36 nmol (cm root)$^{-1}$ h^{-1}). The reason for this difference is unclear.

The decrease in the OA amounts released by roots of P-stressed plants after 4 weeks (Fig. 4D–F) suggests that root exudation was influenced by plant age (Dechassa and Schenk 2004; Badri and Vivanco 2009).

Citric acid has the strongest ability to release soil P (Ryan *et al.* 2001), malic acid is $\sim$10 times less effective than citric acid in mobilizing soil P and succinic acid complexes metal cations only very weakly and, hence, has only a relatively weak ability to release soil P (Nagarajah *et al.* 1970; Jones and Darrah 1995). Therefore, canola, with its relatively rapid exudation of citric acid, is likely to have the greatest ability among the investigated crops to mobilize insoluble soil P. These results provide good evidence to support our Hypothesis (1), but the effect of root exudates on P uptake could not be assessed in this study, because P was only added to the system as aqueous KH_2PO_4. Further investigation is required to reveal the effect of root exudates (OAs) on P mobilization and uptake from soil.

Conclusions

Using a hydroponic culture system and various analytical methods, we have tested our hypotheses that canola, a dicot oilseed crop; potato, a tuber-producing dicot and barley, a monocot, respond differently under various levels of P availability. Our results revealed that the non-mycorrhizal species, canola, showed rapid rates of carboxylic acid exudation, which, together with its greater root length and root surface area, could allow canola to acquire poorly available P forms and explore a larger soil volume. The monocot and mycorrhizal species barley showed a reduction in root length and root surface area as well as low amounts of root exudates; in soil, this would result in a low P-uptake capacity. Potato released only trace amounts of root exudates but produced double the number of root tips under low-P conditions, which would benefit its P uptake in soil. Hence, our study indicates that plants evolved divergent adaptations of root morphology and exudation as a response to low P availability. The results and information generated in this study are valuable for future effective utilization of P and improving the productivity of the three important crops.

Sources of Funding

This study was supported by the core funding of the strategic institute program on 'Opportunities for sustainable use of phosphorus in food production' at the Norwegian Institute of Bioeconomy Research.

Contributions by the Authors

Y.-L.W., N.C. and J.L.C. designed the experiment. Y.-L.W. and M.A. conducted the experiment. All authors contributed to analysis of data, discussions and manuscript writing.

Acknowledgements

We thank Sissel Haugslien, Jan Erik Jacobsen, Monica Fongen and Toril Drabløs Eldhuset for their valuable help with seed collection, phosphorus determinations, WinRHIZO equipment and manuscript comments.

Literature cited

Andersson H, Bergström L, Djodjic F, Ulén B, Kirchmann H. 2013. Topsoil and subsoil properties influence phosphorus leaching from four agricultural soils. *Journal of Environmental Quality* **42**: 455–463.

Anonymous. 1999. Functions of phosphorus in plants. *Better Crops* **83**:6–7.

Badri DV, Vivanco JM. 2009. Regulation and function of root exudates. *Plant, Cell and Environment* **32**:666–681.

Balemi T, Negisho K. 2012. Management of soil phosphorus and plant adaptation mechanisms to phosphorus stress for sustainable crop production: a review. *Journal of Soil Science and Plant Nutrition* **12**:547–562.

Brown LK, George TS, Dupuy LX, White PJ. 2013. A conceptual model of root hair ideotypes for future agricultural environments: what combination of traits should be targeted to cope with limited P availability? *Annals of Botany* **112**:317–330.

Brundrett MC. 2002. Coevolution of roots and mycorrhizas of land plants. *New Phytologist* **154**:275–304.

Brundrett MC. 2009. Mycorrhizal associations and other means of nutrition of vascular plants: understanding the global diversity of host plants by resolving conflicting information and developing reliable means of diagnosis. *Plant and Soil* **320**:37–77.

Cheng L, Tang X, Vance CP, White PJ, Zhang F, Shen J. 2014. Interactions between light intensity and phosphorus nutrition affect the phosphate-mining capacity of white lupin (*Lupinus albus* L.). *Journal of Experimental Botany* **65**:2995–3003.

Dechassa N, Schenk MK. 2004. Exudation of organic anions by roots of cabbage, carrot, and potato as influenced by environmental factors and plant age. *Journal of Plant Nutrition and Soil Science* **167**:623–629.

Desnos T. 2008. Root branching responses to phosphate and nitrate. *Current Opinion in Plant Biology* **11**:82–87.

Erro J, Zamarreño AM, Yvin J-C, Garcia-Mina JM. 2009. Determination of organic acids in tissues and exudates of maize, lupin, and chickpea by high-performance liquid chromatography–tandem mass spectrometry. *Journal of Agricultural and Food Chemistry* **57**:4004–4010.

Fitter A, Williamson L, Linkohr B, Leyser O. 2002. Root system architecture determines fitness in an *Arabidopsis* mutant in competition for immobile phosphate ions but not for nitrate ions. *Proceedings of the Royal Society of London B: Biological Sciences* **269**:2017–2022.

Gahoonia TS, Asmar F, Giese H, Gissel-Nielsen G, Nielsen NE. 2000. Root-released organic acids and phosphorus uptake of two barley cultivars in laboratory and field experiments. *European Journal of Agronomy* **12**:281–289.

Gerke J, Römer W, Jungk A. 1994. The excretion of citric and malic acid by proteoid roots of *Lupinus albus* L.; effects on soil solution concentrations of phosphate, iron, and aluminum in the proteoid rhizosphere in samples of an oxisol and a luvisol. *Zeitschrift für Pflanzenernährung und Bodenkunde* **157**:289–294.

Hammond JP, White PJ. 2008. Sucrose transport in the phloem: integrating root responses to phosphorus starvation. *Journal of Experimental Botany* **59**:93–109.

Hermans C, Hammond JP, White PJ, Verbruggen N. 2006. How do plants respond to nutrient shortage by biomass allocation? *Trends in Plant Science* **11**:610–617.

Hoffland E, Findenegg GR, Nelemans JA. 1989. Solubilization of rock phosphate by rapeseed II. Local root exudation of organic acids as a response to P-starvation. *Plant and Soil* **113**:161–165.

Jones DL, Darrah PR. 1995. Influx and efflux of organic acids across the soil-root interface of *Zea mays* L. and its implications in rhizosphere C flow. *Plant and Soil* **173**:103–109.

Khorassani R, Hettwer U, Ratzinger A, Steingrobe B, Karlovsky P, Claassen N. 2011. Citramalic acid and salicylic acid in sugar beet root exudates solubilize soil phosphorus. *BMC Plant Biology* **11**:121.

Kuijken RCP, Snel JFH, Heddes MM, Bouwmeester HJ, Marcelis LFM. 2015. The importance of a sterile rhizosphere when phenotyping for root exudation. *Plant and Soil* **387**:131–142.

Lambers H, Teste FP. 2013. Interactions between arbuscular mycorrhizal and non-mycorrhizal plants: do non-mycorrhizal species at both extremes of nutrient availability play the same game? *Plant, Cell and Environment* **36**:1911–1915.

Lambers H, Shane MW, Cramer MD, Pearse SJ, Veneklaas EJ. 2006. Root structure and functioning for efficient acquisition of phosphorus: matching morphological and physiological traits. *Annals of Botany* **98**:693–713.

Lambers H, Raven JA, Shaver GR, Smith SE. 2008. Plant nutrient-acquisition strategies change with soil age. *Trends in Ecology and Evolution* **23**:95–103.

Lambers H, Ahmedi I, Berkowitz O, Dunne C, Finnegan PM, Hardy GESJ, Jost R, Laliberté E, Pearse SJ, Teste FP. 2013. Phosphorus nutrition of phosphorus-sensitive Australian native plants: threats to plant communities in a global biodiversity hotspot. *Conservation Physiology* **1**: doi:10.1093/conphys/cot010.

Lambers H, Clode P, Hawkins H, Laliberté E, Oliveira R, Reddell P, Shane M, Stitt M, Weston P. 2015. *Metabolic adaptations of the non-mycotrophic Proteaceae to soil with a low phosphorus availability: phosphorus metabolism in plants in the post-genomic era: from gene to ecosystem*. Oxford, UK: Wiley-Blackwell.

Ligaba A, Shen H, Shibata K, Yamamoto Y, Tanakamaru S, Matsumoto H. 2004. The role of phosphorus in aluminium-induced citrate and malate exudation from rape (*Brassica napus*). *Physiologia Plantarum* **120**:575–584.

López-Bucio J, Nieto-Jacobo MF, Ramírez-Rodríguez V, Herrera-Estrella L. 2000. Organic acid metabolism in plants: from adaptive physiology to transgenic varieties for cultivation in extreme soils. *Plant Science* **160**:1–13.

López-Bucio J, Hernández-Abreu E, Sánchez-Calderón L, Nieto-Jacobo MF, Simpson J, Herrera-Estrella L. 2002. Phosphate availability alters architecture and causes changes in hormone sensitivity in the *Arabidopsis* root system. *Plant Physiology* **129**:244–256.

Lü J, Gao X, Dong Z, Yi J, An L. 2012. Improved phosphorus acquisition by tobacco through transgenic expression of mitochondrial malate dehydrogenase from *Penicillium oxalicum*. *Plant Cell Reports* **31**:49–56.

Lynch JP. 2011. Root phenes for enhanced soil exploration and phosphorus acquisition: tools for future crops. *Plant Physiology* **156**: 1041–1049.

McArthur DA, Knowles NR. 1993*a*. Influence of vesicular-arbuscular mycorrhizal fungi on the response of potato to phosphorus deficiency. *Plant Physiology* **101**:147–160.

McArthur DA, Knowles NR. 1993*b*. Influence of species of vesicular-arbuscular mycorrhizal fungi and phosphorus nutrition on growth, development, and mineral nutrition of potato (*Solanum tuberosum* L.). *Plant Physiology* **102**:771–782.

Nagarajah S, Posner AM, Quirk JP. 1970. Competitive adsorption of phosphate with polygalacturonate and other organic anions on kaolinite and oxide surfaces. *Nature* **228**:83–85.

Neumann G, Römheld V. 2001. The release of root exudates as affected by plant's physiological status. In: Pinton R, Varanini Z, Nannipieri P, eds. *The rhizosphere—biochemistry and organic substances at the soil-plant interface*. New York: Marcel Decker, 41–93.

Neumann G, Bott S, Ohler MA, Mock H-P, Lippmann R, Grosch R, Smalla K. 2014. Root exudation and root development of lettuce (*Lactuca sativa* L. cv. Tizian) as affected by different soils. *Frontiers in Microbiology* **5**:2.

Nielsen KL, Eshel A, Lynch JP. 2001. The effect of phosphorus availability on the carbon economy of contrasting common bean (*Phaseolus vulgaris* L.) genotypes. *Journal of Experimental Botany* **52**:329–339.

Ogner G, Wickstrøm T, Remedios G, Gjelsvik S, Hensel GR, Jacobsen JE, Olsen M, Skretting E, Sørlie B. 1999. *The chemical analysis program of the Norwegian Forest Research Institute 2000*. Ås: Norwegian Forest Research Institute.

Prieto I, Armas C, Pugnaire FI. 2012. Water release through plant roots: new insights into its consequences at the plant and ecosystem level. *New Phytologist* **193**:830–841.

Raghothama KG. 1999. Phosphate acquisition. *Annual Review of Plant Physiology and Plant Molecular Biology* **50**:665–693.

Read DB, Bengough AG, Gregory PJ, Crawford JW, Robinson D, Scrimgeour CM, Young IM, Zhang K, Zhang X. 2003. Plant roots release phospholipid surfactants that modify the physical and chemical properties of soil. *New Phytologist* **157**: 315–326.

Richardson AE, Hocking PJ, Simpson RJ, George TS. 2009. Plant mechanisms to optimise access to soil phosphorus. *Crop and Pasture Science* **60**:124–143.

Ryan PR, Delhaize E, Jones DL. 2001. Function and mechanism of organic anion exudation from plant roots. *Annual Review of Plant Physiology and Plant Molecular Biology* **52**:527–560.

Shen J, Yuan L, Zhang J, Li H, Bai Z, Chen X, Zhang W, Zhang F. 2011. Phosphorus dynamics: from soil to plant. *Plant Physiology* **156**: 997–1005.

Smith SE, Smith FA, Jakobsen I. 2003. Mycorrhizal fungi can dominate phosphate supply to plants irrespective of growth responses. *Plant Physiology* **133**:16–20.

Smith VH, Schindler DW. 2009. Eutrophication science: where do we go from here? *Trends in Ecology and Evolution* **24**: 201–207.

Steingrobe B, Schmid H, Claassen N. 2001. Root production and root mortality of winter barley and its implication with regard to phosphate acquisition. *Plant and Soil* **237**:239–248.

Valentinuzzi F, Cesco S, Tomasi N, Mimmo T. 2015. Influence of different trap solutions on the determination of root exudates in *Lupinus albus* L. *Biology and Fertility of Soils* **51**:757–765.

Wang Y, Xu H, Kou J, Shi L, Zhang C, Xu F. 2013. Dual effects of transgenic *Brassica napus* overexpressing CS gene on tolerances to aluminum toxicity and phosphorus deficiency. *Plant and Soil* **362**:231–246.

Williamson LC, Ribrioux SPCP, Fitter AH, Leyser HMO. 2001. Phosphate availability regulates root system architecture in *Arabidopsis*. *Plant Physiology* **126**:875–882.

Withers PJA, Edwards AC, Foy RH. 2001. Phosphorus cycling in UK agriculture and implications for phosphorus loss from soil. *Soil Use and Management* **17**:139–149.

4

Carbon dioxide stimulation of photosynthesis in *Liquidambar styraciflua* is not sustained during a 12-year field experiment

Jeffrey M. Warren[1*], Anna M. Jensen[1], Belinda E. Medlyn[2], Richard J. Norby[1] and David T. Tissue[3]

[1] Climate Change Science Institute and Environmental Sciences Division, Oak Ridge National Laboratory, Oak Ridge, TN 37831-6301, USA

[2] School of Biological Sciences, Macquarie University, Sydney, NSW 2019, Australia

[3] Hawkesbury Institute for the Environment, University of Western Sydney, Richmond, NSW 2753, Australia

Associate Editor: Jeffrey S. Dukes

Abstract. Elevated atmospheric CO_2 (eCO_2) often increases photosynthetic CO_2 assimilation (A) in field studies of temperate tree species. However, there is evidence that A may decline through time due to biochemical and morphological acclimation, and environmental constraints. Indeed, at the free-air CO_2 enrichment (FACE) study in Oak Ridge, Tennessee, A was increased in 12-year-old sweetgum trees following 2 years of ~40 % enhancement of CO_2. A was re-assessed a decade later to determine if the initial enhancement of photosynthesis by eCO_2 was sustained through time. Measurements were conducted at prevailing CO_2 and temperature on detached, re-hydrated branches using a portable gas exchange system. Photosynthetic CO_2 response curves (A versus the CO_2 concentration in the intercellular air space (C_i); or A–C_i curves) were contrasted with earlier measurements using leaf photosynthesis model equations. Relationships between light-saturated photosynthesis (A_{sat}), maximum electron transport rate (J_{max}), maximum Rubisco activity (V_{cmax}), chlorophyll content and foliar nitrogen (N) were assessed. In 1999, A_{sat} for eCO_2 treatments was 15.4 ± 0.8 $\mu mol\ m^{-2}\ s^{-1}$, 22 % higher than aCO_2 treatments ($P < 0.01$). By 2009, A_{sat} declined to <50 % of 1999 values, and there was no longer a significant effect of eCO_2 ($A_{sat} = 6.9$ or 5.7 ± 0.7 $\mu mol\ m^{-2}\ s^{-1}$ for eCO_2 or aCO_2, respectively). In 1999, there was no treatment effect on area-based foliar N; however, by 2008, N content in eCO_2 foliage was 17 % less than that in aCO_2 foliage. Photosynthetic N-use efficiency (A_{sat} : N) was greater in eCO_2 in 1999 resulting in greater A_{sat} despite similar N content, but the enhanced efficiency in eCO_2 trees was lost as foliar N declined to sub-optimal levels. There was no treatment difference in the declining linear relationships between J_{max} or V_{cmax} with declining N, or in the ratio of $J_{max} : V_{cmax}$ through time. Results suggest that the initial enhancement of photosynthesis to elevated CO_2 will not be sustained through time if N becomes limited.

Keywords: Acclimation; down-regulation; free-air CO_2 enrichment; nitrogen limitation; sweetgum.

* Corresponding author's e-mail address: warrenjm@ornl.gov

Introduction

In 2013, annual atmospheric CO_2 concentration exceeded 396 ppm at Mauna Loa, 2.6 ppm greater than in 2012 and 25 % greater than the initial measurements in 1959 (Keeling *et al.* 2014). Global atmospheric model simulations estimate that this trajectory will continue (IPCC 2013), and thus terrestrial plant communities will remain exposed to increasing CO_2 for the foreseeable future. Atmospheric CO_2 is a potent climate-forcing agent contributing to increased atmospheric temperatures, and ecosystem balance between plant CO_2 uptake through photosynthesis and release through respiratory activity remains a key uncertainty in mechanistic Earth system models that project ecosystem feedbacks to the atmosphere. To address ecosystem response to elevated CO_2 (eCO_2), various plant species have been exposed to air enriched with CO_2 in short-term open- or closed-topped field chambers. Results suggested substantial eCO_2 stimulation of photosynthesis (Luxmoore *et al.* 1993; Gunderson and Wullschleger 1994; Medlyn *et al.* 1999; Norby *et al.* 1999) across various woody species (e.g. in oak, yellow poplar (Gunderson *et al.* 1993), pine (Tissue *et al.* 1997), eucalypts (Ghannoum *et al.* 2010) and tropical seedlings (Ziska *et al.* 1991)). In other studies, eCO_2 induced significant down-regulation and loss of photosynthetic capacity (e.g. in Arctic tundra grasses and shrubs (Tissue and Oechel 1987; Oechel *et al.* 1994), beech (Epron *et al.* 1996) and spruce (Marek *et al.* 1995)), indicating that interaction with soil resources, longer term feedbacks and progressive plant acclimation remained a key uncertainty (Curtis and Wang 1998). New, larger experiments using free-air CO_2 enrichment (FACE) technology (Norby *et al.* 2001) have allowed field crops, grasses and, in particular, woody forest vegetation to be exposed to eCO_2 over many years. Such studies allow distinction of long-term responses from transient responses due to leaf plasticity, stand development or inter-annual environmental variability.

Similar to the response of plants in chambers, an initial increase in net photosynthesis (*A*) and net primary productivity is common across woody plant species exposed to FACE eCO_2 treatments (Norby *et al.* 2005; Ainsworth and Rogers 2007), including in *Liquidambar styraciflua* (Sholtis *et al.* 2004), *Pinus taeda* (Crous and Ellsworth 2004), *Populus × euramericana*, *Populus alba*, *P. nigra* (Bernacchi et *al.* 2003; Liberloo et *al.* 2007), *P. tremuloides* (Noormets *et al.* 2010), *Fagus sylvatica*, *Quercus petraea*, *Carpinus betulus*, *Acer campestre* and *Tilia platyphyllos* (Bader *et al.* 2010). However, longer term eCO_2 field results are mixed, with some experiments reporting feedbacks such as acclimation and photosynthetic down-regulation through time (Rey and Jarvis 1998; Medlyn *et al.* 1999; Kubiske *et al.* 2002; Bernacchi *et al.* 2003; Crous *et al.* 2008) and others not (Bader *et al.* 2010; Darbah *et al.* 2010).

Leaves typically adjust their Rubisco activity or content to the prevailing CO_2 concentration, i.e. down-regulation by elevated CO_2 (Tissue *et al.* 1993; Drake *et al.* 1997; Stitt and Krapp 1999; Rogers and Ellsworth 2002). Since eCO_2 increases the efficiency of Rubisco, *A* can be maintained or enhanced despite reductions in enzyme content, activity or maximum photosynthetic capacity. This implies greater photosynthetic N-use efficiency (PNUE) as displayed across C3 woody plant species exposed to elevated CO_2 (Peterson *et al.* 1999; Calfapietra *et al.* 2007).

Sweetgum (*L. styraciflua*) is a common temperate North American tree species, and is the dominant canopy species at the ORNL-FACE site in TN, USA, and the primary mid-canopy species at the Duke-FACE site in NC, USA. Prior results from these sites indicate a strong and consistent enhancement of light-saturated photosynthesis (A_{sat}) by elevated CO_2; the ~40 % increase in atmospheric CO_2 resulted in 45 % enhancement of A_{sat} at the ORNL-FACE site after 3 years of treatments (Gunderson *et al.* 2002; Sholtis *et al.* 2004) and consistently >50 % enhancement of A_{sat} at the Duke-FACE site through 6 years of treatments (Herrick and Thomas 1999, 2003; Springer *et al.* 2005). After 3 years of treatment at ORNL-FACE there was no change in maximum photosynthetic capacity (A_{max}), maximum electron transport rate (J_{max}) or maximum carboxylation rate (V_{cmax}) when foliage from both treatments was measured at the same CO_2 concentrations (Sholtis *et al.* 2004). Elevated CO_2 did reduce foliar N content and increase soluble carbohydrates and leaf mass per area (LMA, Herrick and Thomas 2003; Sholtis *et al.* 2004)—results consistent across time for sweetgum in both FACE studies. After 6 years of eCO_2 treatment at ORNL-FACE, there was still no significant difference in A_{sat} when both treatments were measured at 400 ppm (Monson *et al.* 2007), indicating little eCO_2 down-regulation of photosynthesis.

After 12 years of eCO_2 treatment at the ORNL-FACE research study, we re-assessed photosynthetic capacity of the dominant sweetgum trees to determine if the early enhancement of photosynthesis during the years 1–3 (Gunderson *et al.* 2002; Sholtis *et al.* 2004) was sustained after an additional 9–10 years of treatment.

Over the 12-year study, the canopy spread upward and there was expansion of the stand diameter distribution as dominant individual trees maintained growth, while suppressed individuals stagnated and began to die. In addition, soil N availability decreased across the site, with the greatest rates of decline in the eCO_2 plots (Garten *et al.* 2011). Decreased site nutrient availability was reflected by persistent inter-annual reductions in foliar N content and net primary production (NPP) (Norby *et al.* 2010), and a shift in carbon (C) allocation belowground to roots (Norby *et al.* 2004). The reductions in NPP through time were greatest for eCO_2 plots despite treatment-specific increases or decreases in canopy leaf area in response to inter-annual variability such as drought (Warren *et al.* 2011*b*), or the substantial reduction in eCO_2 site water use (Warren *et al.* 2011*a*, *b*). In addition to increased soil moisture availability in eCO_2 plots, the increase in root production should lead to an enhanced capacity for soil nutrient extraction. Yet canopy N content continued to decline through time at a greater rate for eCO_2 plots than for aCO_2 plots, which correlated with a greater reduction in NPP for eCO_2 plots (Norby *et al.* 2010).

The main goal of this project was to determine if the initial enhancement of photosynthesis by eCO_2 was sustained through time, and if responses were linked to the progressive decline in site resource availability (Garten *et al.* 2011), foliar N content (Norby *et al.* 2010) and shift in internal plant C allocation (Norby *et al.* 2004). In 2008–09, we re-measured gas exchange in sweetgum leaves and compared the results with data from previous gas exchange campaigns conducted across seasons from 1998 to 2000 (Gunderson *et al.* 2002; Sholtis *et al.* 2004); specifically the mid-summer 1999 dataset was used as it was the most comprehensive dataset and overlapped with the same period sampled in 2008–09 (from late July to early August). We hypothesized that (i) declining N availability would be reflected in reduced photosynthetic capacity through time and (ii) when photosynthesis is N-limited, eCO_2 leaves maintain greater RuBP-regeneration capacity, at lower Rubisco carboxylation rates, compared with aCO_2 leaves (e.g. the slope of $J_{max} : V_{cmax}$ is steeper in aCO_2 leaves). Finally, we evaluated treatment effects on coupling between stomatal conductance and C-assimilation.

Methods

Site description and CO_2 treatments

The study site was a 20-year-old sweetgum (*L. styraciflua* L.) plantation forest in Oak Ridge National Environmental Research Park in eastern TN, USA (35°54′N; 84°20′W). The soil was an Aquic Hapludult with a silty clay-loam texture. A free-air CO_2 enrichment system (Hendrey *et al.* 1999) was installed at the site in four 25-m-diameter plots in 1996. The FACE system regulated the release of CO_2 from vertical PVC pipes located in a ring around each plot based on wind speed, wind direction and in-situ measurements of current CO_2 concentration within the canopy. From 1998 to 2009, CO_2 was released into two FACE rings during each growing season to a target [CO_2] of 560 ppm. Two FACE rings received ambient [CO_2] and one additional ring, without FACE infrastructure, was established to serve as a third control plot. Actual CO_2 enrichment varied through time as ambient concentrations rose incrementally and through modifications in the release regime, which resulted in a mean 40 % increase over ambient CO_2. In 2008, tree height ranged from 10 to 24 m (mean $= 18.2 \pm 3.4$ m (± 1 SD)), median tree diameter was 14.6 ± 4.1 cm (± 1 SD) and peak leaf area index (mid-July) was 4.1 ± 0.2. Mean growing season (April–October) temperature was 19.6 °C in 2008 and 19.1 °C in 2009, and growing season precipitation was 440 mm in 2008 and 511 mm in 2009 (Riggs *et al.* 2010). The site, experimental design and FACE apparatus have been previously described (Norby *et al.* 2001; Warren *et al.* 2011*b*) and results including micrometeorological data have been archived for public use: http://cdiac.ornl.gov/ftp/FACE/ornldata.

Photosynthesis

For each measurement campaign (2008–09), branches were collected early in the morning from mid-canopy (2009) and fully exposed upper canopy (2008, 2009) dominant or co-dominant trees within each treatment plot. Canopy access was achieved via the FACE infrastructure towers, which provided access to trees within ~8 m from the towers. Trees were selected primarily from within the treatment plots (i.e. >2.5 m from the ring edge); however, in ambient plots, several buffer trees were also sampled. On measurement days, 1–2 m long branches were cut from the mid- and upper canopy of selected trees with a pole pruner and quickly placed into plastic bags with wet paper towels to minimize desiccation. Branches were re-cut under water to remove potential embolism induced during removal from the trees. Most *L. styraciflua* vessel length is <0.3 m, therefore >0.5 m of the branch was removed. Prior measurements of gas exchange on severed or attached branches at the site found no differences in measurements over a 2-h period (Tissue *et al.* 2002; Monson *et al.* 2007). Gas exchange was measured using four portable photosynthesis systems (LI-6400XT, LI-COR, Lincoln, NE, USA) during late July 2008 and early August 2009. Measurements were conducted outside under ambient conditions on one or two leaves per branch (total samples $n = 16$–23 (2008); $n = 29$–31 (2009)). Conditions in the gas exchange cuvette were set to approximate ambient outside temperature: relative humidity was ~50–80 %, photosynthetically active radiation was 1800 $\mu mol\ m^{-2}\ s^{-1}$ and CO_2 concentration was initially set at 400 ppm prior to the A–C_i curve measurements (see below). Foliage was retained for the analysis of N, chlorophyll and LMA. Similar raw datasets were available from previous work performed at the site in late July–August 1999 (Gunderson *et al.* 2002; Sholtis *et al.* 2004). At that time, photosynthesis was measured in situ using vertical man lifts. Upper canopy photosynthetic and biochemical datasets were combined across years (1999, 2008, 2009) to investigate shifts in response through time.

In addition to measurement of the light-saturated photosynthetic rate (A_{sat}) at the CO_2 concentration in which the trees were growing, response curves of assimilation versus the CO_2 concentration within the intercellular spaces of the leaf (C_i), or $A-C_i$ curves, were measured in all years. These response curves were used to estimate maximum electron transport rate, J_{max}, and maximum Rubisco activity, V_{cmax}, using a consistent set of leaf photosynthesis model equations (Medlyn *et al.* 2002). In 2008 and 2009, curves were conducted through an initially declining, then increasing, reference CO_2 regime (400, 300, 200, 100, 50, 400, 550, 700, 900, 1200, 1600 ppm). This contrasts slightly with curves conducted in 1999 using a declining CO_2 regime (1500, 1200, 960, 760, 560, 360, 250, 175, 100, 50, 0 ppm). To contrast gas exchange responses at growth CO_2, results were assessed at 400 or 550 ppm for aCO_2 or eCO_2, respectively. Since those CO_2 concentrations were not part of the 1999 campaigns, values at 400 or 550 ppm were interpolated from the linear portion of each response curve ($R^2 > 0.99$) for comparison. Atmospheric CO_2 varied between 1998 and 2008 due to incremental ambient CO_2 increases and FACE performance and management; range (384–405 or 528–560 ppm CO_2) for aCO_2 or eCO_2, respectively. Relationships between J_{max}, V_{cmax}, stomatal conductance (g_s), foliar N, chlorophyll content and CO_2 treatments were assessed.

Foliar biochemistry and LMA

To assess foliar chlorophyll and N concentrations, and LMA, nine leaf discs (9 mm diameter) were collected from the same area of each leaf (avoiding midrib) used for gas exchange measurements (2008–09). For chlorophyll analysis, two leaf discs were immediately placed in scintillation vials containing 5 mL of *N,N*-dimethylformamide (DMF) and extracted at 4 °C in the dark until analysis. Following extraction total chlorophyll, chlorophyll *a* and chlorophyll *b* were determined based on spectroscopy at 647 and 665 nm (Inskeep and Bloom 1985), and results contrasted with previous chlorophyll analysis in 1999 using an ethanol extraction (Sholtis *et al.* 2004). The two extraction solvents have been shown to provide comparable estimates of total chlorophyll in 9 of 11 tree species studied (Minocha *et al.* 2009). However, paired comparisons at our lab with sweetgum found that while DMF was able to fully extract chlorophyll within 1 day, 95 % ethanol at room temperature took up to 7 days for full extraction, and even then yielded only 80 % of the total chlorophyll yielded by DMF. Chlorophyll was thus likely underestimated in the earlier studies, and results are discussed in this context. Foliar N concentration was assessed for oven-dry (70 °C) discs using an elemental analyser (Costech Analytical Technologies, Inc., Valencia, CA, USA). In 1999, N content was measured on whole leaves, but not leaf discs, as used in 2008–09. Thus, comparison of N content across years required an LMA adjustment from whole leaf to leaf disc ($LMA_{disk} = 0.9635(LMA_{total}) + 0.7013$ (mg cm^{-2}); $n = 28$; $R^2 = 0.66$; based on 2008 foliage). All N content is provided on a disc basis.

Statistics

Regression slopes and treatment differences in photosynthesis, chlorophyll content, V_{cmax} and J_{max} were analysed using *t*-tests and analysis of variance techniques. There were no significant plot (ring) effects on chlorophyll, N, J_{max} or V_{cmax} measured in 1999 or later years ($0.1 < P < 0.9$). As such, individual leaves were considered as the (pseudo-replicated) experimental unit (df = 37–60 per treatment per year) as opposed to use of treatment rings (df = 1–2). This is common for analyses of these large, expensive studies where true replication is limited, and results should be considered in this context; i.e. there is a greater chance to detect *spurious* treatment effects (Type I errors) (Hurlbert 1984). Data manipulation and statistical procedures were completed using SAS statistical software (ver. 9.1.3, SAS Institute, Cary, NC, USA) and SigmaPlot (ver. 11.1, Systat Software, Inc., San Jose, CA). Relative differences between treatments and significance levels (*P* values) are presented; $P < 0.05$ indicates statistical significance.

Results

There were significant changes in photosynthetic rates and the photosynthetic response to eCO_2 over time. In 1999, light-saturated photosynthesis (A_{sat}) was significantly greater in eCO_2 than in aCO_2 foliage, but when re-measured in 2008, A_{sat} was similar for the two treatments. In 1999, A_{sat} was 15.4 μmol m^{-2} s^{-1} for eCO_2 foliage, 22 % greater than for aCO_2 foliage ($P < 0.01$; df = 60) (Fig. 1A). In both CO_2 treatments, A_{sat} declined through time to 50 % of the 1999 rates by 2008 and to 45 % of the 1999 rates by 2009. In 2008 and 2009, there was no longer a significant difference in A_{sat} ($P = 0.27$, df = 37 (2008); $P = 0.17$; df = 56 (2009)) (Fig. 1A). There was no significant difference between treatments when A_{sat} was measured at the same CO_2 concentrations, although measured A_{sat} tended to be lower for eCO_2 foliage than for aCO_2 foliage. In 1999, measured A_{sat} was 13 % lower for eCO_2 foliage than for aCO_2 foliage when both were measured at 550 ppm CO_2 ($P = 0.07$; df = 60) and 12 % lower when both were measured at 400 ppm CO_2 ($P = 0.09$) (Fig. 1B). In 2008, measured A_{sat} was 27 %

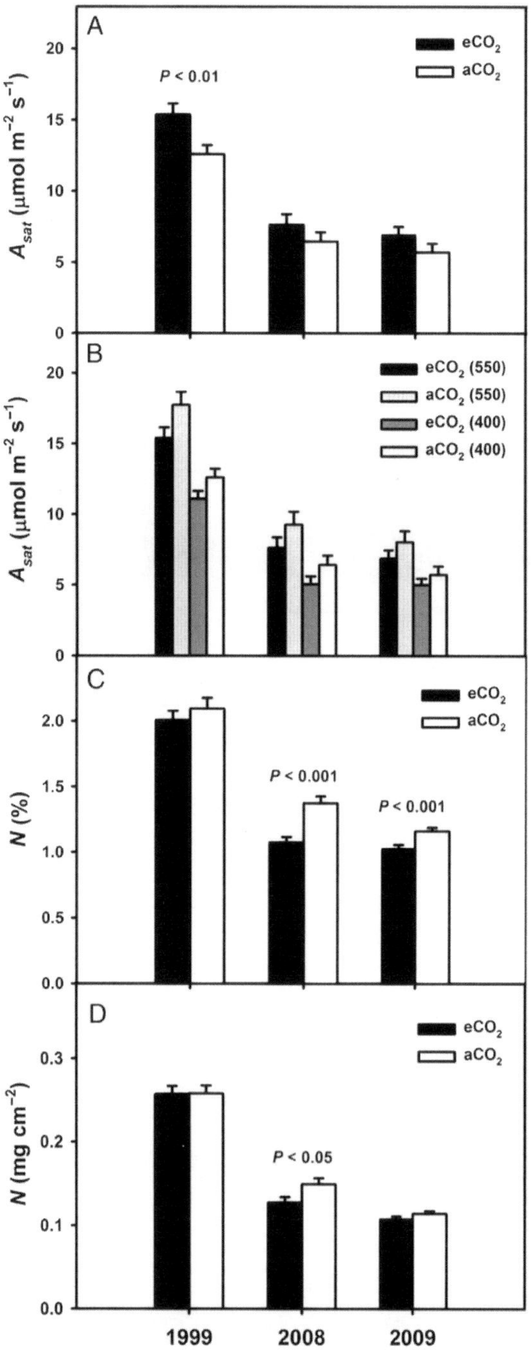

Figure 1. (A) Light-saturated photosynthesis (A_{sat}) at growth CO_2 (400 or 550 ppm for ambient (aCO_2) and elevated (eCO_2) CO_2 treatments, respectively) for mature upper canopy sweetgum foliage in mid-summer 1999, 2008 and 2009 as derived from $A-C_i$ curves. (B) A_{sat} for aCO_2 and eCO_2 foliage at 400 and 550 ppm CO_2. (C) Foliar N content expressed on a mass (%) or (D) leaf area (mg cm^{-2}) basis through time.

lower for eCO_2 foliage than for aCO_2 foliage when both were measured at 550 ppm CO_2 ($P = 0.17$; df $= 37$) and 22 % lower when both were measured at 400 ppm CO_2 ($P = 0.12$). In 2009, measured A_{sat} was 24 % lower for eCO_2 foliage than for aCO_2 foliage when both were measured at 550 ppm CO_2 ($P = 0.24$; df $= 56$) and 23 % lower when both were measured at 400 ppm CO_2 ($P = 0.35$) (Fig. 1B).

Tracking the temporal pattern of A_{sat}, foliar N concentration declined by ~50 % from 1999 to 2009 across both treatments, with values as low as 6.9 or 8.5 mg g^{-1} for eCO_2 or aCO_2, respectively, which was not much higher than the seasonal senescent foliar litter (5.8 mg g^{-1}). Mass-based foliar N content in eCO_2 leaves was significantly lower than for aCO_2 in 2008 and 2009 ($P < 0.001$), but not in 1999 ($P = 0.22$) (Fig. 1C). Elevated CO_2 area-based N content was significantly lower than for aCO_2 in 2008 ($P < 0.05$), but not in 2009 ($P = 0.08$) or in 1999 ($P = 0.47$) (Fig. 1D).

Many of the measured photosynthetic parameters were strongly related to N content, which dropped significantly over the course of the experiment. In 1999, the N content of foliage from both the treatments ranged from ~0.2 to 0.4 mg cm^{-2}, and there was no relationship between foliar N and chlorophyll content (Fig. 2A). When N content dropped <0.2 mg cm^{-2} (2008–09), there was a strong linear relationship with chlorophyll content, with a statistically steeper regression slope for the aCO_2 foliage (Fig. 2A). Mean chlorophyll values were not different between CO_2 treatments in 1999 (0.031 mg cm^{-2}; $P = 0.85$), but were significantly lower for eCO_2 (0.022 mg cm^{-2}) than for aCO_2 (0.028 mg cm^{-2}; $P < 0.001$) foliage in 2008–09. Leaf mass per area also declined through time for both treatments, and was always greater for eCO_2 foliage. In 1999, LMA was 8.6 % greater for eCO_2 foliage (128.9 g m^{-2}) than for aCO_2 foliage (118.7 g m^{-2}; $P < 0.01$). In 2008, LMA was not significantly greater for eCO_2 foliage (115.9 g m^{-2}) than for aCO_2 foliage (107.7 g m^{-2}; $P = 0.10$). In 2009, LMA was 6.6 % greater for eCO_2 foliage (104.5 g m^{-2}) than for aCO_2 foliage (98.0 g m^{-2}; $P < 0.01$).

Nitrogen content also had a significant impact on A_{sat}, particularly at lower values of N. Values of A_{sat} plateaued as N content increased (Fig. 2B). The linear regression slope between A_{sat} and N for 1999–2009 was slightly steeper for eCO_2 than for aCO_2 ($P = 0.055$) slopes and coincided at 0.057 mg cm^{-2}, which is approximately the N content of senescent tissue (regression not shown).

Values of J_{max} and V_{cmax} were not related to N content above ~0.2 mg cm^{-2}, but declined strongly with declining N content (Fig. 3A and B). There was a minimal overlap in N content of leaf samples from early (0.166–0.412 mg cm^{-2}) to late (0.075–0.197 mg cm^{-2}) years, which introduces a potential confounding effect of N × year, and which prohibited further refinement of the ~0.2 mg cm^{-2} N response threshold. Even so, annual measurements throughout the study confirm the N differences as foliage displayed an incremental decline in N content through time (Norby *et al.* 2010). There were

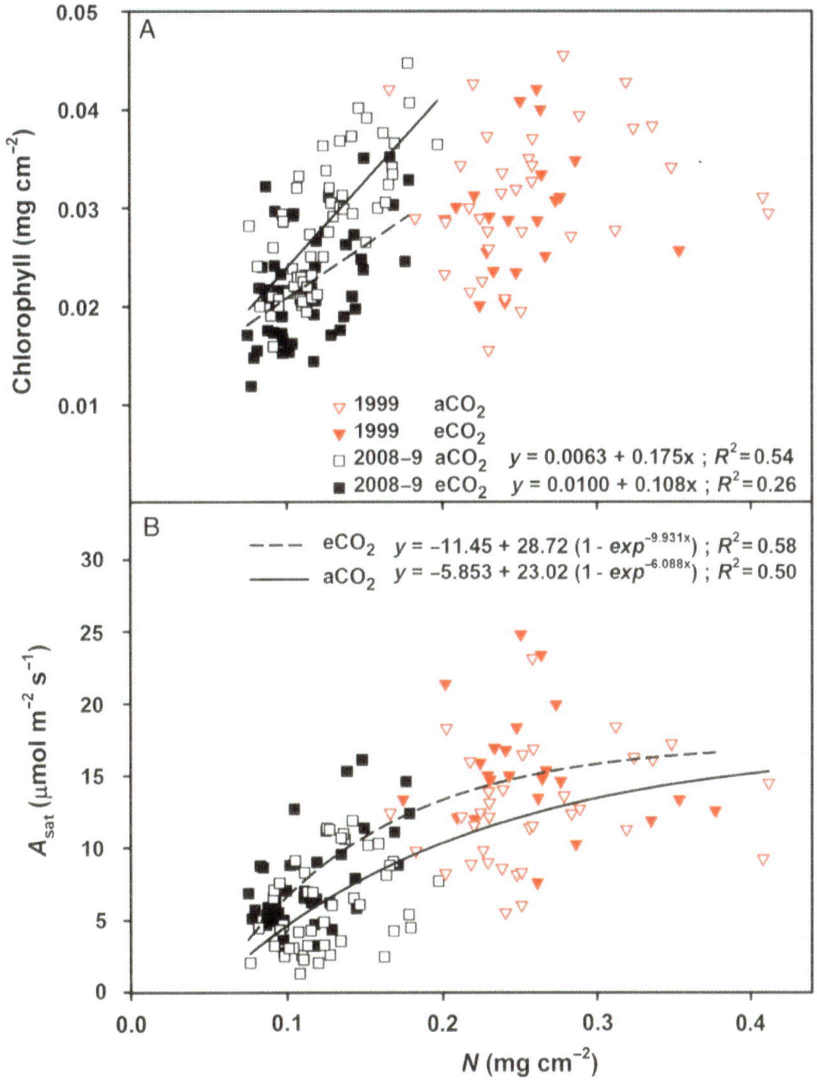

Figure 2. (A) Total foliar chlorophyll (chlorophyll $a + b$) for ambient (aCO_2—solid line) and elevated (eCO_2—dashed line) treatments in mid-summer 1999, 2008 and 2009 in relation to foliar N content, expressed on a leaf-area basis. Regressions were significant for $N < 0.2$ mg cm^{-2} ($R^2 = 0.26$ or 0.54; $P < 0.0001$), and there was a significant treatment effect on slope ($P < 0.05$; $F = 4.09$; df = 107), but little trend was apparent in 1999 when $N > 0.2$ ($R^2 = 0.02$ or 0.08; $P > 0.05$). (B) Light-saturated photosynthesis at growth CO_2 (400 or 550 ppm) in mid-summer across years (1999, 2008, 2009) in relation to foliar N content, expressed on an area basis.

no CO_2 treatment differences in the relationship between N content and J_{max} or N content and V_{cmax} <0.2 mg cm^{-2}. Exponential rise-to-max regressions through all data provided a good fit for J_{max} or V_{cmax} with N content, and again suggested N saturation for the 1999 foliage (Fig. 3C). While there was no CO_2 treatment effect on the ratio of $J_{max}:V_{cmax}$ through time (Fig. 4), the slope of the linear regression of $J_{max}:V_{cmax}$ declined through time across treatments ($P < 0.05$) from ~1.6 in 1999 to ~1.0 in 2009, indicating a decline in electron transport rate with respect to carboxylation rate through time. There was no relationship between $J_{max}:V_{cmax}$ and N content ($R^2 = 0.02$).

Photosynthetic water-use efficiency (WUE, photosynthesis per unit water use; $A_{sat}:g_s$) in 2008–09 was ~35 % greater for eCO_2 foliage than for aCO_2 foliage ($P = 0.02$; Fig. 5). The relationship between A_{sat} and stomatal conductance (g_s) was similar to aCO_2 foliage measured earlier (1998–2000) during different seasons and across years, but not for eCO_2 foliage measured earlier (Fig. 5—regression lines).

Discussion

Photosynthesis was down-regulated after long-term eCO_2 treatments at the ORNL-FACE research site due to a reduction in foliar N content, a symptom indicative of progressive N limitation as the stand developed (Norby *et al.* 2010). Net photosynthesis (*A*) and photosynthetic

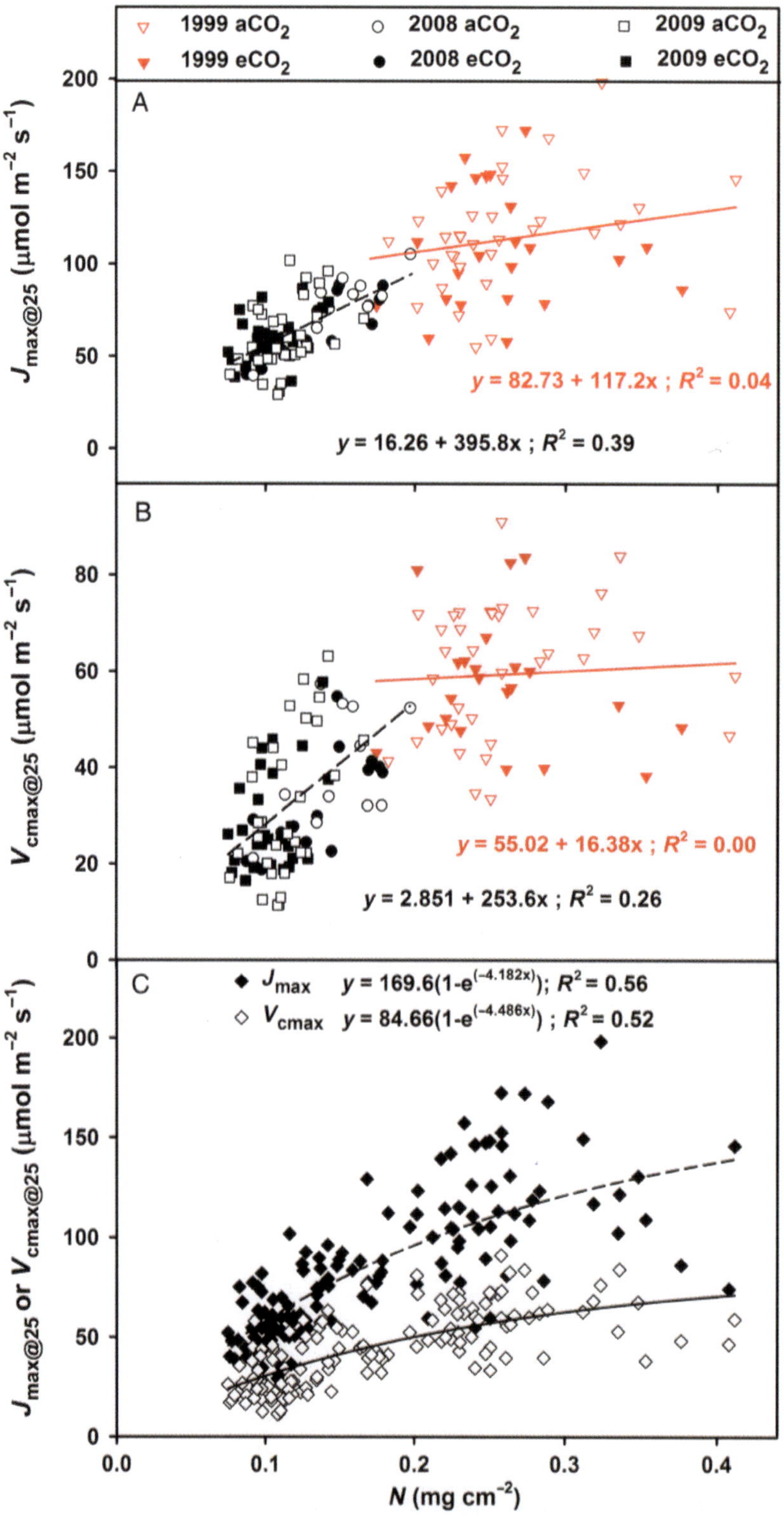

Figure 3. (A) Maximum electron transport rate (J_{max}) and (B) carboxylation rate (V_{cmax}) at 25 °C for ambient (aCO_2) and elevated (eCO_2) CO_2 treatments by year or (C) across all years in relation to foliar N content on an area basis (as leaf discs). Linear regressions by treatment were generally significant in 2008 and 2009 ($R^2 = 0.14$–0.80; $P = 0.07$–0.003) when $N < 0.2$ mg cm^{-2}, but not significant in 1999 ($R^2 = 0$–0.08; $P = 0.98$–0.1). There were no significant treatment effects on the relationships within or across years ($P = 0.22$–0.75). Regression equations are given for combined data across treatments.

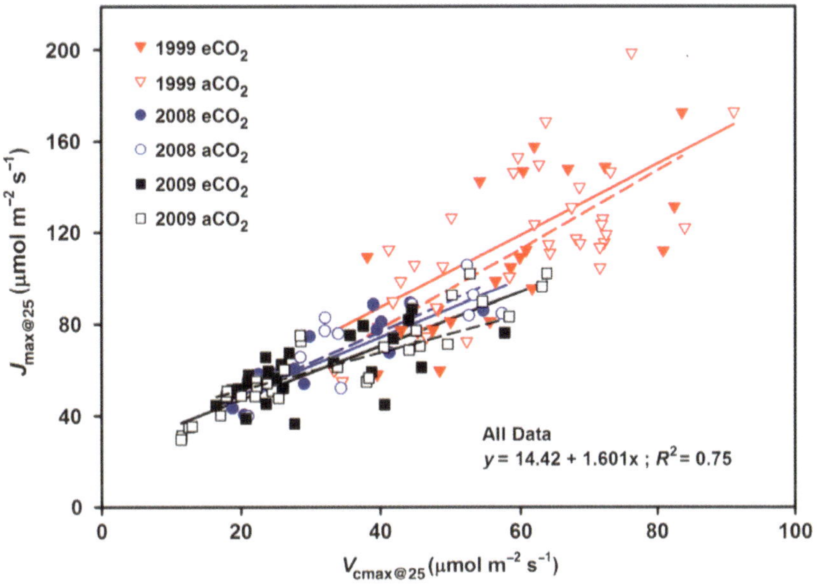

Figure 4. Relationships between maximum carboxylation rate (V_{cmax}) and electron transport rate (J_{max}) standardized to 25 °C for ambient (aCO_2—solid lines) and elevated (eCO_2—dashed lines) CO_2 treatments through time. Regressions were significant for all treatments and years ($R^2 = 0.45$–0.77), but there were no treatment effects on the relationships (in 2009, aCO_2 had a slightly steeper slope, $P = 0.067$).

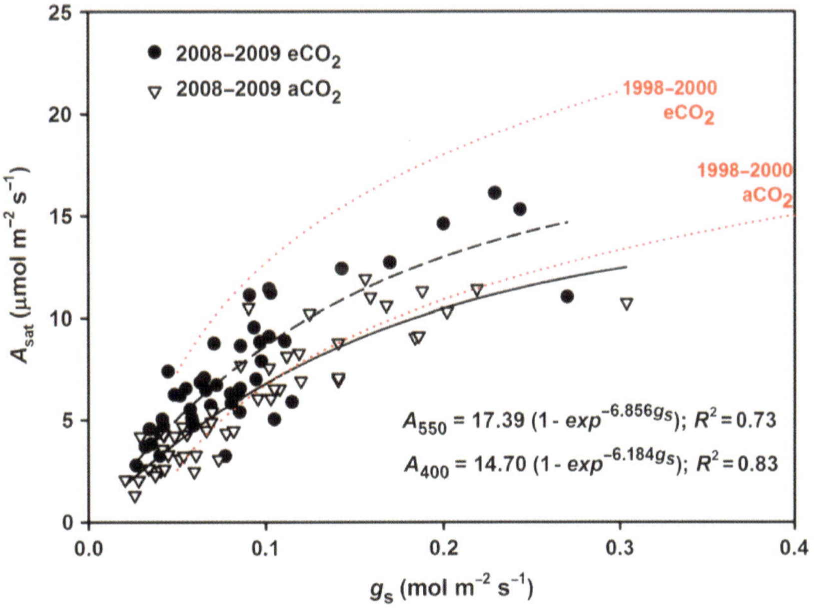

Figure 5. Light-saturated photosynthesis (A) in relation to stomatal conductance (g_s) for mature upper canopy foliage exposed to elevated (eCO_2; dashed line) or ambient (aCO_2; solid line) CO_2 (400 or 550 ppm, respectively) in mid-summer 2008 and 2009. Dotted regression lines represent results of similar measurements conducted across seasons from 1998 to 2000 based on Gunderson *et al.* (2002).

capacity of sweetgum trees declined through time as the stand matured; these responses were correlated with increased nutrient (N, Ca^{2+}, Mg^{2+}) sequestration in long-term biomass or soil organic matter and reduced availability and uptake of N (Finzi *et al.* 2006; Norby *et al.* 2010; Iversen *et al.* 2011; K. Kim *et al.*, unpubl. data). Consistent with our first hypothesis, we found a significant reduction in overall photosynthetic capacity across treatments as leaf N content declined. However, in contrast to our second hypothesis, the reduction in foliar N was so acute that the benefit of eCO_2 in generating enhanced photosynthetic efficiency was lost. As such, the initial CO_2 stimulation (~45 % greater) of net photosynthesis observed after 3 years (Gunderson *et al.* 2002; Sholtis *et al.* 2004) was not sustained after 12 years.

There was no relationship between foliar N and A in 1999, suggesting that foliar N content of upper canopy leaves (2.0–2.1 %) was in excess of photosynthetic requirements during the early years of the experiment. Total canopy N content in aCO_2 plots declined from ~1.8 % prior to the experiment in 1996 to 1.6 % by 2004 (Norby *et al.* 2010). Earlier work in younger 6-year-old sweetgum stands indicated that a leaf N threshold for maximum biomass production was ~1.9 % (Scott *et al.* 2004). In March 2004, part of our sweetgum stand was fertilized (200 kg ha^{-1} urea), which boosted upper canopy leaf N concentrations from ~1.65 % (control) to 2.04 % (fertilized) by late July, and substantially increased canopy leaf area and stem growth (Iversen and Norby 2008). The sweetgum trees at the Duke-FACE site indicated no down-regulation of photosynthesis and no change in tissue N content of upper canopy leaves from 1997 to 2004 (N content >1.5 %) (Herrick and Thomas 1999; Tissue *et al.* 2002; Springer *et al.* 2005; Ellsworth *et al.* 2012). Together, these prior studies support our conclusion that upper canopy photosynthesis at ORNL-FACE was not N limited in 1999, but strongly limited by the end of the experiment when the foliar N content had dropped to 1.03 % (eCO_2) and 1.16 % (aCO_2), which was significantly lower for the eCO_2 treatment.

There was also no relationship between foliar N and chlorophyll in 1999, suggesting that the chlorophyll content was in excess of photosynthetic requirements during the early years of the experiment. However, the chlorophyll values reported for the site in 1999 appear to be too low, suggesting a poor extraction yield occurred with the ethanol solvent as described earlier in the Methods section: 'Foliar biochemistry and leaf mass per area'. Chlorophyll values ranged from ~0.020 to 0.043 mg cm^{-2} in 1999, similar to the values in 2008–09 (0.014–0.041 mg cm^{-2}), despite much lower N content. Actual chlorophyll levels at the ORNL-FACE site in 1999 were thus likely significantly higher than indicated here. Indeed, chlorophyll extracted from similar aged sweetgum trees at the Duke-FACE site that were not N limited indicated late season values from 0.047 to 0.071 mg cm^{-2} (Herrick and Thomas 2001).

The lack of relationship between foliar N content and chlorophyll, J_{max} or V_{cmax} in 1999 suggests that the photosynthetic apparatus in sweetgum was N saturated above ~0.2 mg cm^{-2} and that N had accumulated in excess to meet other N requirements (e.g. growth and storage). A similar pattern of N saturation of V_{cmax} has been suggested for pine at the Duke-FACE site, with saturation of the carboxylation rate at tissue contents >1.4 % N (Palmroth *et al.* 2013). Nitrogen is most often stored as free amino acids or proteins, especially Rubisco, or as inorganic N within the vacuole (Proe and Millard 1994; Warren and Adams 2004). It is likely that reductions in enzyme activation state in eCO_2 foliage can partially uncouple photosynthetic response from foliar N content, as N is shifted into non-active 'storage' components. For example, the activation state of Rubisco has been shown to decline linearly with increasing leaf N in apple (Cheng and Fuchigami 2000). Given that J_{max} and V_{cmax} are often modelled based on N content of the leaf, not on the proportion of N actively used in photosynthesis this may lead to increased uncertainty in estimates of GPP at higher N contents.

While the initial foliar N content of the stand was adequate for photosynthesis, N became progressively more limiting, restricting light harvesting, rates of Rubisco and the electron transport chain. N content for several of the measured leaves in 2008–09 dropped to values as low as 0.075 mg cm^{-2}, which was not much higher than senescent foliage (0.057 mg cm^{-2}). At lower foliar levels of N (0.075–0.2 mg cm^{-2}), eCO_2 leaves utilized N more efficiently (greater PNUE) compared with aCO_2 leaves (Fig. 2B). For leaves operating at greater PNUE, the theoretical photosynthetic N requirement is lower. In our study, this would entail a lower photosynthetic N threshold in eCO_2 leaves. Our results could not determine this threshold, as J_{max} and V_{cmax} became similarly N-restricted below ~0.2 mg cm^{-2}, independent of CO_2 availability.

In mature leaves, the optimum N distribution between RuBP carboxylation/regeneration and electron transport may shift as a result of environmental changes (e.g. CO_2 concentration, nutrient availability or thermal acclimation) such that the processes are co-limiting (e.g. Makino *et al.* 1994; Medlyn 1996; Hikosaka 1997). Theory predicts that the $J_{max}:V_{cmax}$ ratio should increase under eCO_2 (Medlyn 1996), and simulations with pine seedlings suggest that this optimization may vary seasonally (Lewis *et al.* 1996). We found no evidence of a treatment-dependent shift through time (Fig. 4). Similar work at the Duke-FACE site with *P. taeda* also found that there was no change in the slope of $J_{max}:V_{cmax}$ due to eCO_2 treatments (Crous *et al.* 2008). Other observations have found that $J_{max}:V_{cmax}$ can increase slightly (~5 %) with eCO_2 (Medlyn *et al.* 1999; Ainsworth and Long 2007), although this change is generally smaller than predicted by theory (Medlyn 1996).

After long-term eCO_2 treatment at the Duke-FACE site, there was a decline in the slope of J_{max}–N and V_{cmax}–N in pine, indicating eCO_2 down-regulation of overall photosynthetic capacity (Crous *et al.* 2008). In contrast, at the ORNL-FACE site, there was no treatment effect on the relationships between J_{max}–N and V_{cmax}–N in sweetgum. Rather, the declining slope of $J_{max}:V_{cmax}$ through time for all treatments suggests a gradual reallocation of N as plants acclimated to reductions in soil N availability.

The decline in photosynthetic capacity and resultant reduction in eCO_2 enhancement of NPP (Norby *et al.* 2010), coupled with the demonstrated reduction in

N availability for eCO_2 plots (Garten *et al.* 2011), is consistent with the progressive N limitation hypothesis (Comins and McMurtrie 1993; Luo *et al.* 2004). New root production was stimulated in eCO_2 plots (Norby *et al.* 2004) likely as a result of increased internal plant nutrient demand and chemical signalling, including the buildup and translocation of newly fixed sugars. Enhanced root production and exploration at deeper soil depths increased N availability and uptake for eCO_2 trees (Finzi *et al.* 2007; Iversen *et al.* 2011), but N uptake rates were unable to meet demand and resulted in progressive reductions in photosynthesis. Limitation by other elements could also dampen photosynthetic activity, e.g. Mg, which is the central element in chlorophyll and a key cofactor for Rubisco activation. However, base cation uptake (including Mg and Ca) increased through time in eCO_2 trees based on wood composition (K. Kim *et al.*, unpubl. data), likely as a consequence of the enhanced root exploration. Foliar content of other essential mineral nutrients such as P and K did not decline through time in litter collected from 1998 to 2002, nor in wood tissue collected in 2009.

The measurement of gas exchange under fully rehydrated conditions was useful for resolving maximum rates of photosynthesis. Under these hydrated conditions, stomatal conductance remained lower for eCO_2 foliage than for aCO_2 foliage; $eCO_2 : aCO_2$ g_s ranged from 0.75 to 0.81. Similarly, under dry field conditions, whole canopy conductance (g_c) based on sap flow remained lower for eCO_2 foliage; $eCO_2 : aCO_2$ g_c was $\sim$0.7 and declined to $\sim$0.4 as stomatal aperture declined during extreme drought (Warren *et al.* 2011*b*). The relationship between A and g_s did not change much through time for aCO_2 foliage, but was significantly reduced for eCO_2 foliage (Fig. 5). This suggests that the eCO_2 stimulation of photosynthetic WUE declined through time with increasing N limitation. Even so, WUE remained greater for eCO_2 sweetgum trees, and the relationship between stomatal conductance and assimilation was in agreement with predictions of the optimal stomatal model (Medlyn *et al.* 2011), independent of leaf N content (De Kauwe *et al.* 2013). At the Duke-FACE site, partitioning N either to 'photosynthetically active' or 'storage components' based on the N–V_{cmax} threshold allowed for linkage of eCO_2-stimulated WUE to marginal N-use efficiency within the context of optimality theory (Palmroth *et al.* 2013). Results from these studies allow predictive capacity of ecosystem level responses to changes in site resources such as CO_2, water and nutrients.

While shorter-term environmental (e.g. solar radiation, T, CO_2 concentration) regulation of photosynthesis has been studied in detail, knowledge of longer-term shifts in the photosynthetic apparatus in response to eCO_2 has been generally limited to observations at the FACE studies and a few long-term open-top chamber studies (e.g. Erickson *et al.* 2013). Long-term dynamics of soil water or nutrient availability (Finzi *et al.* 2006; Palmroth *et al.* 2013) and growth sink demands (Paul and Pellny 2003; Fatichi *et al.* 2013) provide additional regulation of photosynthetic feedback mechanisms and capacity of C uptake in some ecosystems. These processes are not always well represented in global models that depend on scaling mechanistic photosynthetic responses to the land surface (Smith and Dukes 2013). However, it is encouraging that some optimization models do successfully predict biochemical (e.g. reduced foliar N) and mechanistic (e.g. reduced stomatal conductance) process response to eCO_2 and shifts in resource availability (Dewar *et al.* 2009). Data from the FACE studies, including this one, are being used in a multi-model comparison to assess land surface model structure and variation in predictive capacity, which will result in future model improvement (e.g. De Kauwe *et al.* 2013; Walker *et al.* 2014).

Conclusions

Photosynthesis is the predominant mechanism that removes CO_2 from the atmosphere. Therefore, accurate projections of CO_2 feedbacks to climate forcing rely on the correct mechanistic representation of plant photosynthetic C assimilation capacity. At the ORNL-FACE site, the initial stimulation of carbon assimilation by sweetgum trees exposed to eCO_2 was lost over the 12-year study. The loss of stimulation was due to the declining foliar N content, which has been related to lower soil N availability and uptake. Sweetgum foliar N content below a threshold value ($\sim$0.2 mg N cm^{-2} leaf area) restricted carbon assimilation due to photosynthetic biochemical limitations. This study indicates that atmospheric enrichment of CO_2 may result in an initial 'fertilization' effect, directly increasing photosynthesis and productivity, but this may later be offset by the declining soil N availability, which could completely eliminate the initial positive effects of eCO_2. Overall, we suggest caution in extrapolating shorter-term eCO_2 responses to longer-term ecosystem processes, which are confounded by other controlling factors such as soil nutrient availability, water availability or air temperature that can change through time.

Sources of Funding

This article is based upon work supported by the US Department of Energy, Office of Science, Office of Biological and Environmental Research, under contract DE-AC05-00OR22725.

Contributions by the Authors

J.M.W. designed and conducted the experiment. A.M.J. performed laboratory work. B.E.M. analysed the data.

All authors contributed to the writing and editing of the manuscript.

Acknowledgements

The authors appreciate fieldwork and facility maintenance from Joanne Childs (arbornaut), Cassandra Bruno, Jeff Riggs and Danny Sluss, and data from Colleen Iversen and Carla Gunderson. This manuscript has been authored by UT-Battelle, LLC under Contract No. DE-AC05-00OR22725 with the US Department of Energy. The United States Government retains and the publisher, by accepting the article for publication, acknowledges that the United States Government retains a non-exclusive, paid-up, irrevocable, world-wide license to publish or reproduce the published form of this manuscript, or allow others to do so, for United States Government purposes. The Department of Energy will provide public access to these results of federally sponsored research in accordance with the DOE Public Access Plan (http://energy.gov/downloads/doe-public-access-plan).

Literature Cited

Ainsworth EA, Long SP. 2005. What have we learned from 15 years of free air CO_2 enrichment (FACE)? A meta-analytic review of the responses of photosynthesis, canopy properties and plant production to rising CO_2. *New Phytologist* **165**:351–372.

Ainsworth EA, Rogers A. 2007. The response of photosynthesis and stomatal conductance to rising [CO_2]: mechanisms and environmental interactions. *Plant, Cell and Environment* **30**:258–270.

Bader MK, Siegwolf R, Körner C. 2010. Sustained enhancement of photosynthesis in mature deciduous forest trees after 8 years of free air CO_2 enrichment. *Planta* **232**:1115–1125.

Bernacchi CJ, Calfapietra C, Davey PA, Wittig VE, Scarascia-Mugnozza GE, Raines CA, Long SP. 2003. Photosynthesis and stomatal conductance responses of poplars to free-air CO_2 enrichment (PopFACE) during the first growth cycle and immediately following coppice. *New Phytologist* **159**:609–621.

Calfapietra C, Angelis PD, Gielen B, Lukac M, Moscatelli MC, Avino G, Lagomarsino A, Polle A, Ceulemans R, Mugnozza GS, Hoosbeek MR, Cotrufo MF. 2007. Increased nitrogen-use efficiency of a short-rotation poplar plantation in elevated CO_2 concentration. *Tree Physiology* **27**:1153–1163.

Cheng LL, Fuchigami LH. 2000. Rubisco activation state decreases with increasing nitrogen content in apple leaves. *Journal of Experimental Botany* **51**:1687–1694.

Comins HN, McMurtrie RE. 1993. Long-term response of nutrient-limited forests to CO_2 enrichment; equilibrium behavior of plant–soil models. *Ecological Applications* **3**:666–681.

Crous KY, Ellsworth DS. 2004. Canopy position affects photosynthetic adjustments to long-term elevated CO_2 concentration (FACE) in aging needles in a mature *Pinus taeda* forest. *Tree Physiology* **24**:961–970.

Crous KY, Walters MB, Ellsworth DS. 2008. Elevated CO_2 concentration affects leaf photosynthesis–nitrogen relationships in *Pinus taeda* over nine years in FACE. *Tree Physiology* **28**:607–614.

Curtis PS, Wang X. 1998. A meta-analysis of elevated CO_2 effects on woody plant mass, form, and physiology. *Oecologia* **113**: 299–313.

Darbah JNT, Kubiske ME, Nelson N, Kets K, Riikonen J, Sober A, Rouse L, Karnosky DF. 2010. Will photosynthetic capacity of aspen trees acclimate after long-term exposure to elevated CO_2 and O_3? *Environmental Pollution* **158**:983–991.

De Kauwe MG, Medlyn BE, Zaehle S, Walker AP, Dietze MC, Hickler T, Jain AK, Luo Y, Parton WJ, Prentice IC, Smith B, Thornton PE, Wang S, Wang Y-P, Wårlind D, Weng E, Crous KY, Ellsworth DS, Hanson PJ, Seok KH, Warren JM, Oren R, Norby RJ. 2013. Forest water use and water use efficiency at elevated CO_2: a model-data intercomparison at two contrasting temperate forest FACE sites. *Global Change Biology* **19**:1759–1779.

Dewar RC, Franklin O, Mäkelä A, McMurtrie RE, Valentine HT. 2009. Optimal function explains forest responses to global change. *Bioscience* **59**:127–139.

Drake BG, Gonzalez-Meler MA, Long SP. 1997. More efficient plants: a consequence of rising atmospheric CO_2? *Annual Review of Plant Physiology and Plant Molecular Biology* **48**:609–639.

Ellsworth DS, Thomas R, Crous KY, Palmroth S, Ward E, Maier C, Delucia E, Oren R. 2012. Elevated CO_2 affects photosynthetic responses in canopy pine and subcanopy deciduous trees over 10 years: a synthesis from Duke FACE. *Global Change Biology* **18**: 223–242.

Epron D, Liozon R, Mousseau M. 1996. Effects of elevated CO_2 concentration on leaf characteristics and photosynthetic capacity of beech (*Fagus sylvatica*) during the growing season. *Tree Physiology* **16**:425–432.

Erickson JE, Peresta G, Montovan KJ, Drake BG. 2013. Direct and indirect effects of elevated atmospheric CO_2 on net ecosystem production in a Chesapeake Bay tidal wetland. *Global Change Biology* **19**:3368–3378.

Fatichi S, Leuzinger S, Körner C. 2013. Moving beyond photosynthesis: from carbon source to sink-driven vegetation modeling. *New Phytologist* **201**:1086–1095.

Finzi AC, Moore DJ, DeLucia EH, Licher J, Hofmockel KS, Jackson RB, Kim H-S, Matamala R, McCarthy HR, Oren R, Pippen JS, Schlesinger WH. 2006. Progressive nitrogen limitation of ecosystem processes under elevated CO_2 in a warm-temperate forest. *Ecology* **87**:15–25.

Finzi AC, Norby RJ, Calfapietra C, Gallet-Budynek A, Gielen B, Holmes WE, Hoosbeek MR, Iversen CM, Jackson RB, Kubiske ME, Ledford J, Liberloo M, Oren R, Polle A, Pritchard S, Zak DR, Schlesinger WH, Ceulemans R. 2007. Increases in nitrogen uptake rather than nitrogen-use efficiency support higher rates of temperate forest productivity under elevated CO_2. *Proceedings of the National Academy of Sciences of the USA* **104**: 14014–14019.

Garten CT, Iversen CM, Norby RJ. 2011. Litterfall ^{15}N abundance indicates declining soil nitrogen availability in a free air CO_2-enrichment experiment. *Ecology* **92**:133–139.

Ghannoum O, Phillips NG, Sears MA, Logan BA, Lewis JD, Conroy JP, Tissue DT. 2010. Photosynthetic responses of two eucalypts to industrial-age changes in atmospheric [CO_2] and temperature. *Plant, Cell and Environment* **33**:1671–1681.

Gunderson CA, Wullschleger SD. 1994. Photosynthetic acclimation in trees to rising atmospheric CO_2: a broader perspective. *Photosynthesis Research* **39**:369–388.

Gunderson CA, Norby RJ, Wullschleger SD. 1993. Foliar gas exchange responses of two deciduous hardwoods during 3 years of growth in elevated CO_2: no loss of photosynthetic enhancement. *Plant, Cell and Environment* **16**:797–807.

Gunderson CA, Sholtis JD, Wullschleger SD, Tissue DT, Hanson PJ, Norby RJ. 2002. Environmental and stomatal control of photosynthetic enhancement in the canopy of a sweetgum (*Liquidambar styraciflua* L) plantation during three years of CO_2 enrichment. *Plant, Cell and Environment* **25**:379–393.

Hendrey GR, Ellsworth DS, Lewin KF, Nagy J. 1999. A free-air enrichment system for exposing tall forest vegetation to elevated atmospheric CO_2. *Global Change Biology* **5**:293–309.

Herrick JD, Thomas RB. 1999. Effects of CO_2 enrichment on the photosynthetic light response of sun and shade leaves of canopy sweetgum trees (*Liquidambar styraciflua*) in a forest ecosystem. *Tree Physiology* **19**:779–786.

Herrick JD, Thomas RB. 2001. No photosynthetic down-regulation in sweetgum trees (*Liquidambar styraciflua* L.) after three years of CO_2 enrichment at the Duke Forest FACE experiment. *Plant, Cell and Environment* **24**:53–64.

Herrick JD, Thomas RB. 2003. Leaf senescence and late-season net photosynthesis of sun and shade leaves of overstory sweetgum (*Liquidambar styraciflua*) grown in elevated and ambient carbon dioxide concentrations. *Tree Physiology* **19**:779–786.

Hikosaka K. 1997. Modelling optimal temperature acclimation of the photosynthetic apparatus in C-3 plants with respect to nitrogen use. *Annals of Botany* **80**:721–730.

Hurlbert SH. 1984. Pseudoreplication and the design of ecological field experiments. *Ecological Monographs* **54**:187–211.

Inskeep WP, Bloom PR. 1985. Extinction coefficients of chlorophyll *a* and *b* in *N,N*-dimethylformamide and 80 % acetone. *Plant Physiology* **77**:483–485.

IPCC. 2013. Climate change 2013: the physical science basis. In: Stocker TF, Qin D, Plattner G-K, Tignor M, Allen SK, Boschung J, Nauels A, Xia Y, Bex V, Midgley PM, eds. *Contribution of working group I to the fifth assessment report of the Intergovernmental Panel on Climate Change*. Cambridge, UK and New York, NY, USA: Cambridge University Press, 1029–1136.

Iversen CM, Norby RJ. 2008. Nitrogen limitation in a sweetgum plantation: implications for carbon allocation and storage. *Canadian Journal of Forest Research* **38**:1021–1032.

Iversen CM, Hooker TD, Classen AT, Norby RJ. 2011. Net mineralization of N at deeper soil depths as a potential mechanism for sustained forest production under elevated [CO_2]. *Global Change Biology* **17**:1130–1139.

Keeling RF, Walker SJ, Piper SC, Bollenbacher AF. 2014. Scripps CO_2 Program. Scripps Institution of Oceanography (SIO), University of California, La Jolla, CA, USA. http://scrippsco2.ucsd.edu.

Kubiske ME, Zak DR, Pregitzer KS, Takeuchi Y. 2002. Photosynthetic acclimation of overstory *Populus tremuloides* and understory *Acer saccharum* to elevated atmospheric CO_2 concentration: interactions with shade and soil nitrogen. *Tree Physiology* **22**: 321–329.

Lewis JD, Tissue DT, Strain BR. 1996. Seasonal response of photosynthesis to elevated CO_2 in loblolly pine (*Pinus taeda* L.) over two growing seasons. *Global Change Biology* **2**:103–114.

Liberloo M, Tulva I, Raïm O, Kull O, Ceulemans R. 2007. Photosynthetic stimulation under long-term CO enrichment and fertilization is sustained across a closed *Populus* canopy profile (EUROFACE). *New Phytologist* **173**:537–549.

Luo Y, Su B, Currie WS, Dukes JS, Finzi A, Hartwig U, Hungate B, McMurtrie R, Oren R, Parton WJ, Pataki D, Shaw MR, Zak DR, Field CB. 2004. Progressive nitrogen limitation of ecosystem responses to rising atmospheric carbon dioxide. *Bioscience* **54**: 731–739.

Luxmoore RJ, Wullschleger SD, Hanson PJ. 1993. Forest responses to CO_2 enrichment and climate warming. *Water, Air, and Soil Pollution* **70**:309–323.

Makino A, Nakano H, Mae TS. 1994. Effects of growth temperature on the responses of ribulose-1,5-bisphosphate carboxylase, electron transport components, and sucrose synthesis enzymes to leaf nitrogen in rice, and their relationships to photosynthesis. *Plant Physiology* **105**:1231–1238.

Marek MV, Kalina J, Matouskova M. 1995. Response of photosynthetic carbon assimilation of Norway spruce exposed to long-term elevation of CO_2 concentration. *Photosynthetica* **31**:209–220.

Medlyn BE. 1996. The optimal allocation of nitrogen within the C_3 photosynthetic system at elevated CO_2. *Functional Plant Biology* **23**:593–603.

Medlyn BE, Badeck FW, De Pury DGG, Barton CVM, Broadmeadow M, Ceulemans R, De Angelis P, Forstreuter M, Jach ME, Kellomaki S, Laitat E, Marek M, Philippot S, Rey A, Strassemeyer J, Laitinen K, Liozon R, Portier B, Roberntz P, Wang K, Jstbid PG. 1999. Effects of elevated CO_2 on photosynthesis in European forest species: a meta-analysis of model parameters. *Plant, Cell and Environment* **22**:1475–1495.

Medlyn BE, Dreyer E, Ellsworth DE, Forstreuter M, Harley PC, Kirschbaum MUF, LeRoux X, Loustau D, Montpied P, Strassemeyer J, Walcroft A, Wang KY. 2002. Temperature response of parameters of a biochemically based model of photosynthesis. II. A review of experimental data. *Plant, Cell and Environment* **25**:1167–1179.

Medlyn BE, Duursma RA, Eamus D, Ellsworth DS, Prentice IC, Barton CVM, Crous KY, De Angelis P, Freeman M, Wingate L. 2011. Reconciling the optimal and empirical approaches to modelling stomatal conductance. *Global Change Biology* **17**: 2134–2144.

Minocha R, Martinez G, Lyons B, Long S. 2009. Development of a standardized methodology for quantifying total chlorophyll and carotenoids from foliage of hardwood and conifer tree species. *Canadian Journal of Forest Research* **39**:849–861.

Monson RK, Trahan N, Rosenstiel TN, Veres P, Moore D, Wilkinson M, Norby RJ, Volder A, Tjoelker MG, Briske DD, Karnosky DF, Fall R. 2007. Isoprene emission from terrestrial ecosystems in response to global change: minding the gap between models and observations. *Philosophical Transactions of the Royal Society A* **365**: 1677–1695.

Noormets A, Kull O, Sôber A, Kubiske ME, Karnosky DF. 2010. Elevated CO_2 response of photosynthesis depends on ozone concentration in aspen. *Environmental Pollution* **158**:992–999.

Norby RJ, Wullschleger SD, Gunderson CA, Johnson DW, Ceulemans R. 1999. Tree responses to rising CO_2: implications for the future forest. *Plant, Cell and Environment* **22**:683–714.

Norby RJ, Todd DE, Fults J, Johnson DW. 2001. Allometric determination of tree growth in a CO_2-enriched sweetgum stand. *New Phytologist* **150**:477–487.

Norby RJ, Ledford J, Reilly CD, Miller NE, O'Neill EG. 2004. Fine-root production dominates response of a deciduous forest to atmospheric CO_2 enrichment. *Proceedings of the National Academy of Sciences of the USA* **101**:9689–9693.

Norby RJ, DeLucia EH, Gielen B, Calfapietra C, Giardina CP, King JS, Ledford J, McCarthy HR, Moore DJP, Ceulemans R, De Angelis P, Finzi AC, Karnosky DF, Kubiske ME, Lukac M, Pregitzer KS, Scarascia-Mugnozza GE, Schlesinger WH, Oren R. 2005. Forest response to elevated CO_2 is conserved across a broad range of productivity. *Proceedings of the National Academy of Sciences of the USA* **102**:18052–18056.

Norby RJ, Warren JM, Iversen CM, Medlyn BE, McMurtrie RE. 2010. CO_2 enhancement of forest productivity constrained by limited nitrogen availability. *Proceedings of the National Academy of Sciences of the USA* **107**:19368–19373.

Oechel WC, Cowles S, Grulke N, Hastings SJ, Lawrence B, Prudhomme T, Riechers G, Strain B, Tissue D, Vourlitis GL. 1994. Transient nature of CO_2 fertilization in Arctic tundra. *Nature* **371**:500–503.

Palmroth S, Katul GG, Maier CA, Ward E, Manzoni S, Vico G. 2013. On the complementary relationship between marginal nitrogen and water use efficiencies among *Pinus taeda* leaves grown under ambient and enriched CO_2 environments. *Annals of Botany* **111**:467–477.

Paul MJ, Pellny TK. 2003. Carbon metabolite feedback regulation of leaf photosynthesis and development. *Journal of Experimental Botany* **54**:539–547.

Peterson AG, Ball JT, Luo Y, Field CB, Reich PB, Curtis PS, Griffin KL, Gunderson CA, Norby RJ, Tissue DT, Forstreuter M, Rey A, Vogel CS, Cmeal Participants. 1999. The photosynthesis–leaf nitrogen relationship at ambient and elevated atmospheric carbon dioxide: a meta-analysis. *Global Change Biology* **5**: 331–346.

Proe MF, Millard P. 1994. Relationships between nutrient supply, nitrogen partitioning and growth in young Sitka spruce (*Picea sitchensis*). *Tree Physiology* **14**:75–88.

Rey A, Jarvis PG. 1998. Long-term photosynthetic acclimation to increased atmospheric CO_2 concentration in young birch (*Betula pendula*) trees. *Tree Physiology* **18**:441–450.

Riggs JS, Tharp ML, Norby RJ. 2010. ORNL FACE Weather Data. Carbon Dioxide Information Analysis Center, US Department of Energy, Oak Ridge National Laboratory, Oak Ridge, TN. http://cdiac.ornl.gov.

Rogers A, Ellsworth DS. 2002. Photosynthetic acclimation of *Pinus taeda* (loblolly pine) to long-term growth in elevated pCO_2 (FACE). *Plant, Cell and Environment* **25**:851–858.

Scott DA, Burger JA, Kaczmarek DJ, Kane MB. 2004. Growth and nutrition response of young sweetgum plantations to repeated nitrogen fertilization on two site types. *Biomass and Bioenergy* **27**:313–325.

Sholtis JD, Gunderson CA, Norby RJ, Tissue DT. 2004. Persistent stimulation of photosynthesis by elevated CO_2 in a sweetgum (*Liquidambar styraciflua* L.) forest stand. *New Phytologist* **162**: 343–354.

Smith NG, Dukes JS. 2013. Plant respiration and photosynthesis in global-scale vegetation models: incorporating acclimation to temperature and CO_2. *Global Change Biology* **19**:45–63.

Springer CJ, DeLucia EH, Thomas RB. 2005. Relationships between net photosynthesis and foliar nitrogen concentrations in a loblolly pine forest ecosystem grown in elevated atmospheric carbon dioxide. *Tree Physiology* **25**:385–394.

Stitt M, Krapp A. 1999. The interaction between elevated carbon dioxide and nitrogen nutrition: the physiological and molecular background. *Plant, Cell and Environment* **22**:583–621.

Tissue DT, Oechel WC. 1987. Physiological and growth response of *Eriophorum vaginatum* to field elevated CO_2 and temperature in the Alaskan tussock tundra. *Ecology* **68**:401–410.

Tissue DT, Thomas RB, Strain BR. 1993. Long-term effects of elevated CO_2 and nutrients on photosynthesis and Rubisco in loblolly pine seedlings. *Plant, Cell and Environment* **16**:859–865.

Tissue DT, Thomas RB, Strain BR. 1997. Atmospheric CO_2 enrichment increases growth and photosynthesis of *Pinus taeda*: a 4 year experiment in the field. *Plant, Cell and Environment* **20**: 1123–1134.

Tissue DT, Lewis JD, Wullschleger SD, Amthor JS, Griffin KL, Anderson OR. 2002. Leaf respiration at different canopy positions in sweetgum (*Liquidambar styraciflua*) grown in ambient and elevated concentrations of carbon dioxide in the field. *Tree Physiology* **22**:1157–1166.

Walker AP, Hanson PJ, De Kauwe MG, Medlyn BE, Zaehle S, Asao S, Dietze M, Hickler T, Huntingford C, Iversen CM, Jain A, Lomas M, Luo Y, Mccarthy H, Parton WJ, Prentice IC, Thornton PE, Wang S, Wang Y-P, Wårlind D, Weng E, Warren JM, Woodward FI, Oren R, Norby RJ. 2014. Comprehensive ecosystem model-data synthesis using multiple data sets at two temperate forest free-air CO_2 enrichment experiments: model performance at ambient CO_2 concentration. *Journal of Geophysical Research: Biogeosciences* **119**: 937–964.

Warren CR, Adams MA. 2004. Evergreen trees do not maximize instantaneous photosynthesis. *Trends in Plant Science* **9**: 270–274.

Warren JM, Pötzelsberger E, Wullschleger SD, Thornton PE, Hasenauer H, Norby RJ. 2011*a*. Ecohydrologic impact of reduced stomatal conductance in forests exposed to elevated CO_2. *Ecohydrology* **4**:196–210.

Warren JM, Norby RJ, Wullschleger SD. 2011*b*. Elevated CO_2 enhances premature leaf senescence during extreme climatic events in a temperate forest. *Tree Physiology* **31**:117–130.

Ziska LH, Hogan KP, Smith AP, Drake BG. 1991. Growth and photosynthetic response of nine tropical species with long-term exposure to elevated carbon dioxide. *Oecologia* **86**: 383–389.

5

Climatic niche and flowering and fruiting phenology of an epiphytic plant

Narayani Barve[1,2]*, Craig E. Martin[2] and A. Townsend Peterson[1,2]

[1] Biodiversity Institute, University of Kansas, 1345 Jayhawk Blvd, Lawrence, KS 66045, USA
[2] Department of Ecology and Evolutionary Biology, University of Kansas, Lawrence, KS 66045, USA

Associate Editor: Colin M. Orians

Abstract. Species have geographic distributions constrained by combinations of abiotic factors, biotic factors and dispersal-related factors. Abiotic requirements vary across the life stages for a species; for plant species, a particularly important life stage is when the plant flowers and develops seeds. A previous year-long experiment showed that ambient temperature of 5–35 °C, relative humidity of >50 % and ≤15 consecutive rainless days are crucial abiotic conditions for Spanish moss (*Tillandsia usneoides* L.). Here, we explore whether these optimal physiological intervals relate to the timing of the flowering and fruiting periods of Spanish moss across its range. As Spanish moss has a broad geographic range, we examined herbarium specimens to detect and characterize flowering/fruiting periods for the species across the Americas; we used high-temporal-resolution climatic data to assess the availability of optimal conditions for Spanish moss populations during each population's flowering period. We explored how long populations experience suboptimal conditions and found that most populations experience suboptimal conditions in at least one environmental dimension. Flowering and fruiting periods of Spanish moss populations are either being optimized for one or a few parameters or may be adjusted such that all parameters are suboptimal. Spanish moss populations appear to be constrained most closely by minimum temperature during this period.

Keywords: ERA data; Spanish moss; species distribution; *Tillandsia usneoides* L.

Introduction

Restricted geographic distributions of species are often a consequence of some set of constraints in terms of abiotic requirements, needs in terms of biotic interactions and limitations to dispersal ability (Soberón 2007). All species have a life cycle (be it simple or complex), and each stage in that cycle may have different requirements in terms of climate, soils, topography, other abiotic factors and biotic requirements like food, competitors or mutualisms. Grubb (1977) defined four components of ecological niches of plants: the habitat niche, life-form niche, phenological niche and regeneration niche; much research has examined how regeneration niches may differ in different community assembly processes and how these various niches act in different life stages (Fowler 1988; Tilman 1990; Lavorel and Chesson 1995; Miller-Rushing and Primack 2008). Although several studies have used the regeneration niche concept to explore competition and understand rarity of species at local scales (Engelhardt and Anderson 2011; Ranieri *et al.* 2012), few studies have used the regeneration niche idea to understand species' distributions in terms of their abiotic requirements at geographic scales (Pederson *et al.* 2004; Sweeney *et al.* 2006; Wellenreuther *et al.* 2012).

* Corresponding author's e-mail address: narayani@ku.edu

Phenological stages in plant life cycles comprise critical life stages, in which plants flower, produce seeds, grow or remain dormant (Bond and Midgley 2001; Silvertown 2004). Plants have presumably evolved to flower in seasons and at intervals that ensure maximal reproductive success (Amasino 2010). Considerable research has shown that plants sense and respond in complex ways to environmental cues such as shoots bending towards light and away from gravity (Garner 1933; Lang 1952; Bernier *et al.* 1993; Dennis *et al.* 1996). However, these factors have been investigated chiefly at local scales; at biogeographic scales, the question of whether phenology is optimized or not with respect to physiological responses to abiotic factors like temperature and precipitation remains little investigated (Engelhardt and Anderson 2011; Ranieri *et al.* 2012).

Here, we examine the timing of flowering and fruiting by Spanish moss (*Tillandsia usneoides* L.) populations across the species' broad geographic range in relation to availability of optimal physiological conditions (Barve *et al.* 2014). Physiological measurements have been made previously in year-long field experiments (Martin and Siedow 1981; Martin *et al.* 1981) to estimate intervals of climate-related parameters that are ideal for growth. We used herbarium specimen records of flowering and fruiting Spanish moss to identify population-specific flowering and fruiting periods and tested detailed environmental data for associations with minimum temperature, maximum temperature, relative humidity and rainless days requirements on a univariate basis, building on our earlier analyses of physiological limits in relation to climate across the range of this species (Barve *et al.* 2014). We use these analyses to test whether (i) all four parameters are at optimal physiological values as measured in previous studies during flowering periods and (ii) which physiological parameter(s) is (are) optimized during the flowering periods, if not all are optimized.

Methods

Study organism

Spanish moss (*T. usneoides*) is an epiphytic flowering plant of the family Bromeliaceae, distributed approximately between 38°N and 38°S latitude. It typically grows in warm and humid climates on trees or other supporting structures, such as power cables (Billings 1904; Garth 1964; Callaway *et al.* 2002). Spanish moss occurs over a broad elevational range (0–3300 m), and associations with atmospheric moisture content and temperature vary significantly according to elevation (Gentry and Dodson 1987; Kreft *et al.* 2004). The species does not occur at high elevations, which are apparently too cold for its persistence; indeed, its general natural history suggests that its distribution will prove to be highly constrained by climatic factors (Garth 1964), more or less in line with the 'Hutchinson's dream' scenario of Saupe *et al.* (2012).

Temperature, humidity and drought are known to affect growth and persistence of Spanish moss (Garth 1964; Martin and Siedow 1981; Martin *et al.* 1981; Martin and Schmitt 1989). A year-long field experiment (May 1978–May 1979) was performed by Martin *et al.* (1981) near Elizabethtown, NC, USA (78.594°W, 34.682°N); it found that Spanish moss growth is concentrated in summer months, with winter growth almost negligible. Martin *et al.* (1981) showed that CO_2 uptake was maximal when daytime temperature is 5–35 °C; CO_2 uptake was eliminated at or below 0 °C and at or above 40 °C. Kluge *et al.* (1973) also experimented on Spanish moss, with similar results regarding CO_2 uptake; however, they used greenhouse-grown Spanish moss, and their experiment was carried out in the laboratory under constant temperature and humidity. Martin *et al.* (1985, 1986) assessed North Carolina Spanish moss populations with respect to irradiance effects on morphology and physiology, finding that Spanish moss responds to irradiance by adjusting physiology more than morphology. Garth (1964) showed that Spanish moss cannot survive in Georgia without periodic rainfall, even when water is supplied externally; he found that Spanish moss achieves optimal performance in terms of growth only with $\leq$15 consecutive rainless days. Martin *et al.* (1981) corroborated this latter result, with the additional information that CO_2 uptake is minimal when Spanish moss is wet by rain, suggesting that Spanish moss requires some dry periods for persistence. Overall, then, these experiments identified four parameters that can be analysed at continental extents: minimum temperature $\geq$5 °C (Martin *et al.* 1981), maximum temperature $\leq$35 °C (Martin and Siedow 1981), night-time humidity $\geq$50 % (Martin *et al.* 1981) and $\leq$15 rainless days (Garth 1964).

Input data

We collected information on flowering and fruiting periods of Spanish moss populations by examining herbarium specimens. We photographed 430 specimens in the collections of the Missouri Botanical Garden and 504 specimens from the New York Botanical Garden collections using a 16 megapixel Nikon P510 camera. We took four to five photographs per specimen to capture various details: one of the label to permit capture of associated data, one of the whole specimen and two to three zoomed photographs of flowers or fruits. In addition, we reviewed published floras for flowering dates, although most floras either do not offer sufficient detail about flowering period or do not provide precise locality information. Finally, we downloaded images from various herbaria listed on the Index Herbariorum site (http://sciweb.nybg.

org/science2/IndexHerbariorum.asp) and others (http://herbarium.bio.fsu.edu and http://apps.kew.org/herbcat/navigator.do, 25 September 2015). Flowering and fruiting periods were assumed to be unimodal, so we filled temporal gaps for analyses of optimal physiological conditions. The temporal resolution of flowering and fruiting times was kept at months, so that imprecise date information (e.g. 'April 1914') could be incorporated, and quantity of relevant data maximized.

Information from specimen labels was digitized and stored in a Microsoft Access database. Some labels had geolocations in terms of latitude–longitude coordinates, whereas others had only textual locality information at various administrative levels. In the latter case, geolocations were attached to each record via queries in Google Earth. Overall, we were able to obtain information for 361 sites where both flowering date and geolocation information were available, which we used to profile flowering/fruiting periods at sites across the range of the species.

We examined how physiological thresholds are met (or not) for Spanish moss across the Americas within empirically documented flowering intervals over a 22-year period (January 1989–December 2010) following Barve *et al.* (2014). We used the ERA interim reanalysis climate data developed and supplied by the European Center for Medium-Range Weather Forecasts, which are based on a combination of models and observations, with three-hourly temporal resolution: every second datum is a forecast, whereas the other is a model result. We used only the model result data, thus coarsening the data from three- to six-hourly resolution, but retaining an impressively fine temporal resolution. The data set has a somewhat coarse native spatial resolution of $1.5^\circ \times 1.5^\circ$ or $\sim 165 \times 165$ km grid square resolution at the Equator.

ERA data were downloaded from http://apps.ecmwf.int/datasets/data/interim_full_daily/ for the following parameters: minimum temperature at 2 m, maximum temperature at 2 m, mean temperature at 2 m, dew point temperature at 2 m and precipitation. The data are stored in NetCDF format (http://www.unidata.ucar.edu/packages/netcdf/index.html; Rew and Davis 1990); these data were manipulated and processed via the 'ncdf' package in R (Pierce 2011; R Core Development Team 2012). ERA interim data were processed to identify optimal and suboptimal areas and temporal duration of suboptimal conditions with respect to each physiological variable through time.

Overall, one hundred and thirty-six 1.5° grid squares held at least one Spanish moss record with flowering and fruiting information. As numbers of flowering records were not numerous with respect to so many grid squares, to improve data density, we coarsened the 1.5° grid to 3° grids only to characterize flowering periods, but climate data were kept at the original 1.5° resolution. We generated flowering and fruiting month ranges for each 3° grid square; we assumed single flowering/fruiting months in grid squares in which only single specimens were available, which may be a restrictive assumption in our analyses. We also generated non-flowering month data sets for each grid square for comparison; for example, for a grid square with a flowering/fruiting range of March–May, we generated the remaining 11 possible 3-month sequences for comparison. We identified the average flowering/fruiting month, flowering/fruiting season start and flowering/fruiting season end for each grid square. Average flowering/fruiting month was calculated as a weighted average based on number of flowering or fruiting specimens in each month.

Data analysis

An R script was developed using the raster, ncdf and sp packages (Bivand *et al.* 2008; Pierce 2011; Hijmans and van Etten 2012) to calculate the percentage of time over the 22-year span of the data set that Spanish moss populations experienced optimal conditions with respect to the physiological thresholds described above. For minimum and maximum temperatures, the script checks the value of each variable across four daily observations; a grid square was marked as unsuitable for a day whenever two consecutive observations were outside the limit. For precipitation, whenever all four daily observations were 0 (i.e. no precipitation), it was considered as a day with no precipitation, and all consecutive sets of 15 days were checked; when any 15-day period had no precipitation, the grid square was considered as not suitable. For relative humidity, dew point temperature (T_d) and mean air temperature at 2 m (T_a) were used, and relative humidity was calculated as $Rh = e_s(T_d)/e_s(T_a)$, or the ratio of saturation vapour pressure, e_s, at dew point to that at air temperature, where e_s for any temperature T is given by $e_s(T) = 6.112 \times e^{(17.502 \times T/(240.97 + T))}$ (Stull 1988). We identified grid cells as unsuitable whenever two consecutive observations fell below the humidity threshold. Likewise, we calculated the percentage of time that the grid square spent outside its optimal physiological thresholds within the flowering period for that grid square across the 22-year time span; for comparison, we also generated these percentages for all possible non-flowering periods of similar duration.

We ranked each grid square based on the percentage of time spent outside optimal values in flowering and fruiting periods and each other possible non-flowering period of similar duration. We calculated the rank of each of the observed flowering periods with respect to all other possible periods of the same duration as the number of time periods of non-flowering months that are more suitable.

We used Kolmogorov–Smirnov test to compare distributions of the four variable ranks.

Based on ranks of each grid square for each of the variables, we compared the actual flowering period with the optimum flowering and fruiting periods with respect to those variables. This distance was calculated as a Euclidian distance from an optimal rank of 1 for each of the variables. This Euclidian distance is normalized to a scale of 0–1, such that small distances indicate optimal flowering and fruiting periods for a population, whereas large distances suggest that the population flowers during suboptimal periods. We averaged this distance across all four physiological parameters and mapped these deviations from optimum. We tested for effects of number of herbarium specimen records in each grid square to these optimum distances.

Results

We were able to assemble 361 flowering or fruiting records for Spanish moss across the species' range. Although records concentrated in the US portion of the species' range (159 records, or 44 %), the remaining 202 (56 %) records came from Latin America. Although densities of Latin American points were low at finer spatial resolutions (i.e. most grid squares had single or no flowering-period records), 3° spatial resolution was sufficient to create 83 grid squares, within which we had 1–28 flowering/fruiting records.

Average flowering and fruiting month of Spanish moss populations across the species' range is shown in Fig. 1. The flowering and fruiting periods in eastern Brazil were November–April, while the flowering and fruiting periods in western South America were June–September, with a few exceptions extending to October–November. The flowering period in the USA and Mexico was May–September, with a few exceptions in November–December. Because our identification of flowering and fruiting month(s) was in some sense dependent on numbers of specimens available, we suspect that insufficient data density may be driving the exceptions.

Flowering periods invariably fell in time periods in which least one physiological parameter was optimal in a grid square. The 'optimal' parameter was generally minimum temperature: that is, in 212 of 262 cases, flowering and fruiting periods coincided with months in which minimum temperature was within optimal ranges in at least 70 % of grid square–month combinations. Very few populations experienced minimum temperatures below the 5 °C criterion during their respective flowering and fruiting periods (Fig. 2A). The remainder of Fig. 2 suggests that flowering and fruiting periods depend less critically on parameters like maximum temperature, rainless periods or relative humidity. Ranking months by their optimality for each parameter, a Kolmogorov–Smirnov test revealed that distributions of ranks for minimum temperature were significantly lower than those for the other three factors ($P < 0.0001$, Table 1). The distribution of ranks among grid squares did not differ between maximum temperature and rainless days, whereas ranks of rainless days versus relative humidity showed the latter as significantly more optimal ($P < 0.001$, Table 1).

We identified the optimal month for each pixel across the Americas in terms of each dimension of Spanish moss physiology. Figure 3A shows the optimal flowering and fruiting months for minimum temperature, which centred on July at the northern limit of the distribution, but in January–April at the southwestern distributional limit. However, for maximum temperature, the average expected flowering/fruiting month was February–April at the northern limit of distribution and April–August at the southwestern limit. Similar variation can be seen for relative humidity and rainless days: in short, no pixel had any period in which all four physiological parameters were in optimal states for flowering and fruiting to occur.

To explore how far observed flowering and fruiting months departed from optimal months, we calculated average Euclidean distance in four-dimensional parameter space, ranking months by their suitability, standardizing each dimension to a range of 0–1 (thus creating an index of distance that has rather unclear units but that is useful for visualization) and counting ranks as greater distance from optimal conditions (Fig. 4). Most populations (46 %) showed flowering and fruiting periods with ranked Euclidean distances of ≤0.5. Only a few pixels were under extremely bad conditions and these higher-distance populations were arrayed at the extremes of the distribution (Fig. 4). We tested whether number of available flowering/fruiting specimens affected these latter results (see scatterplot inset in Fig. 4) but found no effect of sample size on distance to optimal month.

Discussion

In overview, we found that Spanish moss populations appear to 'tune' their phenological niches such that they experience optimum minimum temperatures for most of their respective flowering and fruiting periods. Among populations analysed, flowering and fruiting periods of ~76 % of Spanish moss populations experienced optimal minimum temperatures when compared with other time periods throughout the year. Conversely, Spanish moss populations appear to flower and fruit without much consideration of optimality of maximum temperature or relative humidity optimality, though rainless days do have some importance.

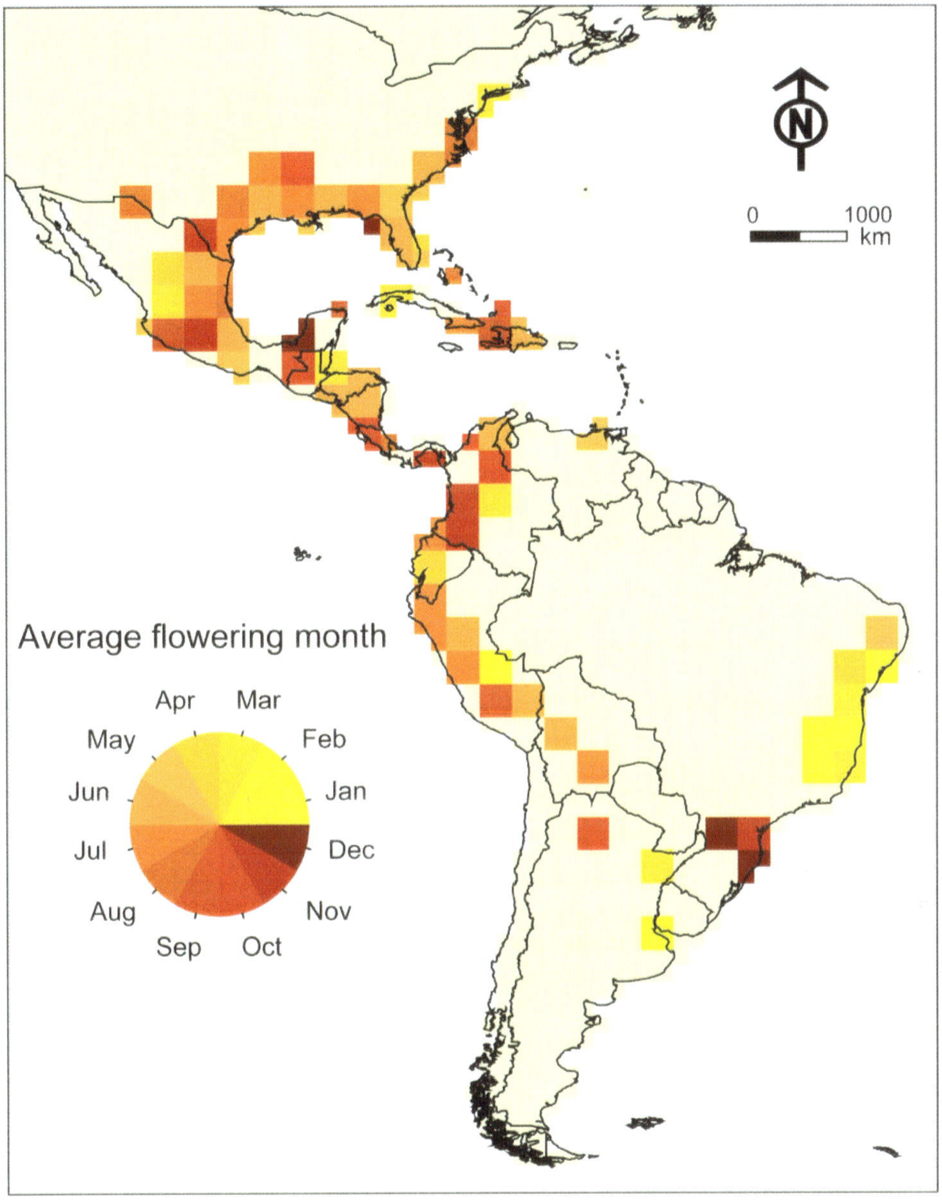

Figure 1. Average flowering month of Spanish moss populations across the Americas calculated as weighted average of flowering or fruiting specimens recorded from each grid square.

Numerous recent studies have documented shifts in flowering and fruiting season as a consequence of climate change (Molau *et al.* 2005; Miller-Rushing and Primack 2008; Telemeco *et al.* 2013). Veriankaitė *et al.* (2010) explored optimum temperatures for flowering and fruiting by comparing air temperatures in climate models with long-term flowering data. However, for this study, we took advantage of known optimum physiological parameters (Martin and Siedow 1981; Martin *et al.* 1981; Martin and Schmitt 1989), so we could explore the degree to which Spanish moss flowering and fruiting periods coincide with months presenting optimal physiological conditions for growth.

Phenological differences are well documented as functions of elevation and latitude (Ruml *et al.* 2011; CaraDonna *et al.* 2014). However, we generated our phenological information from herbarium specimens: few had elevation information, so effects of elevation on flowering phenology cannot be examined particularly in light of the coarse spatial resolution of our weather data. Clearly, as the climate data are coarsened and averaged over broader extents, such details average out in the climate and become invisible to our analyses, as was noted in our previous analyses (Barve *et al.* 2014). Our analyses may also be compromised by our rather coarse characterization of flowering and fruiting periods (i.e. to month) and by our

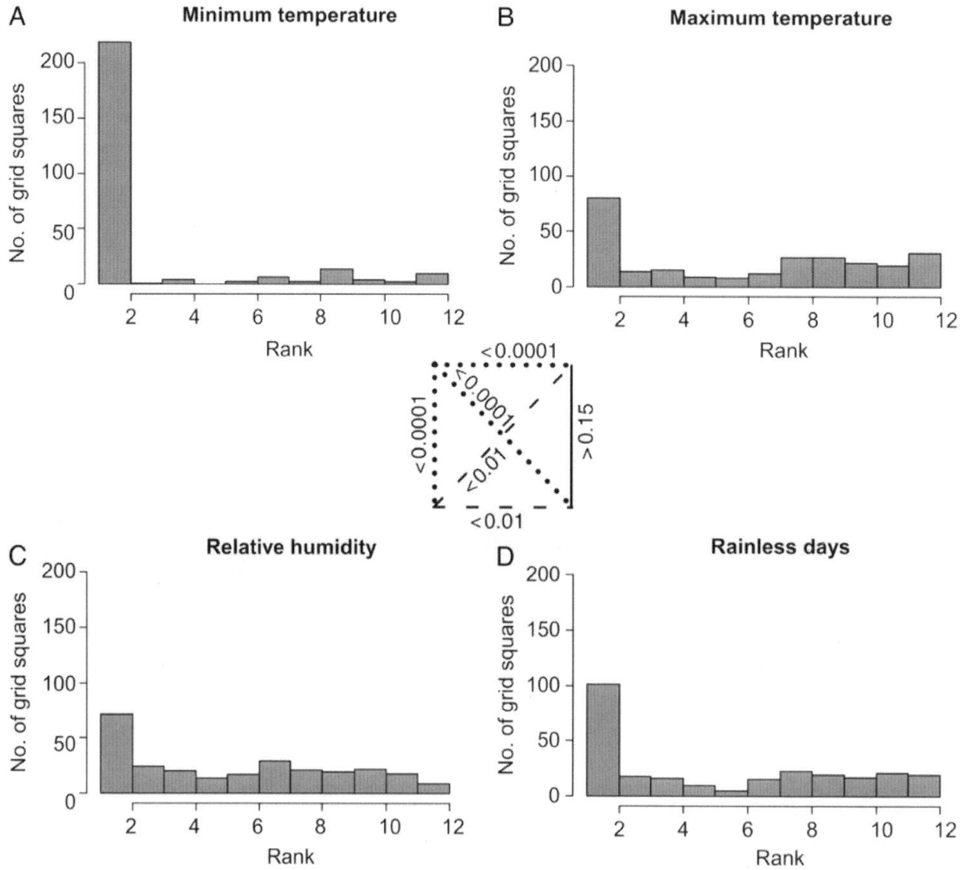

Figure 2. Histogram of ranks based on how long populations in each grid square are outside optimal conditions for each of the four parameters during their flowering and fruiting periods. The box at the centre shows results of Kolmogorov–Smirnov tests for comparison of distributions. Dotted lines indicate highly significant difference, dashed lines significant differences and continuous lines non-significant difference.

Table 1. Result of Kolmogorov–Smirnov test to compare distributions of ranks within four variables, viz, minimum temperature, maximum temperature, rainless days and relative humidity. Upper diagonal are the *P*-values and lower diagonal represents the test statistics.

***P*-value** / **Test statistics**	**Minimum temperature**	**Maximum temperature**	**Rainless days**	**Relative humidity**
Minimum temperature		<0.0001	<0.0001	<0.0001
Maximum temperature	0.5417		>0.1	<0.005
Rainless days	0.4886	0.1023		0.006
Relative humidity	0.6364	0.1402	0.1477	

filling of temporal gaps in flowering periods under the assumption of a single, continuous flowering and fruiting period for each population.

For Spanish moss, we observed that flowering phenology does not generally depend much on maximum temperature. Rather, minimum temperature appears to play a major role (Fig. 2). Comparisons with every other period of similar length in the year for each location suggested that Spanish moss flowering and fruiting periods are moulded such that flowering populations experience optimal minimum temperatures. Hence, an interesting challenge for long-term studies would be to test whether Spanish moss flowering and fruiting advances temporally in relation to rising minimum temperatures, rather than other climate characteristics of warming climates.

In our trade-off maps (Figs 2 and 4), we see that most Spanish moss populations show trade-off distances of 0.5 or less; nonetheless, some populations showed more substantial trade-off distances. Spanish moss populations under such suboptimal conditions likely face challenges to long-term persistence, suggesting that optimality of conditions in flowering period represents a constraint

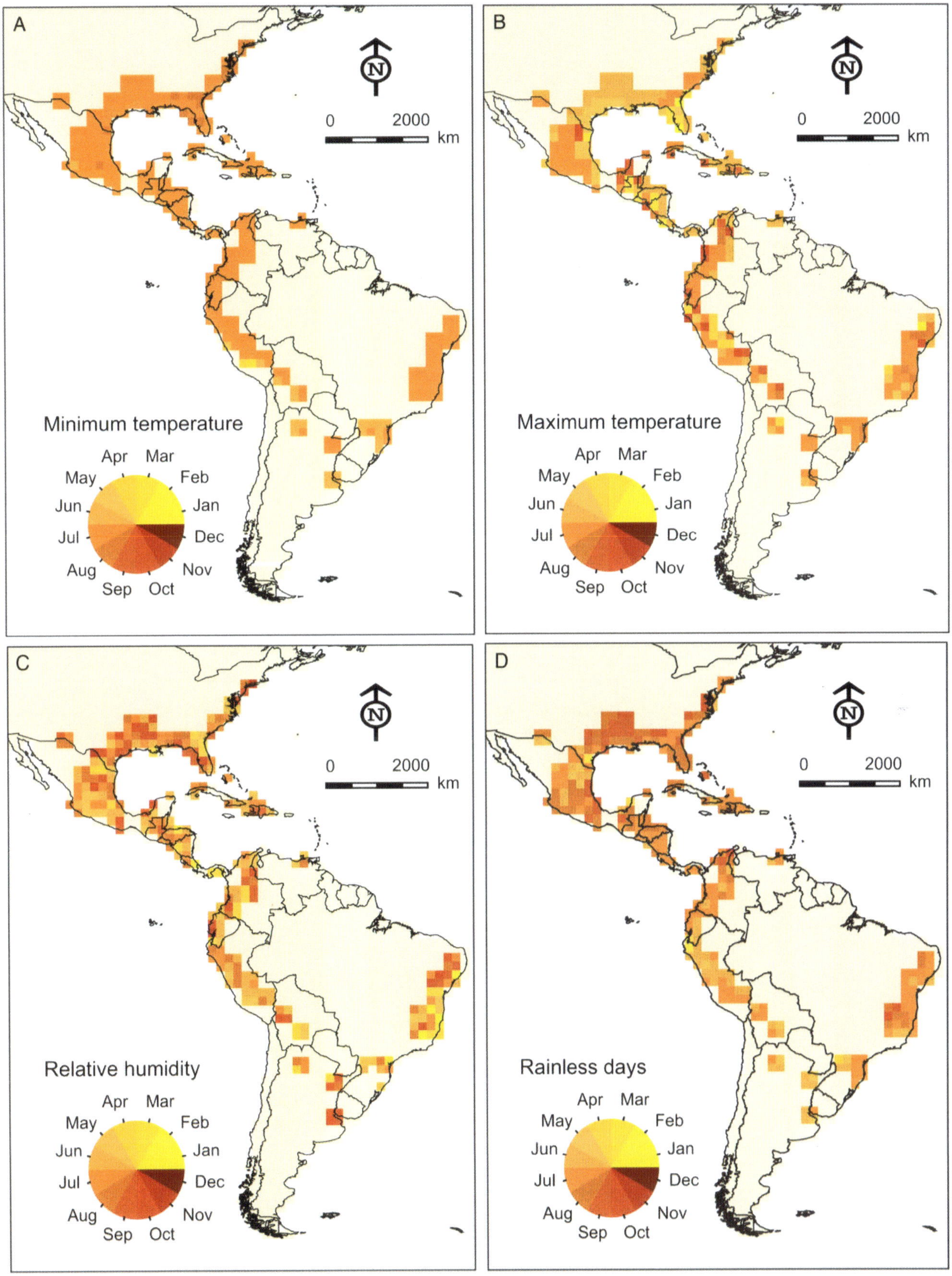

Figure 3. Optimal flowering and fruiting months for Spanish moss populations based on each physiological parameter in isolation.

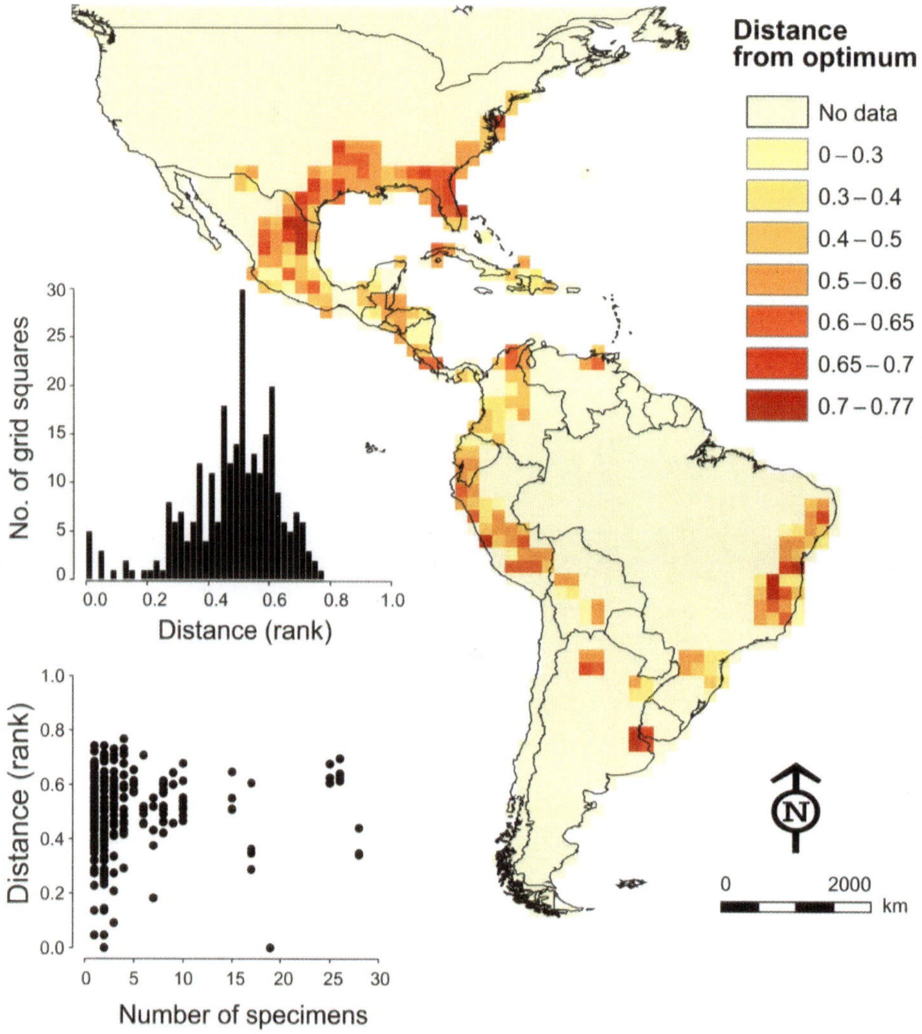

Figure 4. Map of Euclidean distances from observed conditions for Spanish moss populations to the best available across the species' distribution. Inset shows a frequency histogram of distances in grid squares (top) and relationship to numbers of specimens on which distance calculations were based (bottom).

on Spanish moss geographic distributions. Although it is hard to say whether or to what degree climate change will change the geographic distributional potential of Spanish moss, Spanish moss may not flower and produce seeds successfully if climate change takes populations too far from optimal conditions. Even under present-day conditions, our approach can be used to locate where populations of the species will be under particular physiological stress.

Conclusions

We analysed high-temporal-resolution (6-h resolution) climate data over a 22-year span to assess the availability of optimal conditions during flowering and fruiting periods of Spanish moss populations. Our results indicate that Spanish moss populations appear to flower and produce fruit seasonally such that populations experience optimum minimum temperatures. Our finding also shows that the least optimal conditions are experienced by populations along the fringes of the species' distribution. This research is novel in that we used herbarium specimens to assign flowering period to populations, that actual physiological measurements were used to assess optimality of conditions and that high-temporal-resolution weather data were used to provide a near-real-time view of the environmental conditions experienced by the species.

Sources of Funding

This research was supported partly by Microsoft research grant to A.T.P., and N.B. was supported by grants from the Association of Women Geographers and Panorama Fund from the University of Kansas, Biodiversity Institute.

Contributions by the Authors

N.B. and A.T.P. conceived the ideas, N.B. collected and analysed the data, N.B. led the writing and all authors contributed to the writing process.

Acknowledgements

We thank Dr Jim Solomon, Mary McNamara from Missouri Botanical Garden, and Lisa Fruscella and Stella Sylva from New York Botanical Garden for allowing us to access herbarium specimens under their care. We also thank Dr Vijay Barve for assisting in photographing the herbarium specimens. We thank Dr Jorge Soberón and Dr Nathaniel Brunsell for insightful discussion.

Literature Cited

Amasino R. 2010. Seasonal and developmental timing of flowering. *The Plant Journal* **61**:1001–1013.

Barve N, Martin C, Brunsell NA, Peterson AT. 2014. The role of physiological optima in shaping the geographic distribution of Spanish moss. *Global Ecology and Biogeography* **23**:633–645.

Bernier G, Havelange A, Houssa C, Petitjean A, Lejeune P. 1993. Physiological signals that induce flowering. *The Plant Cell* **5**: 1147–1155.

Billings FH. 1904. A study of *Tillandsia usneoides*. *Botanical Gazette* **38**:99–121.

Bivand RS, Pebesma EJ, Gómez-Rubio V. 2008. *Applied spatial data analysis with R*. New York: Springer.

Bond WJ, Midgley JJ. 2001. Ecology of sprouting in woody plants: the persistence niche. *Trends in Ecology and Evolution* **16**:45–51.

Callaway RM, Reinhart KO, Moore GW, Moore DJ, Pennings SC. 2002. Epiphyte host preferences and host traits: mechanisms for species-specific interactions. *Oecologia* **132**:221–230.

CaraDonna PJ, Iler AM, Inouye DW. 2014. Shifts in flowering phenology reshape a subalpine plant community. *Proceedings of the National Academy of Sciences of the USA* **111**:4916–4921.

Dennis ES, Finnegan EJ, Bilodeau P, Chaudhury A, Genger R, Helliwell CA, Sheldon CC, Bagnall DJ, Peacock WJ. 1996. Vernalization and the initiation of flowering. *Seminars in Cell and Developmental Biology* **7**:441–448.

Engelhardt MJ, Anderson RC. 2011. Phenological niche separation from native species increases reproductive success of an invasive species: *Alliaria petiolata* (Brassicaceae)—garlic mustard. *Journal of the Torrey Botanical Society* **138**:418–433.

Fowler NL. 1988. What is a safe site?: Neighbor, litter, germination date, and patch effects. *Ecology* **69**:947–961.

Garner WW. 1933. Comparative responses of long-day and short-day plants to relative length of day and night. *Plant Physiology* **8**: 347–356.

Garth RE. 1964. The ecology of Spanish moss (*Tillandsia usneoides*): its growth and distribution. *Ecology* **45**:470–481.

Gentry AH, Dodson CH. 1987. Diversity and biogeography of Neotropical vascular epiphytes. *Annals of the Missouri Botanical Garden* **74**:205–233.

Grubb PJ. 1977. The maintenance of species-richness in plant communities: the importance of the regeneration niche. *Biological Reviews* **52**:107–145.

Hijmans RJ, van Etten J. 2012. *raster: geographic analysis and modeling with raster data*. http://CRAN.R-project.org/package=raster (10 January 2014).

Kluge M, Lange OL, Eichmann M, Schmid R. 1973. Diurnaler säurerhythmus bei *Tillandsia usneoides*: untersuchungen über Weg des Kohlenstoffs sowie die Abhängigkeit des CO_2-Gaswechsels von Lichtintensität, temperature und Wessergehalt der Pflanze. *Planta* **112**:357–372.

Kreft H, Köster N, Küper W, Nieder J, Barthlott W. 2004. Diversity and biogeography of vascular epiphytes in western Amazonia, Yasuní, Ecuador. *Journal of Biogeography* **31**:1463–1476.

Lang A. 1952. Physiology of flowering. *Annual Review of Plant Physiology* **3**:265–306.

Lavorel S, Chesson P. 1995. How species with different regeneration niches coexist in patchy habitats with local disturbances. *Oikos* **74**:103–114.

Martin CE, Schmitt AK. 1989. Unusual water relations in the CAM atmospheric epiphyte *Tillandsia usneoides* L. (Bromeliaceae). *Botanical Gazette* **150**:1–8.

Martin CE, Siedow JN. 1981. Crassulacean acid metabolism in the epiphyte *Tillandsia usneoides* L. (Spanish moss): responses of CO_2 exchange to controlled environmental conditions. *Plant Physiology* **68**:335–339.

Martin CE, Christensen NL, Strain BR. 1981. Seasonal patterns of growth, tissue acid fluctuations, and $^{14}CO_2$ uptake in the crassulacean acid metabolism epiphyte *Tjllandsia usneoides* L. (Spanish moss). *Oecologia* **49**:322–328.

Martin CE, McLeod KW, Eades CA, Pitzer AF. 1985. Morphological and physiological responses to irradiance in the CAM epiphyte *Tillandsia usneoides* L. (Bromeliaceae). *Botanical Gazette* **146**: 489–494.

Martin CE, Eades CA, Pitner RA. 1986. Effects of irradiance on crassulacean acid metabolism in the epiphyte *Tillandsia usneoides* L. (Bromeliaceae). *Plant Physiology* **80**:23–26.

Miller-Rushing AJ, Primack RB. 2008. Global warming and flowering times in Thoreau's Concord: a community perspective. *Ecology* **89**:332–341.

Molau U, Nordenhäll U, Eriksen B. 2005. Onset of flowering and climate variability in an alpine landscape: a 10-year study from Swedish Lapland. *American Journal of Botany* **92**:422–431.

Pederson N, Cook ER, Jacoby GC, Peteet DM, Griffin KL. 2004. The influence of winter temperatures on the annual radial growth of six northern range margin tree species. *Dendrochronologia* **22**:7–29.

Pierce D. 2011. *ncdf: interface to Unidata netCDF data files*. http:// CRAN.R-project.org/package=ncdf (10 January 2014).

Ranieri BD, Pezzini FF, Garcia QS, Chautems A, França MGC. 2012. Testing the regeneration niche hypothesis with Gesneriaceae (tribe Sinningiae) in Brazil: implications for the conservation of rare species. *Austral Ecology* **37**:125–133.

R Core Development Team. 2012. *R: a language and environment for statistical computing*. Vienna: R Foundation for Statistical Computing.

Rew R, Davis G. 1990. NetCDF: An interface for scientific data access. *IEEE Computer Graphics and Applications* **10**:76–82.

Ruml M, Vulić T, Đorđević B. 2011. The effect of altitude and latitude on the phenology of the plum cv Požegača in Serbia. In: *46th*

Coratian and 6th International Symposium on Agriculture. Croatia: Opatiga, 1052–1055.

Saupe EE, Barve V, Myers CE, Soberón J, Barve N, Hensz CM, Peterson AT, Owens HL, Lira-Noriega A. 2012. Variation in niche and distribution model performance: the need for a priori assessment of key causal factors. *Ecological Modelling* **237–238**:11–22.

Silvertown J. 2004. Plant coexistence and the niche. *Trends in Ecology and Evolution* **19**:605–611.

Soberón J. 2007. Grinnellian and Eltonian niches and geographic distributions of species. *Ecology Letters* **10**:1115–1123.

Stull RB. 1988. *An introduction to boundary layer meteorology*. New York: Springer.

Sweeney AW, Beebe NW, Cooper RD, Bauer JT, Peterson AT. 2006. Environmental factors associated with distribution and range limits of malaria vector *Anopheles farauti* in Australia. *Journal of Medical Entomology* **43**:1068–1075.

Telemeco RS, Abbott KC, Janzen FJ. 2013. Modeling the effects of climate change-induced shifts in reproductive phenology on temperature-dependent traits. *The American Naturalist* **181**: 637–648.

Tilman D. 1990. Constraints and tradeoffs: toward a predictive theory of competition and succession. *Oikos* **58**:3–15.

Veriankaitė L, Šaulienė I, Bukantis A. 2010. The modelling of climate change influence on plant flowering shift in Lithuania. *Zemdirbyste-Agriculture* **97**:41–48.

Wellenreuther M, Larson KW, Svensson EI. 2012. Climatic niche divergence or conservatism? Environmental niches and range limits in ecologically similar damselflies. *Ecology* **93**:1353–1366.

6

Nitrogen addition and harvest frequency rather than initial plant species composition determine vertical structure and light interception in grasslands

Ute Petersen[1,2]* and Johannes Isselstein[1]

[1] Department of Crop Sciences, University of Göttingen, Von-Siebold-Str. 8, D-37075 Göttingen, Germany

[2] Present address: Johann Heinrich von Thünen Institute, Federal Research Institute for Rural Areas, Forestry and Fisheries, Institute of Climate-Smart Agriculture (AK), Bundesallee 50, D-38116 Braunschweig, Germany

Associate Editor: Katharine J.M. Dickinson

Abstract. In biodiversity experiments based on seeded experimental communities, species richness and species composition exert a strong influence on canopy structure and can lead to an improved use of aboveground resources. In this study, we want to explore whether these findings are applicable to agriculturally managed permanent grassland. Vertical layered profiles of biomass, leaf area (LA) and light intensity were measured in a removal-type biodiversity experiment (GrassMan) to compare the canopy structure in grassland vegetation of different plant species composition (called sward types). Additionally, the altered sward types were subjected to four different management regimes by a combination of the factors fertilization (unfertilized, NPK fertilized) and cutting frequency (one late cut or three cuts). In spite of large compositional differences (ratio grasses : non-leguminous forbs : leguminous forbs ranging from 93 : 7 : 0 to 39 : 52 : 9), the vegetation of the same management regime hardly differed in its canopy structure, whereas the different management regimes led to distinct vertical profiles in the vegetation. However, the allocation of biomass in response to cutting and fertilization differed among the sward types. Vegetation dominated by grasses was denser and had more LA when fertilized compared with vegetation rich in dicots which merely grew taller. In functionally more diverse vegetation, light interception was not increased compared with vegetation consisting of more than 90 % of grasses in terms of biomass. Management had a much stronger influence on structure and light interception than plant species composition in this grassland experiment.

Keywords: Canopy structure; dicots; light interception; monocots; permanent grassland; plant species composition.

Introduction

Canopy structure, the 3D arrangement of all aboveground plant components in space, is quite diverse in grasslands, despite their relatively low canopy height compared with woody vegetation (Körner 1994). There are not only species with horizontally arranged leaves and species with steep-angled leaves; there are also small and tall species leading to several distinct growth forms (Liira and Zobel 2000).

The determinants of canopy structure in permanent grasslands have been the subject of many scientific studies. Most of these studies have included some kind of

* Corresponding author's e-mail address: Ute_Petersen@gmx.net

classification according to the growth form of the different plants (e.g. Liira and Zobel 2000). The density, total plant height and vertical distribution of leaf area (LA) and biomass in a grassland canopy are influenced by water and nutrient content of the soil as well as by the frequency and type of biomass removal (pasture vs. meadow) (Werger *et al.* 1986; Mitchley and Willems 1995; Liira and Zobel 2000; Lienin and Kleyer 2012). Further, research on the leaf area index (LAI) of single species has been conducted (Sheehy and Cooper 1973; Sheehy and Peacock 1977; Haase *et al.* 1999; Cheng *et al.* 2009) with early studies by Brougham (1958) on LAI and light interception, emphasizing the importance of different leaf angles in a grassland canopy.

Within the still expanding research area on investigating the relationship between biodiversity and ecosystem processes, vertical structure and LAI in grasslands have generally been analysed in relation to species richness. On the one hand, the vertical structure has been shown to exert a strong influence on biodiversity due to its impact on light availability (Mitchley and Willems 1995; Lamb 2008), whilst on the other hand, species richness and species composition can also influence vertical structure. Indeed, Spehn *et al.* (2000), Fridley (2003), Lorentzen *et al.* (2008) and Vojtech *et al.* (2008) all found a positive effect of species richness on LAI and biomass, and concluded that resource partitioning of light and complementary use of aboveground space plays a role in canopy development, especially in environments where nutrients are not limiting. Since space *per se* is not a resource plants actively compete for (Wilson *et al.* 2007), the term 'complementary use of space' here in this article refers to the arrangement of leaves in space to indicate the effectiveness of light interception. However, the previously mentioned authors (Spehn *et al.* 2000; Fridley 2003; Lorentzen *et al.* 2008; Vojtech *et al.* 2008) used seeded, experimental plant communities and the applicability of results from these young, immature plant communities on real-world ecosystems is currently debated (Schmid 2002; Thompson *et al.* 2005; Hector *et al.* 2007; Mokany *et al.* 2008; Jiang *et al.* 2009). One criticism is that many of these types of experiments are not managed according to common agricultural practices in managed European grasslands (several cuts/grazing events per year and appropriate fertilization) (Wrage *et al.* 2011), thus lowering the transferability of results obtained from studies of experimental plant communities to actual managed grasslands. In addition, biodiversity experiments are generally designed with only a few species used (up to five, e.g. Vojtech *et al.* 2008; Nyfeler *et al.* 2009) or their species richness levels follow a logarithmic scale (1, 2, 4, 8, 16, 32 species, e.g. Spehn *et al.* 2000; Roscher *et al.* 2004; Tilman *et al.* 2012). Hence, only two of these experimental species richness levels (8 and 16 species) are comparable with species richness in managed grasslands in Europe where species richness ranges from 7 to 25 species per square metre (Pywell *et al.* 2007; Scimone *et al.* 2007). Therefore, the applicability of the experimental results to managed grasslands in Europe is questionable. Another approach, and a reasonable complement to biodiversity experiments, is removal experiments, where one part of the vegetation (most of the time one or several functional groups) is removed manually or by herbicide application (Díaz *et al.* 2003; McLaren and Turkington 2010; Petersen *et al.* 2012).

In 2008, we set-up a removal experiment and manipulated permanent grassland by herbicide application to obtain three different types of grassland vegetation (hereafter referred to as sward types); vegetation containing, in terms of biomass, <5 % dicots (-Dic); vegetation with a ratio monocots : dicots = 50 : 50 (-Mon) and untreated control vegetation (Co; for further details see Petersen *et al.* 2012 and the Methods section). With our experiment, we tried to close the gap between low (<10) and high (>30) species numbers per experimental unit. In the first year after the set-up, the grassland vegetation of the experimental plots had a species richness ranging from 7 to 16/m^2 or 16–28 per experimental unit (here 225 m^2). Roy (2001) outlined in his meta-analysis of biodiversity–ecosystem functioning studies that the largest effects were found at species numbers up to four. Clearly, we have more than four species in this experiment and therefore expect species richness effects on canopy structure and light interception to be small. However, we predict that over and above species richness functional and compositional differences amongst the sward types due to different growth forms are important factors to be taken into account. We hypothesize that the results of the seeded experiments mentioned above (Spehn *et al.* 2000; Fridley 2003; Lorentzen *et al.* 2008; Vojtech *et al.* 2008) are transferrable to managed grassland. Hence, the following two hypotheses should apply to the vegetation of our experimental plots: (a) depending on the fertilization and cutting regime, the species composition of the vegetation determines its vertical structure and LAI and (b) coexistence of different growth forms (i.e. different leaf angles) in a functionally more diverse grassland vegetation leads to better light interception compared with grass-dominated vegetation, because different growth forms lead to a more efficient arrangement of leaves for light interception.

Methods

Biodiversity experiment

The grassland management (GrassMan) experiment is a removal-type biodiversity experiment incorporating different management intensities. It was established in 2008 on permanent grassland between Silberborn and Neuhaus in the Solling Uplands, Germany (51°44′53″N, 9°32′42″E, 490 m above sea level (a.s.l.)). Prior to the set-up of the experiment, the grassland had been used as summer pasture for cattle. The area has a mean annual temperature of 6.9 °C and mean annual precipitation of 1028 mm (German Meteorological Service, DWD 1960–91, station Silberborn-Holzminden, 440 m a.s.l.). The soil type of the experimental area was classified as Inceptisol, more specifically as Haplic Cambisol (Keuter *et al.* 2013) and its vegetation was a nutrient-poor *Lolio-Cynosuretum* Br.-Bl. et De Leeuw 1936.

GrassMan is a three-factorial experiment (Table 1). The three sward types were obtained by a single herbicide application in July 2008 targeting dicotyledonous and monocotyledonous species, respectively. For the scheduled harvests and yield determination in 2009 we used a Haldrup® forage combine harvester with a cutting height of 7 cm.

The combination of all factors and levels resulted in 12 experimental treatments, which were replicated six times. The treatment names were generated by a combination of the abbreviation of experimental factor levels in the order sward–utilization–nutrients, e.g. Co3x = control sward, cut three times, no fertilization. The two factors simulating the intensity of agricultural management, i.e. fertilization and cutting, will later be referred to as management regimes. To account for potential spatial gradients, the 72 experimental plots were arranged to form a Latin Rectangle with the upper part bordering a forest sloping gently (<5 % slope) towards the lower rows. Two columns form one block. A detailed description of the experiment can be found in Petersen *et al.* (2012).

Table 1. Experimental factors and treatment levels of the GrassMan experiment. The acronyms of the different treatments are generated by combination of the factor level abbreviations in the order sward–utilization–nutrients, e.g. -Dic1x = dicot-reduced sward, cut once, no fertilization. *N fertilizer: calcium ammonium nitrate N27, P&K fertilizer: Thomaskali® (8 % P_2O_5, 15 % K_2O, 20 % CaO).

Factor	Level	Abbreviation
Sward type	1.1 untreated control sward	Co
	1.2 dicots reduced	-Dic
	1.3 monocots reduced	-Mon
Utilization	2.1 1 cut (July)	1
	2.2 3 cuts (May, July, September)	3
Nutrients	3.1 no fertilization	*X*
	3.2 180/30/100 kg NPK ha^{-1} $year^{-1}$*	NPK

Biomass and measurements for canopy structure

To describe the canopy structure, we used the vertical layer-wise distribution of biomass within the vegetation. Here, biomass comprised the total aboveground biomass of a defined area, not just the part which was harvested for agricultural purposes (yield). The spatial arrangement of biomass was analysed by stratified clipping (Monsi and Saeki 1953; Sheehy and Cooper 1973), i.e. harvesting the biomass in 10 cm thick layers along the canopy. Prior to each of the scheduled harvests in 2009 (May, July and September 2009), we pegged three randomly selected 15 × 15 cm quadrats per plot (=subsample). The generative growing points, i.e. the buds of all dicotyledonous species, were counted before cutting the subsample at the ground level. The cut bunch of grass was separated into 10 cm layers starting at the ground level, i.e. separated into volumes of 15 × 15 × 10 cm except for the lowest layer which was 7 cm thick, corresponding to the cutting height of the forage harvester to ensure comparability of total biomass and harvested yield at the different harvest dates. The material of each layer was placed in a plastic bag and frozen before further processing. The three subsamples of each plot corresponding to the same layer were combined to one sample and sorted to the three functional groups [grasses, non-leguminous forbs and leguminous forbs (later called legumes)] and dead plant material (=standing litter). The sorted material was dried at 105 °C for 48 h and weighed. Additionally, the number of grass tillers and vegetative growing points of clover in the lowest layer were counted. Both vegetative and generative growing points will hereafter be referred to as buds.

From these data, we calculated the harvest index, i.e. the ratio of economic yield to biological yield or total aboveground biomass. This index originates from crop yield calculations, e.g. grain mass versus total biomass of cereals (Donald 1962). Moreover, we determined the centre of gravity (*c*) per plot according to Spehn *et al.* (2000) as a measure of vertical biomass density.

$$c = \frac{1}{m_t} \times \sum_{l=1}^{n} (m_l \times h_l)$$

where m_l is the biomass of one layer, h_l the corresponding average height of this layer (here 10 cm or 7 cm) and m_t the total biomass of the profile. The centre of gravity indicates the canopy height (in cm), around which most of the biomass is concentrated. The centre of gravity is raised in very tall canopies compared with short canopies or in canopies of equal height where the biomass is distributed more evenly across the different canopy layers, i.e. where the vertical profile has a convex rather than a concave shape.

Leaf area index and light measurements

The LAI was determined at the time of peak standing biomass at the second harvest (first harvest in the 1-cut treatments) in July. The LAI of the subsamples was not measured directly but calculated by multiplying the specific leaf area (SLA) of each biomass fraction (stems and leaves of grasses, non-leguminous forbs and legumes) by the corresponding biomasses harvested by the stratified clipping. For the determination of SLA, a mixed sample of fresh plant material from each block separated according to cutting regime was taken, put into a cool box and transferred to the laboratory immediately. Legumes were collected separately from plots belonging to the same utilization × nutrient combination, since preceding tests had shown that in contrast to non-leguminous forbs and grasses, the SLA of the legumes (consisting of more than 90 % *Trifolium repens* L.) was influenced not only by cutting regime but also by nutrient supply. We followed the protocol for determining SLA by Garnier *et al.* (2001) and kept the material damp and cool before scanning 20–30 subsamples of stems and leaves of each functional group using the software WinRhizo (Version 2007 d, Regent Instruments, Inc., Quebec, Canada). The scanned samples were dried at 105 °C for 48 h before weighing. From a regression between dry biomass and scanned leaf area, SLA was calculated. Owing to the large heterogeneity of stem diameters in legumes and non-leguminous forbs, the goodness of fit of the mass–area regression was more variable for all forbs than for grasses (adjusted R^2 between 0.68 and 0.95 for forbs, from 0.85 to 0.99 for grasses). Additionally, the stem data of non-leguminous forbs from one block were missing completely and would have had to be estimated from the other data. In all plots containing a certain amount of forbs, this would have led to a larger variability in estimated LA than in plots with more than 90 % grasses. To eliminate these biased values, we decided to exclude all LA calculated from stem biomass from further analyses and only used LA from leaf blades. Thus, the leaf area ratio (LAR, leaf area per sample dry weight, $mm^2\ mg^{-1}$) was not based on the total leaf area (Ryser and Lambers 1995) but on the leaf area of laminae, likewise the LAI (called LAI leaves from here on).

Prior to each harvest, the light interception was measured using the SunScan canopy analysis system (Typ SS1; Delta-T Devices Ltd, Cambridge, UK) with a portable reference sensor directly above the canopy of each plot. The photosynthetically active photon flux density (PPFD) was measured in vertical profiles from different heights within and above the canopy (0, 7, 17, 27 cm and so on up to 117 cm in the tallest canopies) each with three replicates per plot, corresponding to the biomass layers in the subsamples cut afterwards.

The light extinction coefficient *k* was calculated following the Lambert–Beer extinction law $I = I_0 \times e^{-k\mathrm{LAI}}$ by using the determined LAI of the leaves and light intensity at the ground level (Saeki 1959; Spehn *et al.* 2000), where I = PPFD at the ground level and I_0 = PPFD above the canopy measured by the reference sensor. The extinction coefficient *k* does not only describe how readily the incoming light is intercepted (= absorbed or reflected back); it also gives a hint to the distribution of leaf angles within the canopy (Geyger 1977). The more vertically orientated the leaf blades are, the lower is their extinction coefficient and vice versa. The species richness and the functional diversity, i.e. the proportions of grasses, non-leguminous forbs and legumes, of the vegetation had been determined by vegetation relevés (9 m^2) in May and August 2009. Since the three sward types Co, –Dic and –Mon differed significantly in their proportions of functional groups with grass contents of 70, 92 and 45 %, respectively, and with almost no legumes present in the –Dic-swards (Petersen *et al.* 2012), the sward types were taken as surrogates for functional diversity in this study with functional diversity increasing from –Dic via Co-swards to –Mon-swards.

Statistical analysis

Univariate statistics were calculated in R version 2.12.2 (R Development Core Team 2011). Since in an analysis of variance (ANOVA), the order of explaining variables is important (see example given below), row and block were included first of all factors into all (general) linear models (GLM/LM) we applied on our data. With this order of factors, we were able to extract all environmental variability before assessing treatment effects. The effects of the experimental factors sward, utilization and nutrients on biomass, centre of gravity, harvest index, light interception and extinction, LAI and LAR were tested using a three-way ANOVA including all treatment factors and their interactions. For example harvest index = row + block + sward × nutrients × utilization (data from July harvest) and harvest index = row + block + sward × nutrients (May and September data). For biomass, LAI and LAR we further fitted alternative models including species richness instead of sward type, to compare the effects of species richness (one aspect of species composition) on our subplots with those obtained on seeded experimental plots. The models were simplified by dropping insignificant interactions and factors (apart from row and block) following the instructions given in Zuur *et al.* (2009) using the Aikaike Information criterion (AIC) to detect the best fitting models. For comparison of means of several factor levels, we used Tukey's 'Honest Significant Difference' method (Miller 1981) with a confidence level of 0.95. To ensure better comparability of the layer-wise distribution of LA and biomass, these data were expressed as the proportion of the total LA or biomass. We applied general linear models with the binomial error distribution on these proportional data. We compared the proportions of biomass in a given canopy layer in each of the sward types belonging to the same cutting and fertilization level. All layers were compared up to the ones where 80 % of total biomass or LA was reached. This was the layer 37–47 cm for biomass and 27–37 cm for LA. The models included the experimental factors or, to compare the means of single treatments, the 12 treatments themselves. Overdispersion larger than 1.5 (i.e. the variance in a GLM with Poisson distributed errors was at least 1.5 times larger than the mean) was corrected by multiplication of standard errors with the overdispersion parameter Φ, using a quasi-GLM model or a model with negative binomial distribution of errors (Zuur *et al.* 2009). Normal distribution and homogeneity of the residuals were inspected using QQ plots and conditional boxplots of the residuals against all factor levels as suggested by Anscombe (1973) and Zuur *et al.* (2010). Heterogeneity of variances was corrected using general least square (GLS) models including the 'varIdent' variance structure (different variance per factor level, this R-function is found in the 'nlme' package (Pinheiro *et al.* 2009). Necessary data transformations are indicated in the results.

Results

Canopy composition and structure

The botanical and functional composition of the three sward types showed a distinct variation in the buds and tillers per square metre (Table 2). Since there were almost no legume buds in the –Dic-swards, the variance of the depending variable 'legume bud number' was close to zero in this sward type. Consequently, this sward type had to be excluded from all legume analyses. The three-way ANOVAs and linear contrasts (data not shown) underlined the effects of the herbicide treatment, i.e. the sward types, on bud and tiller numbers. At all harvest dates, the bud numbers of legumes and non-leguminous forbs averaged across all cutting and fertilization levels differed significantly among the sward types ($t_{(20,\ \alpha)} \geq 2.853$, $P \leq 0.0098$, linear contrasts in LM and GLS), except for the May harvest, when legume bud numbers were almost equal in the Co- and –Mon-swards. The impact of fertilization on bud numbers increased over the course of the year. No effect was found in May and July, while in September the bud numbers of legumes had been reduced by fertilization ($z_{(11,\ \alpha)} = 2.568$, $P = 0.010$, linear contrasts in GLM with negative binomial error distribution) and non-leguminous forb bud numbers were significantly lower in fertilized plots ($t_{(20,\ \alpha)} = 2.789$, $P = 0.011$, linear contrasts in LM). The number of grass tillers in May, however, was increased by fertilization ($t_{(22,\ \alpha)} = 2.401$, $P = 0.025$, LM, response variable raised to the power of 0.25). Significant sward effects were only found in May and September (ANOVA, $F_{May(2,\ 22)} = 21.62$, $F_{Sept(2,\ 19)} = 14.11$, $P < 0.001$).

Total biomass of the July canopy profiles belonging to the same management regime did not differ among the sward types (means and standard deviations displayed in Fig. 1). Likewise, species richness had no effect on biomass when

Table 2. Vegetative and generative buds/number of tillers per m^2 in the experimental treatments at each harvest 2009. Mean and standard deviation for grasses, median and median deviation for non-leguminous forbs and legumes, $n = 6$. For treatment abbreviation, refer to Table 1.

	Grass	Non-leguminous forbs	Legumes
May 2009			
Co3x	10 173 ± 3071	504 ± 222	238 ± 148
Co3NPK	11 291 ± 2922	593 ± 74	326 ± 230
–Dic3x	13 111 ± 5150	170 ± 111	0 ± 0
–Dic3NPK	18 304 ± 4632	200 ± 148	0 ± 0
–Mon3x	7948 ± 2532	1171 ± 164	200 ± 111
–Mon3NPK	8605 ± 3044	897 ± 89	252 ± 111
July 2009			
Co1x	5563 ± 3504	511 ± 126	82 ± 67
Co1NPK	4203 ± 2733	474 ± 200	45 ± 45
Co3x	8763 ± 3636	771 ± 133	126 ± 52
Co3NPK	7607 ± 5882	407 ± 89	111 ± 111
–Dic1x	5600 ± 2878	134 ± 112	0 ± 0
–Dic1NPK	6360 ± 4627	148 ± 104	0 ± 0
–Dic3x	6425 ± 996	289 ± 89	0 ± 0
–Dic3NPK	10 286 ± 5258	134 ± 111	0 ± 0
–Mon1x	5869 ± 2463	875 ± 327	348 ± 185
–Mon1NPK	2768 ± 1075	1252 ± 415	119 ± 89
–Mon3x	4615 ± 1569	971 ± 319	274 ± 59
–Mon3NPK	11 477 ± 3810	704 ± 156	82 ± 52
Sept 2009			
Co3x	9353 ± 1191	778 ± 163	141 ± 104
Co3NPK	10 062 ± 5411	289 ± 119	156 ± 82
–Dic3x	13 953 ± 3973	223 ± 44	0 ± 0
–Dic3NPK	11 916 ± 6993	96 ± 96	0 ± 0
–Mon3x	8479 ± 2762	1304 ± 282	400 ± 111
–Mon3NPK	8721 ± 4448	608 ± 363	341 ± 282

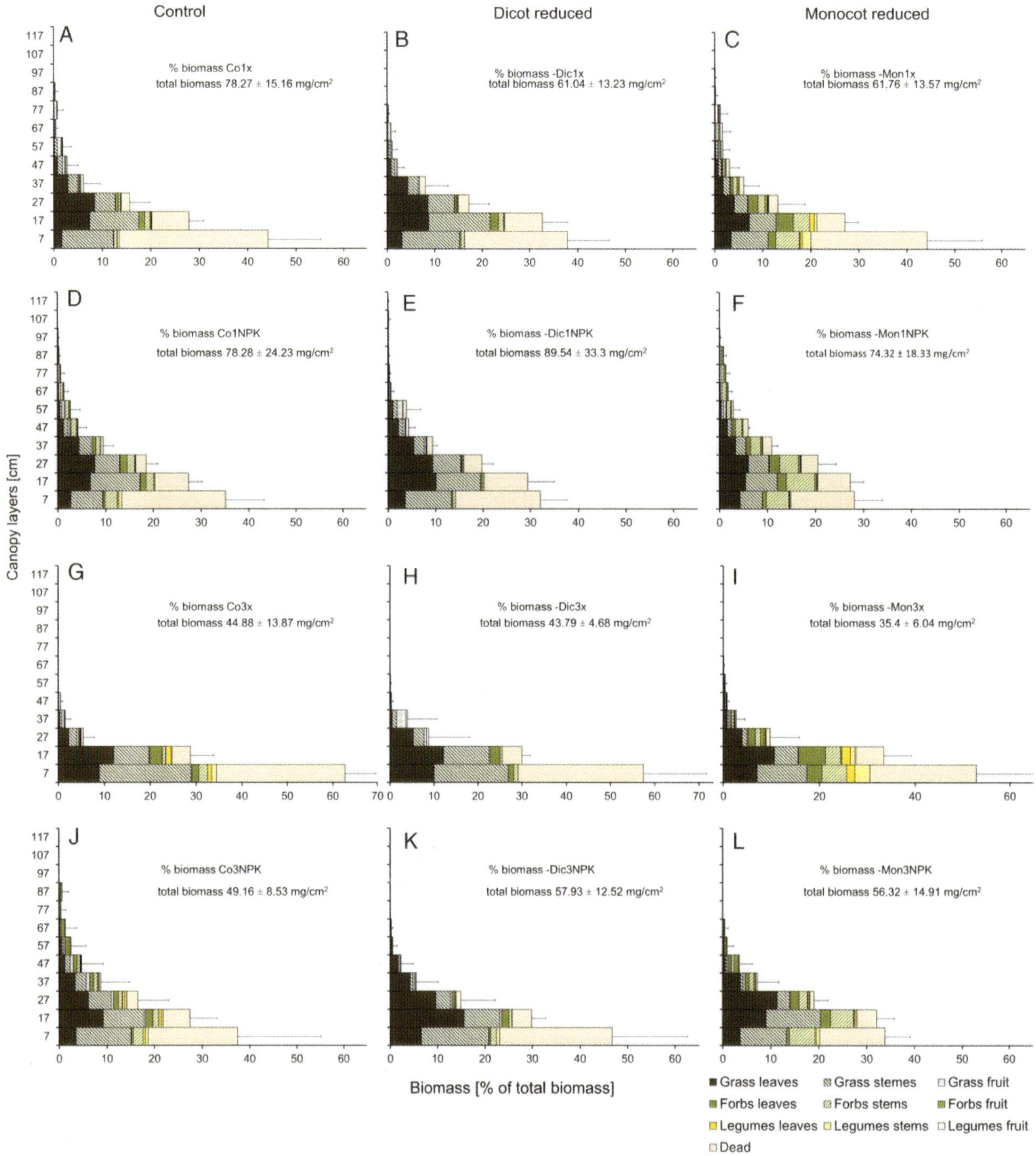

Figure 1. Vertical distribution of the different functional fractions of biomass for all 12 experimental treatments (July 2009). Given are means of total biomass, biomass of each fraction per layer and standard deviations of each layer, $n = 6$. For abbreviations of treatments, refer to Table 1.

included as first experimental factor in a GLS with different variances per fertilization level. However, the two manipulated sward types gained significantly more biomass from fertilization than the Co-swards (interaction sward × nutrients, $t_{(51,\ \alpha)} = 2.051$, $P = 0.045$ for –Dic and $t_{(51,\ \alpha)} = 2.345$, $P = 0.023$ for –Mon-swards, respectively, GLS allowing for different variance in each cut × nutrient combination). Averaged across all management treatments, the –Mon-swards had produced the least biomass ($t_{(51,\ \alpha)} = 2.49$, $P = 0.015$).

All three experimental factors influenced the vertical structure of the vegetation in the subplots, as demonstrated by the layered diagrams in Fig. 1. The outer shape of the layered diagrams (a line from top to bottom connecting the outer edges of the horizontal bars) was defined by the management. The biomass diagrams of unfertilized plots had a concave shape, the fertilized ones a more convex shape, especially close to the ground. The tallest canopy height was determined by the cutting regime. Significant sward effects were only found within the same management regime. Most of the structural differences, i.e. significant differences in biomass proportions per 10 cm layer due to the factor 'sward type', were observed between -Mon (Fig. 1C, F, I and L) and -Dic-swards (Fig. 1B, E, H and K) and they were confined to two of the medium layers (7–17 and 27–37 cm). Differences were either found in fertilized plots belonging to different sward types or when comparing the change in vertical structure between different management regimes (fertilized vs. unfertilized, cut once vs. cut thrice). Under fertilization, both Co- and -Mon-swards had increased their biomass proportions in layer 27–37 cm significantly more than the -Dic-swards, by 7.2 % (Co) and 4.9 % (-Mon) compared with 1.4 % in the -Dic-swards (sward × nutrient interaction in Co-swards cut thrice, $t_{(16,\ \alpha)} = 2.208$, $P = 0.004$ and -Mon-swards cut once $t_{(16,\ \alpha)} = 2.758$, $P = 0.006$). Compared with the plots cut once, the proportion of biomass belonging to layer 7–17 cm in the three-cut regime increased significantly more in the -Mon-swards (5.3 %) than in the -Dic-swards (1.1 %) (significant sward × cut interaction, $t_{(45,\ \alpha)} = 2.321$, $P = 0.020$). The -Mon-swards cut once had allocated significantly less biomass, on average 24 %, to layer 7–17 cm than the -Dic-swards ($t_{(18,\ \alpha)} = -2.118$, $P = 0.034$, Fig. 1B, C, E and F), where around 30 % of total biomass belonged to this layer. Above all, fertilization led to significant relocation of biomass from the lowest into higher (>17 cm) canopy layers in Co- and -Mon-swards, whereas the -Dic-swards hardly changed their vertical biomass distribution under nutrient enrichment.

All diagrams (Fig. 1A–L) show that a large proportion of the total biomass was concentrated in the lowest layer and remained as stubble after the harvest, which led to harvest indices from 36 ± 5.6 up to 73.5 ± 6.7. Still, the harvest indices in July were clearly larger than in May (between 17.9 ± 2.6 and 27.8 ± 1.5) and September (ranging from 33.7 ± 11.2 to 40 ± 13.8). The sward type was not responsible for any variation in harvest indices, they were mainly influenced by fertilization (ANOVA with arcsine-square root transformation, $F_{(1,\ 24)} = 7.31$, $P = 0.012$ in May, $F_{(1,\ 50)} = 35.76$, $P < 0.001$ in July) and, in July, by the cutting frequency ($F_{(1,\ 50)} = 25.10$, $P < 0.001$). The third regrowth in September had similar harvest indices across all experimental treatments.

As already indicated in the shapes of the layered diagrams, the location of the centre of gravity was raised by fertilization (linear contrasts in LM with log transformation, $t_{(52,\ \alpha)} = 6.180$, $P < 0.001$) and lowered by a higher cutting frequency ($t_{(52,\ \alpha)} = 6.827$, $P < 0.001$). The centre of gravity was located between 7.5 ± 1.0 cm (Co3x) and 19.2 ± 1.3 cm (-Mon1NPK) above the ground. In the -Mon-swards the centre of gravity was located ~1 cm further up in the canopy than in the other two sward types, although this effect was not significant.

The amount of standing litter within the vegetation (Fig. 2) mainly depended on age and nutrient status of the plots. The high cutting frequency with an early first cut (plots belonging to the three-cut regime) and fertilization significantly decreased the ratio of dead to green plant material. However, the amount of grasses present in the vegetation also tended to influence the amount of litter (analysis of deviance of a linear model with negative binomial error distribution $P = 0.07$). The -Mon-swards, containing the least grasses also had the lowest amounts of standing litter.

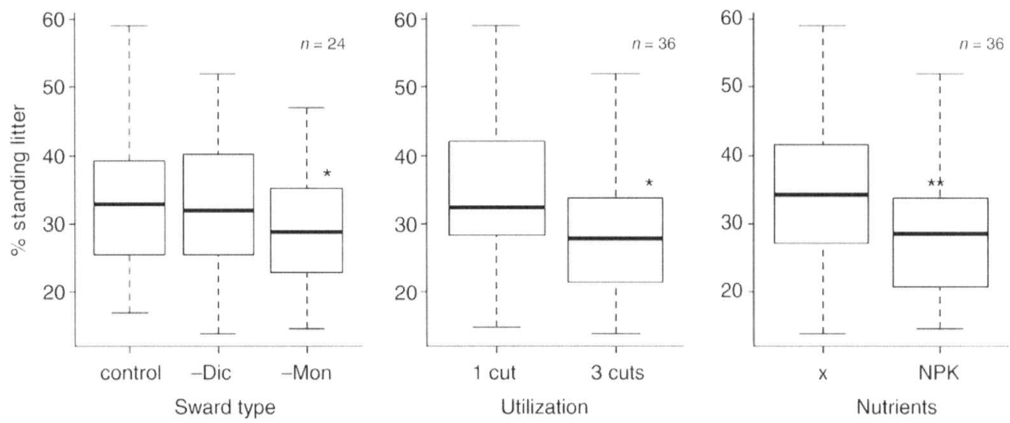

Figure 2. Proportion of standing litter based on total biomass found in the canopy profiles shortly before the harvest in July 2009 depending on experimental factors. The boxes show the median and first and third quartiles. For abbreviations of treatments, refer to Table 1. Asterisks denote significant differences of factor levels compared with the reference level (sward control, utilization 1/year, fertilization x) in a GLM with negative binomial error distribution (link: logit). **$P < 0.01$, *$P < 0.05$.

Leaf area index and light distribution

The total LAI of the vegetation was neither influenced by the sward type nor the species richness within the same management regime. Just the ability to transform applied fertilizer into leaf area was significantly higher in the –Dic-swards than in the Co-swards (interaction sward × nutrients, $t_{(49,\ \alpha)} = 2.517$, $P = 0.015$, GLS allowing for different variances per fertilization level). The LA profiles had a concave–convex shape across all treatments due to a concentration of LA in canopy layers from 7 to 27 cm. The only significant sward influences on the vertical LA distribution were found in fertilized experimental treatments. In the fertilized plots cut three times, we found a significantly higher proportion of LA in higher canopy layers in the –Mon than in the –Dic-swards ($t_{(43,\ \alpha)} = 2.047$, $P = 0.050$). The lowest layer close to the ground had the lowest proportion of LA in the fertilized Co-swards of cut once, significantly lower compared with the –Mon-swards ($t_{(4,\ \alpha)} = 2.812$, $P = 0.048$).

A combination of values of light extinction and LA showed more pronounced differences between the three sward types (Fig. 3). The extinction curves all followed a distinct shape depending on cutting frequency. In the

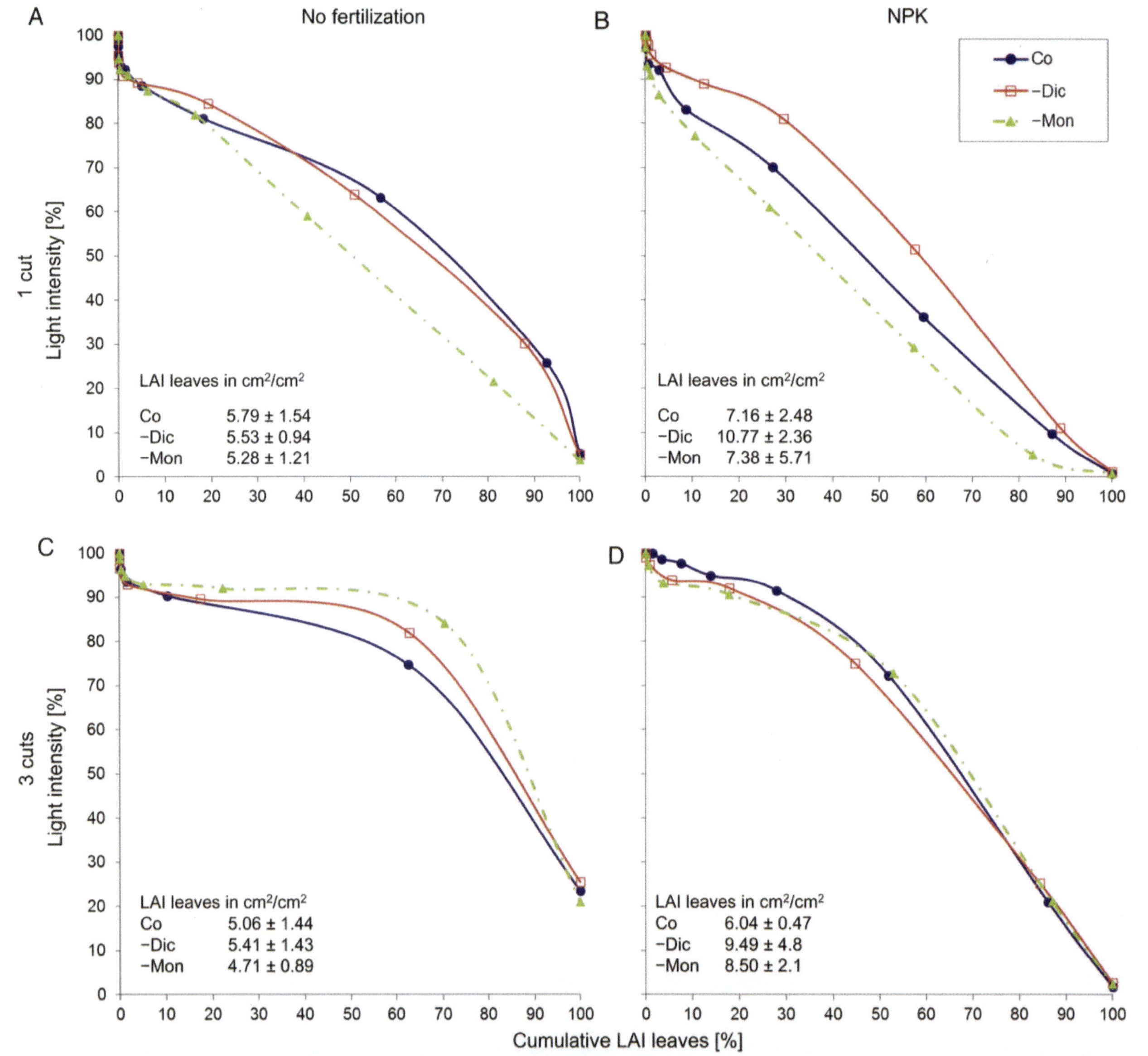

Figure 3. Means of LAI (leaves) and light intensity in relation to cumulative LAI (leaves) in the experimental treatments in July 2009 grouped by management regime (*A* = 1x, *B* = 1NPK, *C* = 3x, *D* = 3NPK). The sloping curves show the light intensity from above the canopy (0 % LAI) in 10 cm steps down to the ground (100 % LAI), *n* = 6.

mature vegetation cut once, it took much less LA to intercept a certain amount of light compared with the vegetation of the plots in the three-cut regime. The −Mon-swards showed the strongest structural changes under management variation. In the plots cut three times without fertilization (Fig. 3C), the vegetation of the −Mon-swards tended to intercept the least light with 50 % of its LAI compared with the vegetation of the other sward types. In contrast, in the fertilized plots cut once (Fig. 3B), the vegetation of the −Mon-swards intercepted the most light (significantly more than the other two sward types, $P \leq 0.03$, linear contrasts in LM).

The extinction coefficient was lowered by a high cutting frequency (Fig. 4). In both Co- and −Mon-swards, it increased due to fertilization, whereas the −Dic-swards showed a similar light extinction in both nutrient regimes (sward × nutrient interaction in ANOVA, $F_{(2, 49)} = 3.30$, $P = 0.045$). In just a few of the fertilized Co-swards, the leaves grew mainly in wide angles (>45° between stem and leaf), indicated by an extinction coefficient k larger than 1. The vegetation in all other treatments showed an equal distribution of leaf angles apart from the vegetation in the −Dic-swards, which had mainly vertically orientated leaves (leaf

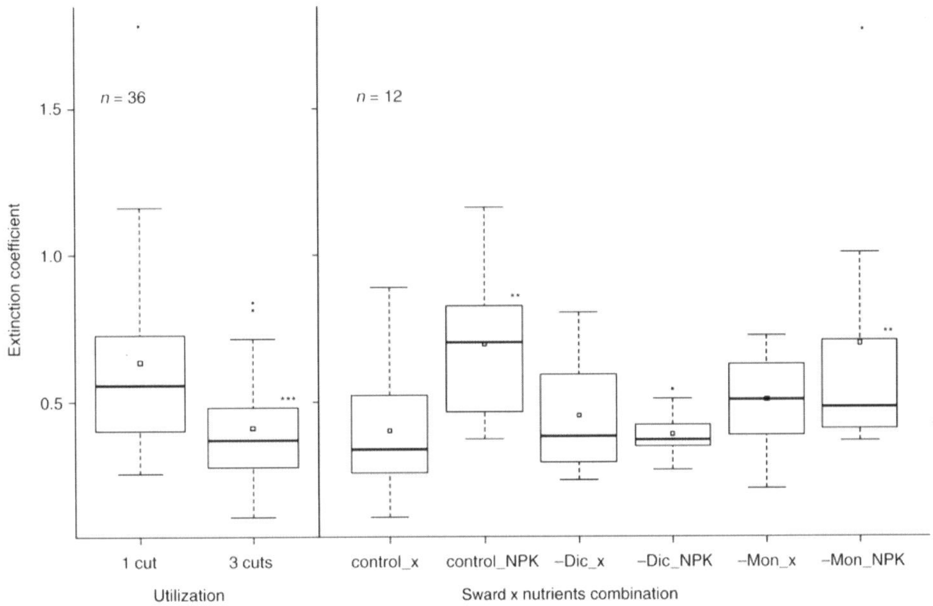

Figure 4. Extinction coefficient in the different experimental treatments in July 2009. Asterisks denote significant differences of factor levels compared with the reference level (sward control, utilization 1/year, control_ x). Interaction sward × nutrients ($F_{(2, 49)} = 3.30$, $P = 0.045$).

Table 3. Median and median deviations of light transmission (I/I_0) per treatment at ground level in July 2009 with corresponding ANOVA table (adjusted for the factors row and block, data log transformed, $n = 6$).

	No fertilization (*x*)		**Fertilized (NPK)**	
	1 cut	**3 cuts**	**1 cut**	**3 cuts**
Control	0.05 ± 0.03	0.24 ± 0.04	0.01 ± 0.003	0.02 ± 0.01
−Dic	0.05 ± 0.03	0.26 ± 0.09	0.01 ± 0.002	0.03 ± 0.02
−Mon	0.04 ± 0.01	0.21 ± 0.08	0.01 ± 0.003	0.02 ± 0.02
	Df	**MSq**	***F*-value**	***P***
Sward type	2	0.847	1.72	0.187
Utilization	1	30.296	61.43	<0.001
Nutrients	1	59.570	120.78	<0.001
Residuals	56	0.535		

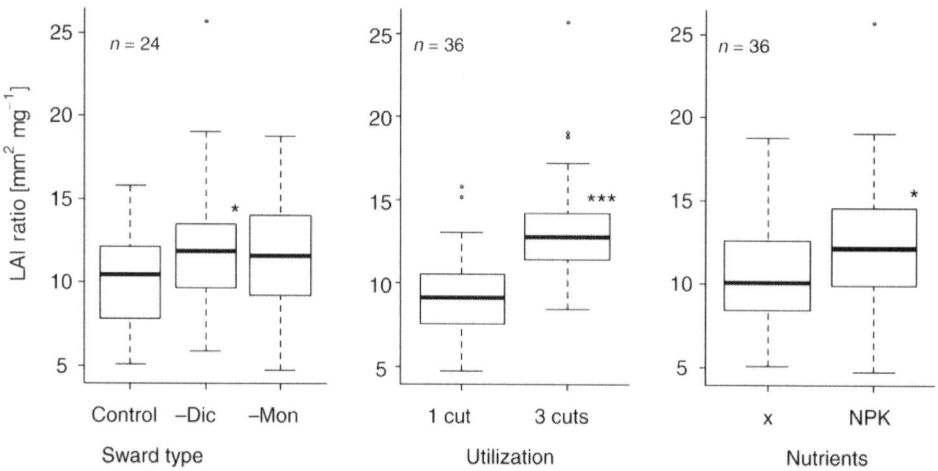

Figure 5. Leaf area ratio (median, first and third quartiles) per experimental factor in July 2009. For abbreviations of factors, refer to Table 1. Asterisks denote significant differences of factor levels compared with the reference level (sward control, utilization 1 cut, nutrients x) in a linear model (with log transformation of the response variable for comparison of treatments). $***P < 0.001$, $**P < 0.01$, $*P < 0.05$.

angle $<45°$, $0.3 < k < 0.7$). In spite of these different extinction coefficients, the light reaching the ground was similar in all sward types; it was only influenced by management regime (Table 3).

The LAR was influenced mainly by management, but also by the amount of grasses present in a sward (Fig. 5). Overall, the –Dic-swards had a higher LAR than the Co-swards. The comparison of the single treatments, however, did not show any sward or species richness effects. The highest LAR was found in the fertilized –Dic-swards cut three times.

Discussion

In this study we examined the canopy structure of permanent grassland under different management regimes with contrasting plant species composition of species from the same plant community. In accordance with many other publications, we consider the biomass or LA profiles as main indicators for vertical structure (e.g. Fliervoet and Werger 1984, 1985; Werger *et al.* 1986; Mitchley and Willems 1995). To complement the profiles, the following canopy variables or indices were used: centre of gravity, extinction coefficient representing the prevalent leaf angles, light extinction (as in Spehn *et al.* 2000), LAR and the total LAI (as in Fliervoet 1987). We aimed to compare our results with those obtained in seeded grassland vegetation commonly used in biodiversity experiments.

The species composition, described by the numbers of growing points of each functional group, was not stable over the whole season. Pavlů *et al.* (2006) also encountered fluctuating numbers of grass tillers, legume buds and non-leguminous forb plants when comparing vegetation density in mown and continuously grazed pastures. In our experiment, nutrient addition, especially in combination with defoliation, stimulated the formation of grass tillers while reducing the numbers of buds in non-leguminous forbs and legumes. Management is known for strongly influencing species composition and density in permanent grassland (Prins and Van Burg 1979; Oomes and Mooi 1981; Klimek *et al.* 2007), leading to a new equilibrium in plant species composition. Since we were especially interested in structural differences brought about by species composition, we had restricted our analyses to one growing season.

In spite of the differences in species composition and the different growth forms, the sward types hardly showed any structural differences within one management regime. As also reported by Werger *et al.* (1986), who analysed different permanent grassland types in the Netherlands, the canopy form (from concave via linear up to convex) was largely determined by the nutrient status and the productivity of a plot. Nutrient addition generally elongated the canopy and raised the centre of gravity. However, the reaction to fertilization differed significantly among the sward types. The canopies of –Mon-swards grew taller after fertilization than –Dic-swards which remained largely unchanged in their vertical structure. This shift of biomass into upper canopy layers can be explained by the increased presence of tall species with leafy stems. This trait combination is typical for competitive species, which increase their biomass after nutrient addition and low levels of disturbance in natural grassland communities (Liira and Zobel 2000; Gross *et al.* 2007). In the –Mon-swards the total amount of these tall species was considerably higher than in the −Dic-swards (see Petersen *et al.* 2012) since they hosted apart from tall grasses also tall dicots with leafy stems (e.g. *Rumex acetosa* L., *Achillea millefolium* L.). Consequently, the biomass shift was much more pronounced compared with the −Dic-swards. Spehn *et al.* (2000) likewise observed a raised centre of gravity in functionally more diverse vegetation in seeded experimental plots.

The harvest index was not improved by functional diversity; it was only influenced by management. Especially the young (cut in May) or regularly cut short vegetation (in September, the vegetation only had a canopy height of around 20–25 cm) had not developed a distinct structure, which consequently could not have influenced the harvest index; a result which was also found by Liira and Zobel (2000). Only tall, mature grassland could develop a clear vertical structure they reported. All young or recently cut grassland vegetation is too short to develop a canopy with distinct layers, i.e. in all layers there is roughly the same amount of biomass. This was in line with our observations. The tillering of grasses stimulated by the first cut in May led to a dense canopy layer close to the ground which lowered the harvest index in the regrowth in July especially in −Dic-swards compared with the other sward types.

The amount of standing litter in the vegetation strongly increased due to a late cut without fertilization (1x regime) is also encountered by Marriott *et al.* (2005) on grassland vegetation in Scotland. However, the functional composition of the vegetation also influenced the amount of standing litter. The more grasses that were present; the more dead biomass was found. This correlation was also found by Fliervoet and Werger (1985) for Calthion grasslands in the Netherlands. Since the −Dic-swards were quite dense and consisted almost completely of grasses (=leafy stem species), they had the most standing litter. The fact that by the beginning of July most of the −Dic-swards were not standing upright any more increased the light deficiency and stimulated the death of leaves close to the ground.

The reason for the significant LA changes in grass-rich vegetation after fertilization lies in the different growth physiology of grasses compared with dicots. If they are fertilized, grasses increase their leaf area a lot more than dicots (Körner 1994). The LAI ratio underpinned these findings. The more grasses the vegetation contained, the higher was its LAR.

In this study, hypothesis (a) was only partly supported. We did find a strong influence of management on the vertical structure, but species composition effects were negligible. The location of light interception and hence the shape of the light intensity curves were mainly defined by the cutting regime. The different light extinction coefficients of the sward types were obviously due to differences in leaf angles. Light extinction is influenced by the vegetation's density but mainly by the angles of leaves with grassy vegetation having a lower extinction coefficient ($0.3 < k < 0.7$) than vegetation containing a higher percentage of dicots ($k > 1.0$, Geyger 1977). Fertilization and a late cut increased leaf size and biomass density, visible by a stronger light interception in upper canopy layers in −Mon-swards. Since fertilization did not change the leaves' angles, the −Dic-swards had a lower extinction coefficient in spite of a larger LA, corroborated by Brougham's (1958) findings. Even though most of the flat-lying plants (= horizontal leaves from ~17 cm downwards) were found in −Dic-swards, this angle effect was still visible. Fliervoet and Werger (1984) also mentioned the inclination and vertical distribution of leaves as the main factor determining the light interception in the permanent grassland they analysed.

The absolute light interception was similar in all plots of the same management regime. The hypothesized positive effect of species or functional diversity on light interception, as Spehn *et al.* (2000) found in seeded grassland vegetation, could not be confirmed. One reason for this discrepancy could be the different age of the vegetation. The grassland vegetation had been sown 2 years previous to the experiment and especially the vegetation consisting only of grass species had not even reached full ground cover. The vegetation in the GrassMan experiment was much older and had recovered completely from the herbicide treatment in the previous year in terms of cover and biomass (Petersen *et al.* 2013). Even though we had a contrasting species composition in terms of monocot to dicot ratio, these differences did not lead to structural or functional differences. The vegetation in plots cut three times may have been disturbed too often to build up a distinct structure as described for managed permanent grassland by Liira and Zobel (2000).

The fact that the light interception efficiency of our grassland ecosystem was not influenced by compositional differences [hypothesis (b)] is possibly due to its relatively high species number compared with experimental, seeded grassland. As Schwartz *et al.* (2000) and Waide *et al.* (1999) pointed out, most of the ecosystem processes are saturating at low species numbers; Roy (2001) had summarized in his meta-analysis that, in many studies, five plant species were enough to reach 90 % of the net primary productivity of the examined vegetation type. Since the vegetation in our experiment consisted of at least seven species per square metre, it was not at all surprising that the light interception, the ecosystem function analysed here, was not limited by the species number. A mixture of just two productive grass species achieves a denser filling of the aboveground space and hence a higher light interception compared with a monoculture as shown by Vojtech *et al.* (2008). Consequently, the efficient light interception in the -Dic-swards consisting of at least five grass species and small amounts of two or more dicot species is not surprising. In addition, the broad range of structural and functional diversity among monocots and dicots in this study may have been more important for other ecosystem functions (e.g. nutrient acquisition) than for light interception. Griffin *et al.* (2009) suggest that functional diversity can even have adverse effects on different ecosystem functions.

Conclusions

Above all, management had a much stronger influence on structure and light interception than species composition in this grassland experiment. Compositional and functional differences turned out to be important only if different management intensities in terms of fertilization and cutting frequency were compared. Then different physiological and morphological traits of dicots and monocots became visible, leading to a slightly different structural development of the vegetation, whereas no effects of species richness were found. Hence, the strong effects due to species composition found in experimental, seeded grassland vegetation cannot be transferred directly to managed permanent grassland. Obviously, even in plots with a low number of species ($7/m^2$) the structural and functional diversity was high enough to ensure an optimal arrangement of leaves in space to fully capture the incoming light.

Sources of Funding

This study, funded by the Ministry of Science and Culture of Lower Saxony and the 'Niedersächsisches Vorab', was part of the Cluster of Excellence 'Functional Biodiversity Research'.

Contributions by the Authors

J.I. conceived the study design. U.P. performed the experiment, analysed the data and wrote the manuscript, which was then edited by J.I.

Acknowledgements

We thank the technical staff of the Department of Crop Sciences at the University of Göttingen and the experimental farm Relliehausen for help with experimental set-up and maintenance and two anonymous reviewers for their helpful comments on the manuscript.

Literature Cited

Anscombe FJ. 1973. Graphs in statistical analysis. *The American Statistician* **27**:17–21.

Brougham R. 1958. Interception of light by the foliage of pure and mixed stands of pasture plants. *Australian Journal of Agricultural Research* **9**:39–52.

Cheng X, Luo Y, Su B, Verburg PSJ, Hui D, Obrist D, Arnone JA III, Johnson DW, Evans RD. 2009. Responses of net ecosystem CO_2 exchange to nitrogen fertilization in experimentally manipulated grassland ecosystems. *Agricultural and Forest Meteorology* **149**: 1956–1963.

Díaz S, Symstad AJ, Chapin FS III, Wardle DA, Huenneke LF. 2003. Functional diversity revealed by removal experiments. *Trends in Ecology and Evolution* **18**:140–146.

Donald CM. 1962. In search of yield. *Journal of the Australian Institute of Agricultural Science* **28**:171–178.

Fliervoet LM. 1987. Characterization of the canopy structure of Dutch grasslands. *Vegetatio* **70**:105–117.

Fliervoet LM, Werger MJA. 1984. Canopy structure and microclimate of two wet grassland communities. *New Phytologist* **96**:115–130.

Fliervoet LM, Werger MJA. 1985. Vegetation structure and microclimate of three Dutch Calthion palustris communities under different climatic conditions. *Vegetatio* **59**:159–169.

Fridley JD. 2003. Diversity effects on production in different light and fertility environments: an experiment with communities of annual plants. *Journal of Ecology* **91**:396–406.

Garnier E, Shipley B, Roumet C, Laurent G. 2001. A standardized protocol for the determination of specific leaf area and leaf dry matter content. *Functional Ecology* **15**:688–695.

Geyger E. 1977. Leaf area and productivity in grasslands. In: Krause W, ed. *Application of vegetation science to grassland husbandry. Handbook of vegetation science vol 13*. Dordrecht: Springer, 499–520.

Griffin JN, Méndez V, Johnson AF, Jenkins SR, Foggo A. 2009. Functional diversity predicts overyielding effect of species combination on primary productivity. *Oikos* **118**:37–44.

Gross N, Suding KN, Lavorel S. 2007. Leaf dry matter content and lateral spread predict response to land use change for six subalpine grassland species. *Journal of Vegetation Science* **18**:289–300.

Haase P, Pugnaire FI, Clark SC, Incoll LD. 1999. Environmental control of canopy dynamics and photosynthetic rate in the evergreen tussock grass *Stipa tenacissima*. *Plant Ecology* **145**:327–339.

Hector A, Joshi J, Scherer-Lorenzen M, Schmid B, Spehn EM, Wacker L, Weilenmann M, Bazeley-White E, Beierkuhnlein C, Caldeira MC, Dimitrakopoulos PG, Finn JA, Huss-Danell K, Jumpponen A, Leadley PW, Loreau M, Mulder CPH, Neßhöver C, Palmborg C, Read DJ, Siamantziouras ASD, Terry AC, Troumbis AY. 2007. Biodiversity and ecosystem functioning: reconciling the results of experimental and observational studies. *Functional Ecology* **21**:998–1002.

Jiang L, Wan S, Li L. 2009. Species diversity and productivity: why do results of diversity-manipulation experiments differ from natural patterns? *Journal of Ecology* **97**:603–608.

Keuter A, Hoeft I, Veldkamp E, Corre MD. 2013. Nitrogen response efficiency of a managed and phytodiverse temperate grassland. *Plant and Soil* **364**:193–206.

Klimek S, Richter gen. Kemmermann A, Hofmann M, Isselstein J. 2007. Plant species richness and composition in managed grasslands: the relative importance of field management and environmental factors. *Biological Conservation* **134**:559–570.

Körner C. 1994. Scaling from species to vegetation: the usefulness of functional groups. In: Schulze ED, Mooney HA, eds. *Biodiversity and ecosystem function*. Heidelberg: Springer, 117–140.

Lamb EG. 2008. Direct and indirect control of grassland community structure by litter, resources, and biomass. *Ecology* **89**:216–225.

Lienin P, Kleyer M. 2012. Plant trait responses to the environment and effects on ecosystem properties. *Basic and Applied Ecology* **13**: 301–311.

Liira J, Zobel K. 2000. Vertical structure of a species-rich grassland canopy, treated with additional illumination, fertilization and mowing. *Plant Ecology* **146**:185–195.

Lorentzen S, Roscher C, Schumacher J, Schulze E-D, Schmid B. 2008. Species richness and identity affect the use of aboveground space in experimental grasslands. *Perspectives in Plant Ecology, Evolution and Systematics* **10**:73–87.

Marriott CA, Bolton GR, Fisher JM, Hood K. 2005. Short-term changes in soil nutrients and vegetation biomass and nutrient content following the introduction of extensive management in upland sown swards in Scotland, UK. *Agriculture Ecosystems and Environment* **106**:331–344.

McLaren JR, Turkington R. 2010. Ecosystem properties determined by plant functional group identity. *Journal of Ecology* **98**: 459–469.

Miller RG. 1981. *Simultaneous statistical inference*. New York: Springer.

Mitchley J, Willems JH. 1995. Vertical canopy structure of Dutch chalk grasslands in relation to their management. *Vegetatio* **117**: 17–27.

Mokany K, Ash J, Roxburgh S. 2008. Functional identity is more important than diversity in influencing ecosystem processes in a temperate native grassland. *Journal of Ecology* **96**:884–893.

Monsi M, Saeki T. 1953. Über den Lichtfaktor in den Pflanzengesellschaften und seine Bedeutung für die Stoffproduktion. *Japanese Journal of Botany* **14**:22–52.

Nyfeler D, Huguenin-Elie O, Suter M, Frossard E, Connolly J, Lüscher A. 2009. Strong mixture effects among four species in fertilized agricultural grassland led to persistent and consistent transgressive overyielding. *Journal of Applied Ecology* **46**:683–691.

Oomes MJM, Mooi H. 1981. The effect of cutting and fertilizing on the floristic composition and production of an *Arrhenatherion elatioris* grassland. *Vegetatio* **47**:233–239.

Pavlů V, Hejcman M, Pavlů L, Gaisler J, Hejcmanová-Nežerková P, Meneses L. 2006. Changes in plant densities in a mesic species-rich grassland after imposing different grazing management treatments. *Grass and Forage Science* **61**:42–51.

Petersen U, Wrage N, Köhler L, Leuschner C, Isselstein J. 2012. Manipulating the species composition of permanent grasslands—A new approach to biodiversity experiments. *Basic and Applied Ecology* **13**:1–9.

Petersen U, Wrage-Mönnig N, Isselstein J. 2013. Effects of herbicide application to control sward composition in different management variants. *International Journal of Biodiversity Science, Ecosystem Services and Management* **9**:155–165.

Pinheiro J, Bates D, DebRoy S, Sarkar D, the R Core team. 2009. *nlme: linear and nonlinear mixed effects models*. R package version 3.1-96. http://CRAN.R-project.org/package=nlme (18 June 2013).

Prins WH, Van Burg PFJ. 1979. The seasonal response of grassland to nitrogen at different levels of nitrogen pretreatment 1. Experiments during 1972 and 1973. Netherlands Nitrogen Technical Bulletin No. 11, 1–46.

Pywell RF, Bullock JM, Tallowin JB, Walker KJ, Warman EA, Masters G. 2007. Enhancing diversity of species-poor grasslands: an experimental assessment of multiple constraints. *Journal of Applied Ecology* **44**:81–94.

R Development Core Team. 2011. *R: a language and environment for statistical computing. Computer program*. Vienna: R Foundation for Statistical Computing.

Roscher C, Schumacher J, Baade J, Wilcke W, Gleixner G, Weisser WW, Schmid B, Schulze ED. 2004. The role of biodiversity for element cycling and trophic interactions: an experimental approach in a grassland community. *Basic and Applied Ecology* **5**:107–121.

Roy J. 2001. How does biodiversity control primary productivity? In: Roy J, Saugier B, Mooney HA, eds. *Terrestrial global productivity*. San Diego: Academic Press.

Ryser P, Lambers H. 1995. Root and leaf attributes accounting for the performance of fast- and slow-growing grasses at different nutrient supply. *Plant and Soil* **170**:251–265.

Saeki T. 1959. Interrelationships between leaf amount, light distribution and total photosynthesis in a plant community. *Botanical Magazine Tokyo* **73**:56–63.

Schmid B. 2002. The species richness-productivity controversy. *Trends in Ecology and Evolution* **17**:113–114.

Schwartz MW, Brigham CA, Hoeksema JD, Lyons KG, Mills MH, van Mantgem PJ. 2000. Linking biodiversity to ecosystem function: implications for conservation ecology. *Oecologia* **122**:297–305.

Scimone M, Rook AJ, Garel JP, Sahin N. 2007. Effects of livestock breed and grazing intensity on grazing systems: 3. Effects on diversity of vegetation. *Grass and Forage Science* **62**:172–184.

Sheehy JE, Cooper JP. 1973. Light interception, photosynthetic activity, and crop growth rate in canopies of six temperate forage grasses. *Journal of Applied Ecology* **10**:239–250.

Sheehy JE, Peacock JM. 1977. Microclimate, canopy structure and photosynthesis in canopies of three contrasting temperate forage grasses: I. Canopy structure and growth. *Annals of Botany* **41**:567–578.

Spehn EM, Joshi J, Schmid B, Diemer M, Körner C. 2000. Above-ground resource use increases with plant species richness in experimental grassland ecosystems. *Functional Ecology* **14**:326–337.

Thompson K, Askew AP, Grime JP, Dunnett NP, Willis AJ. 2005. Biodiversity, ecosystem function and plant traits in mature and immature plant communities. *Functional Ecology* **19**:355–358.

Tilman D, Reich PB, Isbell F. 2012. Biodiversity impacts ecosystem productivity as much as resources, disturbance, or herbivory. *Proceedings of the National Academy of Sciences of the USA* **109**: 10394–10397.

Vojtech E, Loreau M, Yachi S, Spehn EM, Hector A. 2008. Light partitioning in experimental grass communities. *Oikos* **117**:1351–1361.

Waide RB, Willig MR, Steiner CF, Mittelbach G, Gough L, Dodson SI, Juday GP, Parmenter R. 1999. The relationship between productivity and species richness. *Annual Review of Ecology and Systematics* **30**:257–300.

Werger MJA, Dusink EM, Fliervoet LM. 1986. Types of phytomass—and leaf area index profiles in grassland vegetation. *Vegetatio* **65**:39–45.

Wilson JB, Steel JB, Steel S-LK. 2007. Do plants ever compete for space? *Folia Geobotanica*, **42**:431–436.

Wrage N, Strodthoff J, Cuchillo MH, Isselstein J, Kayser M. 2011. Phytodiversity of temperate permanent grasslands: ecosystem ser-

7

Composition of fungal soil communities varies with plant abundance and geographic origin

Vanessa Reininger[1,2,3*], Laura B. Martinez-Garcia[1,4], Laura Sanderson[1] and Pedro M. Antunes[1]

[1] Department of Biology, Algoma University, Sault Ste. Marie, ON, P6A 2G4 Canada

[2] Present address: Agroscope, Institute for Plant Production Sciences IPS, Schloss 1, 8820 Wädenswil, Switzerland

[3] Present address: ETH Zurich, Institute of Integrative Biology, Forest Pathology and Dendrology, Universitätstrasse 16, 8092 Zurich, Switzerland

[4] Present address: Department of Soil Quality, Wageningen University, 6700 AA, Wageningen, The Netherlands

Associate Editor: Jean H. Burns

Abstract. Interactions of belowground fungal communities with exotic and native plant species may be important drivers of plant community structure in invaded grasslands. However, field surveys linking plant community structure with belowground fungal communities are missing. We investigated whether a selected number of abundant and relatively rare plants, either native or exotic, from an old-field site associate with different fungal communities. We also assessed whether these plants showed different symbiotic relationships with soil biota through their roots. We characterized the plant community and collected roots to investigate fungal communities using 454 pyrosequencing and assessed arbuscular mycorrhizal colonization and enemy-induced lesions. Differences in fungal communities were considered based on the assessment of α- and β diversity depending on plant 'abundance' and 'origin'. Plant abundance and origin determined the fungal community. Fungal richness was higher for native abundant as opposed to relatively rare native plant species. However, this was not observed for exotics of contrasting abundance. Regardless of their origin, β diversity was higher for rare than for abundant species. Abundant exotics in the community, which happen to be grasses, were the least mycorrhizal whereas rare natives were most susceptible to enemy attack. Our results suggest that compared with exotics, the relative abundance of remnant native plant species in our old-field site is still linked to the structure of belowground fungal communities. In contrast, exotic species may act as a disturbing agent contributing towards the homogenization of soil fungal communities, potentially changing feedback interactions.

Keywords: Arbuscular mycorrhizal fungi; community ecology; α diversity; β diversity; 454 pyrosequencing.

Introduction

One of the major goals in ecology is to understand why some plant species in natural communities are more abundant than others (Hubbell 2001; Chave 2004; McGill *et al.* 2007; Dumbrell *et al.* 2010). Many studies focusing on plant community ecology have provided insight into community assembly through testing plant–plant competitive interactions and plant–abiotic interactions such as nutrient and light availability (Tilman 1985; Thompson 1987; Wilby and Brown 2001; Miller *et al.* 2005; Dybzinski and Tilman 2007; Bever *et al.* 2012). There is increasing evidence that belowground plant–microbe interactions

* Corresponding author's e-mail address: Vanessa.Reininger@env.ethz.ch

are important to explain plant community dynamics (Bever *et al.* 1997, 2010, 2012; Reynolds and Haubensak 2009). Many studies have already been conducted focusing on belowground microbes and their interactions with the host plant (Johnson *et al.* 1997; Jumpponen and Trappe 1998; Reynolds *et al.* 2003; Bonfante and Anca 2009; Compant *et al.* 2010; Porras-Alfaro and Bayman 2011; Reininger *et al.* 2012; Reininger and Sieber 2012, 2013). However, little is known about the impact of plant community structure on the belowground soil microbial community (Klironomos 2003; Reynolds *et al.* 2003; Bagchi *et al.* 2010; Bever *et al.* 2010, 2012) especially with regard to plant communities which are progressively invaded by exotic plant species (Levine 2000; Klironomos 2002). According to the Wild Species Report issued by the Government of Canada, extant natural plant communities in Canada now include a large portion of exotic species, comprising 24 % of all the vascular plants (Wild Species 2010, http://publications.gc.ca/collections/collection_2011/ec/CW70-7-2010-eng.pdf, November 2014). Due to establishment of exotic plant species, plant community structure can change and consequently, interactions with soil microbial communities may change accordingly. By comparing the associations between co-occurring native and exotic plant species and their respective belowground fungal communities, we may be better able to understand and predict plant community dynamics in invaded ecosystems.

According to Klironomos (2002), plant species relative abundance may partially be explained by interactions between plants and soil microbes. Plant–soil feedbacks are defined as the effect that plant species exert on soil biotic and abiotic factors, which in turn can feedback into growth and fitness effects on subsequent conspecific or heterospecific plants (Van der Putten *et al.* 2013). In native plants, feedbacks are subjected to co-evolution between symbionts. Exotic species invading a new range will also establish interactions with the soil biota but by virtue of the host plant's novelty in the system, those interactions may not yet have experienced co-evolutionary processes. As a result, exotics may associate indiscriminately with soil microbial communities and function as a non-host for some plant pathogens, i.e. they show a symptomless infection (Nürnberger and Lipka 2005; Schulz and Boyle 2005). These fungal symbionts, accumulating in exotic species without necessarily causing population declines, may in turn negatively affect native plants to which they are pathogenic (Roy and Handley 2012; Flory and Clay 2013; Van der Putten *et al.* 2013).

Exotic plants are assumed to leave their enemies behind in their home geographic range (Enemy Release Hypothesis) and this could explain the differential in feedback in their new range relative to natives, which are more affected by a negative feedback (Keane and Crawley 2002). However, $\sim$50 % of the studies do not support the Enemy Release Hypothesis (Agrawal *et al.* 2005; Jeschke *et al.* 2012). These contrasting hypotheses highlight the uncertainty surrounding microbial communities associating with native and exotic plant species in natural communities and how alterations imposed by exotics may affect the soil biota community and by extension, the entire plant community.

Next-Generation Sequencing technologies such as 454 pyrosequencing using universal primers provide an opportunity to survey the soil microbial diversity associated with plant species in natural communities (Buée *et al.* 2009; Costa and Antunes 2012). These surveys contribute to improve our understanding of the biodiversity of microorganisms that co-exist and interact with plant roots across spatial scales and possibly influence plants' relative abundance and the dynamics between native and exotic plant species. Specifically, fungi are key drivers of primary productivity and plant community structure due to their diverse roles as mycorrhizal fungi, endophytes, decomposers and pathogens. They directly interact with plant roots resulting in functional outcomes that range from mutualism to parasitism (Schulz and Boyle 2005; Sinclair and Howard 2005; Smith and Read 2008). Better knowledge of fungal community structure within plant roots among different plant species is a first step towards understanding how fungi influence plant community dynamics.

We surveyed the fungal diversity present in the roots of four native and four exotic plant species from a natural old-field plant community. Within each of these two groups, two species were highly abundant in the community and the other two were relatively rare. α Diversity was used to measure the fungal diversity in each plant species whereas β diversity was a measure of variation in fungal composition among groups (abundant versus rare and native versus exotic) (Anderson *et al.* 2011). We also determined levels of arbuscular mycorrhizal colonization and lesions inflicted by enemies to plant roots.

We predicted that fungal communities associating with plant roots differ depending on plant species 'abundance' and 'origin'. More specifically, abundant natives would associate with more similar fungal communities than rare natives as they are more often surrounded by conspecifics that are likely to harbor the same fungal community. However, exotics were expected to associate with a more variable fungal community as they are not yet restricted to a particular (i.e. co-evolved) fungal community. Abundant and rare exotic plant species were expected to show the same variable pattern as they have not had the same amount of time to interact with the belowground fungal communities and establish species-specific relationships in comparison to the natives.

Methods

Study site and plant community diversity

The study site, our long-term ecological research site (LTERS), is a 100 × 100 m flat homogeneous (i.e. plant cover and edaphic conditions) old agricultural field at the Ontario Ministry of Natural Resources Research (OMNR) Arboretum in Sault Ste. Marie, Ontario (46°32′34.1″N, 84°27′29.4″W). The field site has not been cultivated for six decades and has not been mowed since 2009. The soil is a silty loam and soil analysis conducted in 2011 were as follows ($n = 6$): 22.6 ± 1.06 % clay, 53.1 ± 0.27 % silt, 24.4 ± 1.07 % sand, 0.13 ± 0.001 % total N, 1.81 ± 0.02 % total C, 11.4 ± 0.07 mg kg^{-1} Bray available P, 360.7 ± 2.55 mg kg^{-1} Ca, 45.0 ± 0.23 mg kg^{-1} K, 38.9 ± 0.32 Mg and pH = 5.74 ± 0.01. The field was divided into two hundred and eighty-nine 1 × 1 m plots on a 17 × 17 plot grid. Each plot was surrounded by a 1 m wide buffer zone. One hundred plots were selected randomly for assessment of plant species richness and abundance. Species richness was assessed in the second week of each month from May to August 2012. Plots were thoroughly examined and all species present were recorded. Percent occurrence of species was calculated by counting the number of plots (i.e. the 100 randomly assigned plots) containing each species. This provided an estimate of species' abundance (i.e. percent occurrence) in our old-field site (Fig. 1).

The survey comprised four exotic and four native plant species (Table 1). In each category, i.e. exotic and native, two species were abundant and two were relatively rare (termed abundant and rare hereon) at this old-field site (Fig. 1). The species were chosen according to their natural relative abundance at the old-field site and, as such, the lack of a balanced phylogenetic design was unavoidable. Exotic grasses dominate the site and we selected two of the most dominant for our exotic

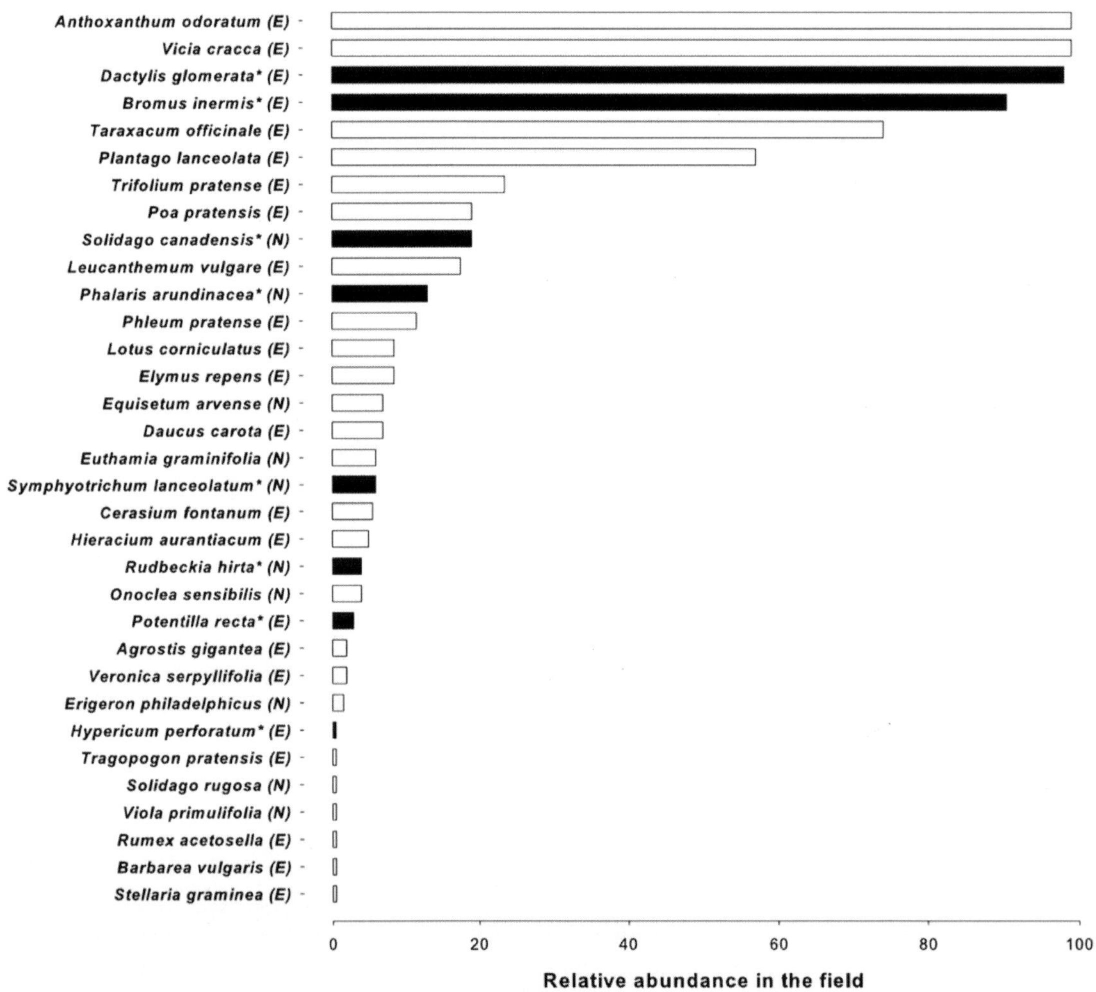

Figure 1. Plant community at the old-field site in Sault Ste. Marie, ON, Canada. Relative abundance was calculated as the number of plots in which each species was present out of 100 plots (1 × 1 m). Black bars and species names with an asterisk indicate the species used in this study. E, exotic origin; N, native origin.

Table 1. Plant species from our LTERS investigated in this study.

Native rare	Native abundant	Exotic rare	Exotic abundant
R. hirta	*S. canadensis*	*H. perforatum*	*B. inermis*
S. lanceolatum	*P. arundinacea*	*P. recta*	*D. glomerata*

abundant category (i.e. *Dactylis glomerata* and *Bromus inermis*). For the native-abundant category we selected the most abundant natives at the site (i.e. *Solidago canadensis* and *Phalaris arundinacea*). Rare (i.e. relatively rarer) species were chosen based on the presence of at least four replicates in the field. *Potentilla recta* and *Hypericum perforatum* were the selected rare exotic species and *Symphyotrichum lanceolatum* and *Rudbeckia hirta* the rare natives. Our main goal was not to control for phylogenetic relatedness or functional group but instead to survey and compare plant species according to their relative abundance in this community. Our eight selected species represent 24 % of all the species in the community. Across the entire old-field site four individuals per species were sampled for roots and to minimize potential spatial variability, sets of replicates (one individual per plant species) were sampled in close proximity to each other in a smaller area wherever possible. To account for this possible spatial variation we included the factor 'replicate' in our statistical model. The choice of four replicates is consistent with other studies investigating belowground fungal community interactions with plants (Johnson *et al.* 2004; Martínez-García *et al.* 2011).

454 Pyrosequencing of fungal communities

To determine the fungal community in the selected species, plants were collected including their whole-root system using a spade within 2 days in the second half of August 2013. They were placed into Ziploc bags and cooled immediately in a cold box and later at 4 °C until further processing. Plants were stored at 4 °C no longer than 24 h. Root systems were washed in the laboratory with cold tap water to remove the soil and to separate any root fragments that were not connected to the main system and which could belong to a different plant.

DNA was extracted from six technical replicates per individual plant to ensure having a representative amount of DNA. For each technical replicate twenty-one 0.5-cm-long root pieces were excised. For very thick roots, the amount of root material was adjusted accordingly to avoid overloading the DNA extraction column. Roots were surface sterilized for 30 s in 30 % hydrogen peroxide and the reaction was stopped using 70 % EtOH. They were then frozen at −80 °C and freeze-dried for 4 days (Labconco FreeZone 2.5, USA). To verify proper surface sterilization three 0.5-cm root pieces of each individual plant were excised and surface sterilized as described above and incubated at room temperature on potato dextrose agar plates. Surface sterilization of the plant root systems was successful as fungi started growing out of the roots only after several days at 20 °C. This means organisms attached to the surface were excluded but fungi within the roots were still actively growing.

DNA was extracted using the DNeasy 96 Plant Kit (Qiagen, Toronto, ON, Canada) according to the manufacturer's protocol. The extracted DNA was cleaned and purified using the Genomic DNA Clean & Concentrator™ Kit (Zymo Research, USA). After DNA purification, the six technical replicates coming from one single root system were pooled together leading to 32 samples in total (eight plant species × four replicates). The amount of DNA was measured using a BioTek™ Epoch™ Microplate Spectrophotometer (Winooski, VT, USA).

Primers fITS7 as a forward (5′-GTGARTCATCGAATCTTTG-3′) and ITS4 as a reverse (5′-TCCTCCGCTTATTGATATGC-3′) were chosen to sequence samples using 454 pyrosequencing (Ihrmark *et al.* 2012). We specifically chose the new highly specific fungal primer fITS7 because we were interested in amplifying the entire fungal community present within the plant roots. In combination with ITS4 this pair specifically amplifies the ITS 2 region and produces shorter amplicons compared with the ITS1f/ITS4 primer pair. We expected that as we sequenced an entire fungal community any bias resulting from different sequence lengths would be minimized by amplifying the shorter ITS 2 region instead of the entire internal transcribed spacer (ITS) region. According to a BLASTN search, our selected primer pair should match 68 % of the Glomeromycota (Ihrmark *et al.* 2012). Samples were sent to Genome Québec (Montréal, Canada) and purified and diluted 1 : 4 using a MultiScreen PCR filtration plate (Millipore, Canada). PCR with primers fITS7 and ITS4 was conducted in 25 μL reaction volume containing 2.5 μL purified DNA, 2.5 μL 10× buffer, 1.5 mM $MgCl_2$, 0.2 μM dNTPs, 0.4 μM of each primer and 5 U μL^{-1} HotStarTaq DNA Polymerase (Qiagen) as final concentrations. PCR running conditions were as follows: initial denaturation at 96 °C for 15 min followed by 35 cycles of denaturation at 96 °C for 30 s, annealing at 52 °C for 30 s, extension at 72 °C for 1 min and a final extension step at 72 °C for 10 min. Samples were tagged using 10-bp multiplex identifiers (MID Tags) in a separate barcoding PCR reaction. The tagged, equimolar amplicons were sequenced on a Roche 454 GS FLX instrument using the FLX+ chemistry at Genome Québec. A total of 327 867 single reads were produced in a one-fourth region of a 454 plate. Sequence data were submitted to NCBI (accession no. SRP043133).

Symbiotic relationships

To investigate symbiotic (i.e. functional) relationships between mutualistic and antagonistic belowground biota and our target species, plants were collected at our old-field site in mid-August 2012, with the exception of *R. hirta*, which could not be found on that occasion. Four individual plants were collected for each species using a spade (except for *H. perforatum* for which we could only sample three individuals). In the laboratory, root samples were carefully washed and divided into two sub-samples. One sub-sample was used to quantify the percentage of root colonization by arbuscular mycorrhizal fungi (AMF) using the gridline intersect method by McGonigle *et al.* (1990). The roots were stained using 0.03 % Chlorazol-black E and a minimum of 100 intersections per replicate were examined. The other sub-sample was used to quantify the incidence of disease and enemy attack by visually estimating damage (i.e. the number of lesions per root) caused by pathogens, parasites and herbivores. This method is identical to the root staining and quantification method used for AMF, excluding the staining step. This approach provides a consistent estimate of plant tolerance against pathogens and other enemies (Mitchell 2003; Schnitzer *et al.* 2011). For both AMF and lesion quantification, *H. perforatum* samples had to undergo an additional bleaching step using alkaline hydrogen peroxide to remove root pigmentation to allow visualization.

Bioinformatics and statistical analysis

Bioinformatics were performed at Genome Québec using Qiime (version 1.7.0) the pipeline for data processing. In total, 327 867 reads were generated, assessed for their quality and clustered. Reads with a Phred quality score $\geq$30, $<$3 undetermined bases and $<$4 bases having a Phred quality score $<$20 passed the quality control step. All reads were cut to a final length of 220 bases. Reads were clustered at 100 % ID, followed by 99 % and finally by 97 %. Chimera clusters were removed during this step. Sequences were blasted against the ITS2 database of the UNITE ITS database (Kõljalg *et al.* 2013), trimmed to the ITS2 region. An Operational Taxonomic Units (OTU) table was generated representing fungi only and samples were rarefied to 1000 reads per sample. The Qiime pipeline (version 1.7.0) was used to calculate the rarefraction curves. After rarefaction, singletons and doubletons were removed and the resulting OTU table was used for further analysis.

To determine whether fungal communities differed between the factors 'abundance' (i.e. rare and abundant plant species) and 'origin' (i.e. native and exotic plant species), a Permutational MANOVA was conducted using function 'adonis' from the *vegan* package (Oksanen *et al.* 2013) implemented in the software R (R Core Team 2013). The effect of plant abundance and origin was calculated by measuring the effect of each factor after accounting for the variance explained by the other factor (variance partitioning (Schmid *et al.* 2002)). This was done after first accounting for our four replicates (i.e. spatial variability) and we also took into consideration the two different plant functional groups (i.e. monocotyledons and dicotyledons). We first ran a model including 'plant species' as a third factor and, as expected, plant species were a significant predictor of fungal community composition. By this factor, 42 % of the variance was explained. However, since our main goal was to survey general patterns of fungal diversity rather than species-specific effects, we considered species as 'plant functional group' thereby gaining degrees of freedom in our two factor model.

To calculate the variance explained by the factor 'abundance' (full model including interaction) we used the following formula:

$$\text{Dissimilarity matrix} \sim \text{replicate} + \text{plant functional group} + \text{origin} + \text{abundance} + \text{origin:abundance} \quad (1)$$

To calculate the variance explained by the factor 'origin' (full model including interaction) we used the following formula:

$$\text{Dissimilarity matrix} \sim \text{replicate} + \text{plant functional group} + \text{abundance} + \text{origin} + \text{abundance:origin} \quad (2)$$

The plant functional group was kept in the model since it explained 9.21 % of the total variance. By keeping it, the variance from this factor could be deducted from our variables of interest ('abundance' and 'origin'). As dissimilarity measure, Bray–Curtis was used to account for the composition of the fungal communities as well as to exclude joint absences. α Diversity was tested for each sample using the function 'rarefy' from the *vegan* package (Oksanen *et al.* 2013) in R (R Core Team 2013) and an ANOVA was run to test for significant differences depending on the factors 'abundance' and 'origin'. A distance matrix was created using the *vegan* package (Oksanen *et al.* 2013) in R (R Core Team 2013) to estimate β diversity using the function 'betadisper' and to draw principle coordinates analysis (PCoA) plots.

Another ANOVA was run to test whether AMF root colonization and lesions varied according to 'abundance' and 'origin'. To meet the requirements of the statistical tests, variables were arcsine transformed. Means among treatments were compared using Tukey's honestly significant difference (HSD) test ($P < 0.05$). These statistical analyses were performed with the software StatSoft,

Inc. (2013), STATISTICA (data analysis software system), version 12. www.statsoft.com.

Results

Fungal community patterns

We detected 164 fungal OTUs across all samples. The phylum Ascomycota comprised 138 OTUs (84 % of fungal OTUs) whereas Basidiomycota only comprised 23 OTUs (14 % of fungal OTUs), the remaining three OTUs (1.8 % of fungal OTUs) could only be assigned to Kingdom Fungi (Table 2). We were unable to recover AM fungi as we expected based on our primer choice. Fungal richness of rarefied and non-rarefied samples for each plant species is provided in Table 3. The two plant species *H. perforatum* and *P. arundinacea* had the highest number of fungal OTUs (Table 3) and the Helotiales was the most common fungal order across all plant species (Fig. 2 and **Supporting Information—Table S1** for detailed taxonomic results).

Permutational MANOVA showed significant fungal community dissimilarities for 'abundance' ($F_{1,29} = 5.194$, $R^2 = 0.140$, $P = 0.001$) and 'origin' ($F_{1,29} = 4.590$, $R^2 = 0.120$, $P = 0.001$) after accounting for both the spatial variation (factor 'replicate') in our field site ($F_{3,29} = 1.322$ with $R^2 = 0.103$, $P = 0.161$ (1) and, $R^2 = 0.103$, $P = 0.140$ (2)) and our two plant functional groups (monocotyledons and dicotyledons) which explained 9.21 % of the total variance ($F_{1,29} = 3.756$ with $R^2 = 0.092$, $P = 0.005$ (1) and $R^2 = 0.092$, $P = 0.002$ (2)). However, the interaction of 'abundance' and 'origin' was not significant ($F_{1,29} = 1.235$, $R^2 = 0.034$, $P = 0.255$) in either of the two models and therefore, we continued with the analysis using the reduced models (i.e. by excluding the interaction but keeping the factor 'replicate') (Table 4). The PCoA plots were in concert with the results from the permutational MANOVA pointing towards the structuring of fungal communities as a function of 'abundance' and 'origin' (Fig. 3). Figure 3C shows that our 'plant functional groups' did not share the same fungal communities but those were rather dependent on origin except for *H. perforatum* and *S. canadensis*.

α Diversity, i.e. the number of OTUs per sample, was not affected by the factors 'abundance' ($F_{1,26} = 1.108$, $P = 0.302$) or 'origin' ($F_{1,26} = 0.163$, $P = 0.690$) but their interaction was significant ($F_{1,26} = 6.908$, $P = 0.014$) (Fig. 4A). Fungal β diversity varied significantly depending on plant species' relative 'abundance' ($F_{1,28} = 5.77$, $P = 0.017$) but not on their 'origin' (Fig. 4B).

Overall, abundant plant species tended to associate with a more similar fungal community compared with rare plants (Fig. 3A). This was also supported by β diversity (Fig. 4B). In addition, PCoA indicated that compared with native plants, fungal community composition among exotics was more similar (Fig. 3B).

Table 2. Taxonomic fungal groups detected in the roots of all plant species in the filtered and rarefied OTU table excluding singletons and doubletons.

Taxonomic group	No. of OTUs	OTUs in %	No. of reads	Reads in %
Ascomycetes	138	84	26 754	95.5
Helotiales	57	41.3	17 964	67.1
Chaetothyriales	25	18.1	1291	4.8
Pleosporales	17	12.3	1522	5.7
Capnodiales	9	6.5	1522	5.7
Sordariales	7	5.1	809	3
Hypocreales	3	2.2	129	0.5
Xylariales	2	1.4	195	0.7
Chaetosphaeriales	2	1.4	121	0.5
Dothideales	2	1.4	50	0.2
Hysteriales	1	0.7	52	0.2
Basidiomycetes	23	14	1258	4.5
Agaricales	6	26.1	198	15.7
Auriculariales	5	21.7	102	8.1
Sebacinales	3	13	630	50.1
Cantharellales	2	8.7	285	22.7
Polyporales	1	4.3	8	0.6
Undetermined (fungi)	3	1.8	3134	11.2

Table 3. Fungal richness per plant species for non-rarefied and rarefied samples.

Plant species	Non-rarefied	Rarefied
R. hirta	45.26127	27.25000
S. lanceolatum	78.96880	49.00000
S. canadensis	75.55551	44.33333
P. arundinacea	107.10158	64.75000
H. perforatum	108.59286	55.25000
P. recta	63.01892	40.33333
B. inermis	69.51156	43.50000
D. glomerata	64.02056	39.25000

Symbiotic relationship patterns

Species 'abundance' as well as the interaction between 'abundance' and 'origin' significantly determined AMF colonization by hyphae in our selected plant groups

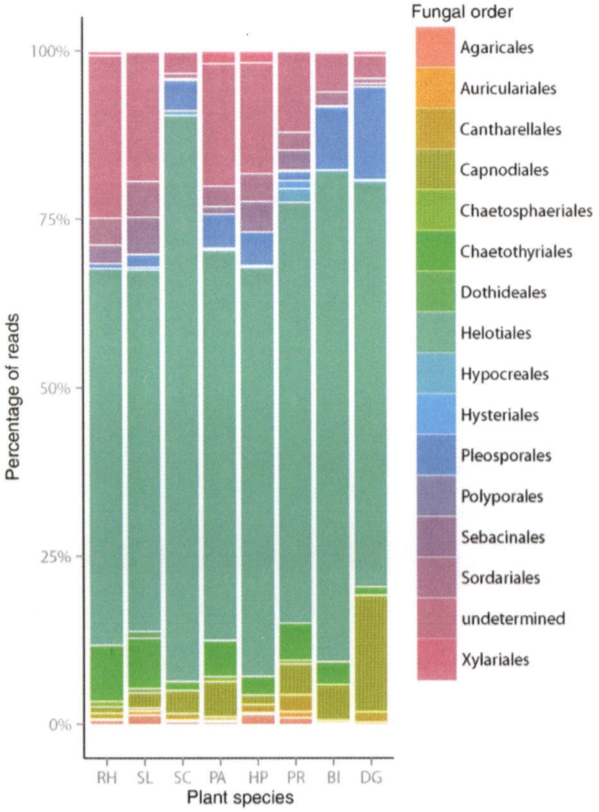

Figure 2. Rarified abundance of fungal sequence reads discriminated by fungal order for each plant species. 'Undetermined' means OTUs not determined to the level 'Order' but at least to the kingdom of fungi. RH, *R. hirta*; SL, *S. lanceolatum*; SC, *S. canadensis*; PA, *P. arundianacea*; HP, *H. perforatum*; PR, *P. recta*; BI, *B. inermis*; DG, *D. glomerata*.

Table 4. Factors and their interaction referring to community dissimilarity and resulting from the Permutational MANOVA for the response variables 'abundance' and 'origin'; *Significant at $\alpha = 0.05$.

	F-value	R^2	*P*-value
Calculating factor 'abundance'			
Abundance	5.194	0.14	0.001*
Replicate	1.322	0.103	0.161
Plant functional group	3.756	0.092	0.005*
Abundance × origin	1.235	0.034	0.255
Calculating factor 'origin'			
Origin	4.59	0.12	0.001*
Replicate	1.322	0.103	0.14
Plant functional group	3.756	0.092	0.002*
Abundance × origin	1.235	0.034	0.255

($F_{1,23} = 11.548$, $P = 0.002$ and $F_{1,23} = 6.972$, $P = 0.015$, respectively). Rare exotic plants had the largest amount of AMF colonization and abundant exotics the least. In contrast, hyphal colonization did not vary significantly with plant relative abundance for native plants (Fig. 5A). Arbuscular colonization showed the same trends but the results were not statistically significant due to a high variance for the rare exotic species, which were underrepresented (Fig. 5B). We found a significant effect of 'origin' as well as 'origin' × 'abundance' for vesicular colonization ($F_{1,23} = 7.663$, $P = 0.011$ and $F_{1,23} = 6.937$, $P = 0.015$, respectively). However, this was in contrast with hyphal colonization. The rare native species had the greatest number of vesicles and rare exotics the least (Fig. 5C).

'Origin' was a significant factor determining the susceptibility of plant species to enemy attack belowground ($F_{1,22} = 17.811$, $P \leq 0.001$), and despite the lower replication of rare plants due to the absence of *R. hirta*, we detected a marginally significant effect of 'abundance' ($F_{1,22} = 4.148$, $P = 0.054$). In addition, there was a marginally significant effect for the interaction ($F_{1,22} = 3.273$, $P = 0.084$). Rare natives were the most susceptible group to enemy attack whereas for exotics, their lesion number was the same regardless of their relative abundance in the community (Fig. 5D).

Discussion

The main objective of this study was to investigate patterns connected with plant and belowground fungal community assembly. More specifically, we determined whether plant species of contrasting relative abundance and geographic origin in a natural old-field community typical of eastern North America associated differently with the belowground fungal community. Our main goal was to primarily survey plants based on their relative abundance. As such, the lack of a balanced phylogenetic design was an unavoidable feature of the study system. Our choice of plant species provides insight into how natural plant diversity may contribute to influence the belowground fungal community of a typical old-field grassland in eastern North America. Nevertheless, in addition to plant abundance (abundant and rare) and plant origin (native and exotic) we accounted for the proportion of the variance explained by plant functional groups (monocotyledons and dicotyledons). Our results were consistent with our prediction that fungal communities associated differently with plant species depending on the plants' relative 'abundance' and 'origin'.

Plant species 'abundance' was one of the two factors influencing fungal communities in plant roots suggesting that soil communities may also play a role in determining the relative abundance of our plant species in the community. This is consistent with reports emphasizing the importance of soil microbial communities in structuring plant communities (Klironomos 2002; Bever *et al.* 2010; Pendergast *et al.* 2013).

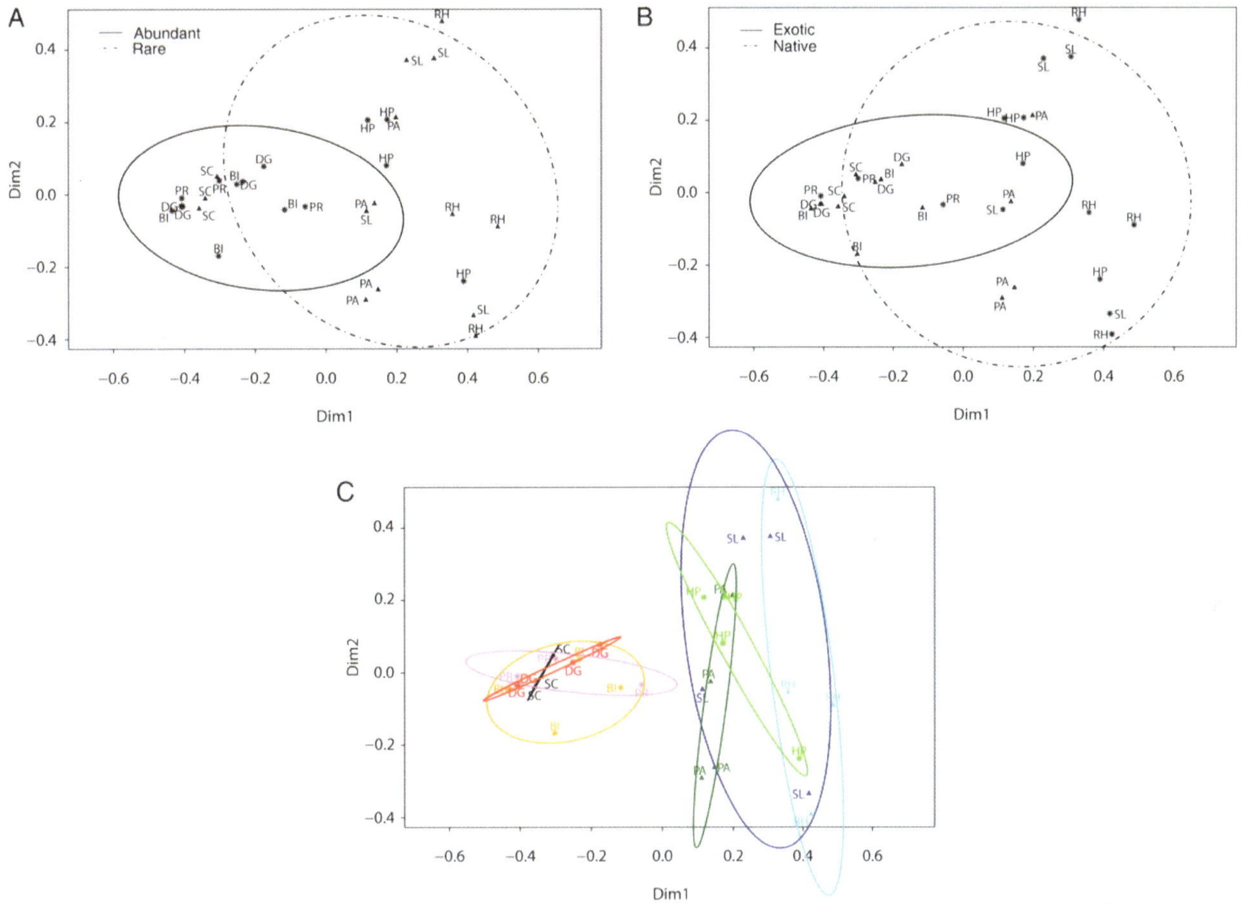

Figure 3. Principal coordinate analysis plots discriminating between belowground root-associated fungal communities according to (A) abundance, (B) origin and (C) plant species. In (A and C) exotic plant species are labeled with a star and native species with triangles and in (B) abundant plant species are labeled with a triangle and rare species with a star. Confidence interval = 80 % for (A and B) and 75 % for (C). RH, *R. hirta*; SL, *S. lanceolatum*; SC, *S. canadensis*; PA, *P. arundianacea*; HP, *H. perforatum*; PR, *P. recta*; BI, *B. inermis*; DG, *D. glomerata*.

Partly consistent with our prediction, we found β diversity to be significantly higher for rare than for abundant plant species, indicating that fungal communities among abundant plant species were more similar compared with those present in rare plant species (Fig. 4B). This result was additionally supported by the PCoA plot (Fig. 3A). However, the interaction between 'abundance' and 'origin' was not significant contrary to our prediction that patterns of fungal community diversity were connected to both plant abundance and origin. β Diversity was not significantly different between exotic and native species. β Diversity of rare plants may be influenced by the fungal community of neighboring plant species which are likely to be composed of different plant species. Conversely, abundant plants are more often surrounded by conspecifics and therefore, share a more similar fungal community than rare plant species. Alternatively, our four rare species (Table 1) may be more selective in their fungal associations, explaining the higher overall β diversity but lower α diversity. Another scenario could be that rare species occupy a greater number of microsites to avoid competition as they are presumably rather weak competitors compared with abundant plant species. Due to these different environments rare plants cannot be very selective in choosing their fungal communities. This in turn might influence their β diversity which was rather high compared with abundant species. Since rarity in communities may be driven by negative feedback (Klironomos 2002) based on our results it is tempting to speculate that rare native species may associate with putatively more antagonistic fungi (Bever *et al.* 2012; Van der Putten *et al.* 2013). This was further supported by our root lesion data, showing that root enemies affected mostly rare native plants (Fig. 5B). Additionally, a negative feedback has been shown to be prevalent in grasslands maintaining a high plant diversity (Kulmatiski *et al.* 2008).

Fungal richness (α diversity) in abundant native plants was higher than in rare natives but for exotics we did not observe a significant difference (Fig. 4A). Perhaps such higher fungal diversity in the roots of native plants

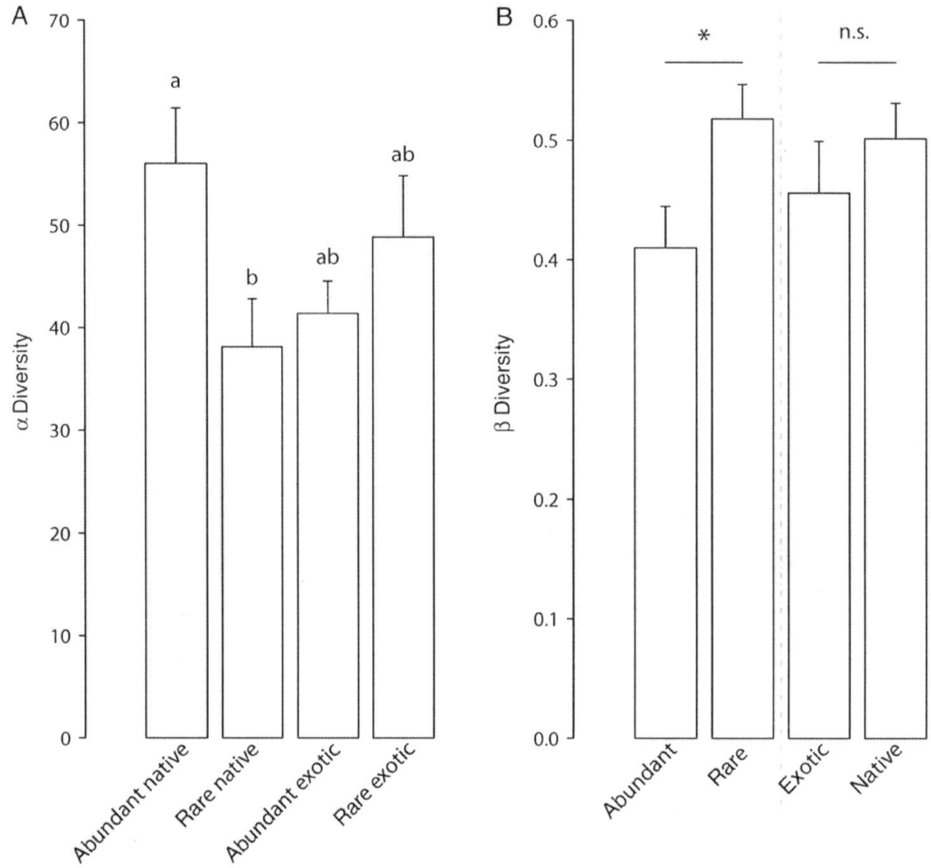

Figure 4. α (A) and β (B) diversity of fungal communities for the factors 'abundance' and 'origin' in plant roots from a natural old-field community. α Diversity showed a significant interaction between 'abundance' and 'origin' ($P = 0.014$). Standard errors are indicated by error bars. Means in a bar sharing the same letter are not significantly different according to Tukey's HSD at a significance level of $P \leq 0.05$.

could be important to explain their abundance. Ecosystems and communities may be considered to be more stable if biodiversity is higher since productive species can compensate for other species that may not perform as well (Naeem and Li 1997). This may also be linked to the success of individual plant species. For instance, higher genotypic diversity of root-associated dark septate endophytic fungi has been shown to increase their host plants' growth rate (Reininger *et al.* 2012). Perhaps the microbial communities associated with abundant native plant species may be even more conducive to positive feedback (Van der Putten *et al.* 2013).

As predicted, fungal communities associating with exotic species were different from fungal communities associating with native species. However, we found no differences in α- and β diversity between abundant exotic and rare exotic species, and exotic and native species, respectively. This suggests that the relative abundance of exotics in the community may be independent from their fungal communities. Since exotic species may have escaped from their co-evolved soil enemies (Keane and Crawley 2002; Agrawal *et al.* 2005), they might be uncoupled from the requirement to associate with a specific belowground microbial community. Despite the importance of fungi as drivers of plant community structure (Van der Heijden *et al.* 1998; Reynolds *et al.* 2003) only very few studies have compared the fungal communities associating with exotics versus natives in a natural community (Bongard *et al.* 2013). Consistent with our results, Nelson and Karp (2013) found that distinct communities of pathogenic fungi associate with native and exotic haplotypes of *Phragmites australis*, suggesting that exotic plants may associate with a different fungal community than native species.

Our molecular methodological approach detected fungi belonging to the Ascomycota, which is the largest clade of fungi (Kirk *et al.* 2001; Wang *et al.* 2006), and Basidiomycota. More specifically, the most represented order across all plant species regardless of their 'abundance' and 'origin' were the Helotiales belonging to the Ascomycota (Fig. 2). This very diverse and large fungal order includes ~2300 species which take several ecological functional roles ranging from ericoid mycorrhiza, endophytes, saprobes and pathogens (Webster and Weber 2007). Therefore, it is not surprising that we found so many OTUs belonging to this

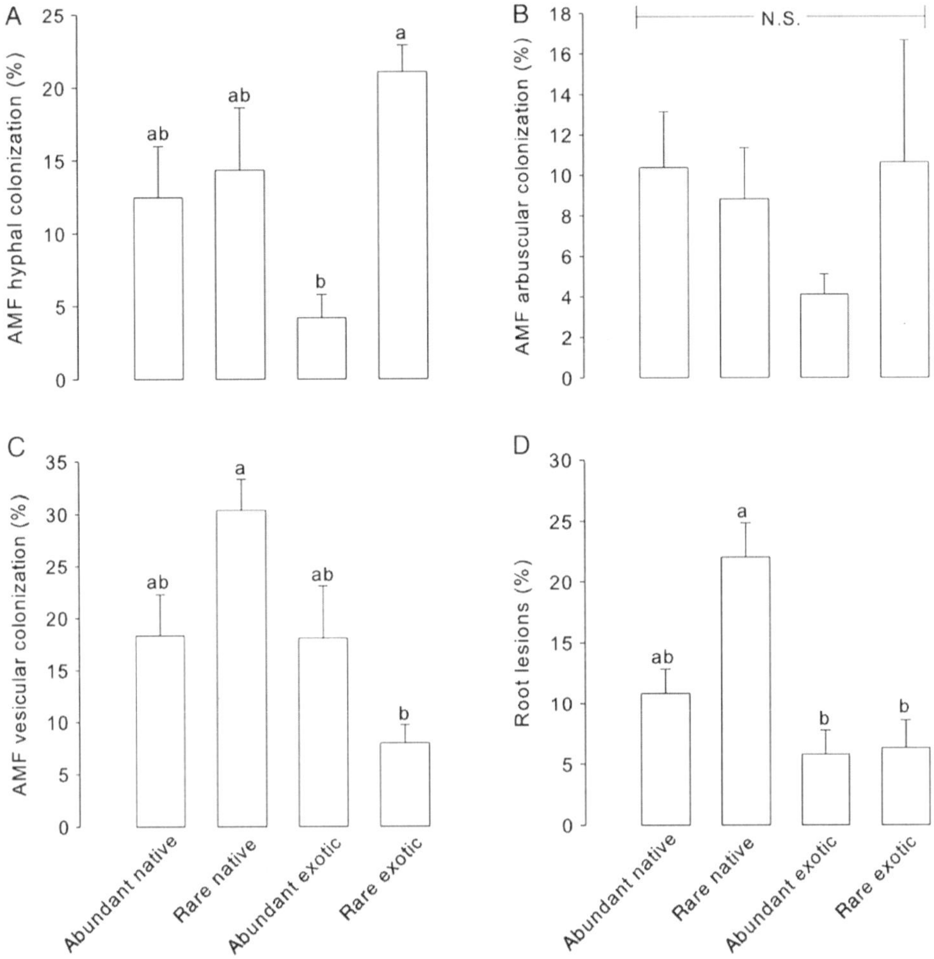

Figure 5. Arbuscular mycorrhizal fungal colonization by hyphae (A), arbuscules (B) and vesicles (C) and incidence of lesions (D) in the roots of native and exotic plants of contrasting relative abundance in a ruderal natural plant community. Means in a bar sharing the same letter are not significantly different according to Tukey's HSD at a significance level of $P \leq 0.05$.

large and diverse fungal order. Since OTUs were not determined to species level, we were unable to make any statement about their particular function in our old-field site.

The ITS region has been shown a high degree of success for the broad detection of fungi (Buée *et al.* 2009) and was recently proposed to serve as the primary fungal barcode marker (Schoch *et al.* 2012). Even though the primer pair fITS7/ITS4 has been tested positively to match Glomeromycota (Ihrmark *et al.* 2012), we failed to do so even though AMF were certainly present in the roots of our plant species (Fig. 5A). This is in contrast with a BLASTN search at NCBI with fITS7, which matched 68 % of the Glomeromycota 5.8S sequences (Ihrmark *et al.* 2012). Even though we selected this primer pair taking into consideration the shorter sequence length, it is known that the choice of ITS could bias against certain groups such as Early Diverging Lineages, Ascomycota and Glomeromycota (see Schoch *et al.* 2012). For the latter group, Stockinger *et al.* (2010) proposed a larger barcode, combining the ITS and LSU regions. Furthermore, due to the fact that in our study the primers were degenerate, there is an increasing chance of getting unspecific results by running too many PCR cycles (Ihrmark *et al.* 2012). This primer bias can be expected as a result of a high number of PCR cycles (35) and the low GC content of the Glomeromycota (Stockinger *et al.* 2010; Gianinazzi-Pearson *et al.* 2012; Kauserud *et al.* 2012; Pinto and Raskin 2012). Nevertheless, all our samples were treated equally and even though we were unable to capture the full breadth of the fungal Kingdom, our 454 pyrosequencing approach was able to consistently detect a broad diversity of taxa among our plant groups. This is in accordance with Ihrmark *et al.* (2012) as our primers are supposed to better preserve the quantitative composition of the template between-sample differences in fungal community composition. We also need to consider that fungal community composition shifts seasonally (Thoms and Gleixner 2013; Voříšková *et al.* 2014). The sampling was undertaken mid-summer to ensure that fungal communities had sufficient time to re-establish after the winter. However, our study represents a snapshot of the fungal community at a certain time. As such it is entirely possible that the outcome of the study might be different in another season and it would be interesting to contrast our results with those collected in late fall or early spring. Furthermore, our study should be seen as a first coarse survey of fungal communities across a selected number of plants of contrasting abundance and origin in a community. Future work should include more species in this and other communities and take into consideration phylogenetic relationships.

We were able to evaluate patterns of root interactions with mutualistic and antagonistic soil biota by estimating AM fungal colonization and lesions in plant roots, respectively (Fig. 5). Arbuscular mycorrhizal fungi are considered widespread among plants and represent perhaps the most important mutualism globally (Van der Heijden *et al.* 1998; Wang and Qiu 2006; Parniske 2008; Öpik *et al.* 2010; Kiers *et al.* 2011). We found that AMF barely colonized the most abundant exotic plants (i.e. <5 %), which could be considered as invasive plants in our old-field site (Fig. 1). These plants also happened to be grasses and some subfamilies in the Poaceae (e.g. subfamily Pooideae to which *B. inermis* belongs) are known to have low mycorrhizal responsiveness and colonization (Reinhart *et al.* 2012). The opposite was true for rare exotic species (Fig. 5A). Taken together, our results are consistent with the framework proposed by Pringle *et al.* (2009) suggesting, despite the very limited data available, that at least for North America, exotic invasive plants (i.e. in this particular case, mostly grasses in old-field systems) may not associate or depend on AMF. Taking this into consideration, our results are consistent with studies showing that abundant cool-season grasses have low mycorrhizal dependency (i.e. their biomass responses are not correlated with AMF colonization (Reinhart *et al.* 2012; Treseder 2013) whereas exotic forbs can establish strong associations with AMF (Wilson and Hartnett 1998; Lekberg *et al.* 2013).

Our results showed that rare native plants harbored more vesicles, which are energy reserve structures and whose production may be triggered in response to stress (Smith and Read 2008) (Fig. 5C). They were also significantly more affected by root enemies than exotic plants present in the community (Fig. 5D). This confirms and further provides a mechanism (i.e. increased specialist plant biotic interactions) for the result of a previous greenhouse feedback study with soil biota from an old-field plant community in which negative feedback was stronger on rare native plants (Klironomos 2002). It is possible that some of the fungal groups detected using our sequencing approach may be responsible for some of the observed root lesions present in higher number on native rare plants. However, considering that fungal function is highly context dependent (e.g. host dependent), future work should use a more targeted approach that can provide higher taxonomic resolution and capture functional processes, to determine the extent to which changes in fungal diversity associating with roots are either beneficial or detrimental to particular plants species or guilds.

To our knowledge, this is the first study investigating linkages between belowground fungal community structure and abundant and rare plant species in a natural grassland community, including plant species originating from different geographic origins. Future research connected to this field should use molecular methodologies capable to increase the level of taxonomic resolution. In addition, a major goal should be the establishment of links between taxonomic and functional diversity driving plant communities.

Conclusions

In conclusion, our results are the first linking natural plant abundance and geographic origin with fungal community structure. Overall, plant relative abundance but not origin determined fungal β diversity. Fungal α diversity varied only for native species of contrasting abundance; not for exotics. Taken together, this suggests the structuring of fungal communities even in the remnant native plants of an invaded community. Conversely, exotic plants appear to act as a disturbing (homogenizing) agent of the fungal community possibly changing feedback interactions across the entire plant community (e.g. mutualist disruption). These patterns may be linked to invasion meltdown processes and need further investigation.

Sources of Funding

This study was funded by a Research Chair from the Ontario Ministry of Natural Resources and by a Natural Sciences and Engineering Research Council of Canada (NSERC) Discovery Grant awarded to P.M.A.

Contributions by the Authors

P.M.A. originally formulated the idea, P.M.A., V.R. and L.S. developed original idea, V.R. and L.S. developed methodology, V.R., L.S. and P.M.A. conducted fieldwork, V.R. generated sequencing data and molecular analyses, V.R., L.B.M.-G. and P.M.A. performed statistical analyses, and V.R., L.B.M.-G., P.M.A. and L.S. wrote the manuscript.

Acknowledgements

We thank Jacqueline Ayotte for technical support in the field. We also thank the Ontario Forestry Research Institute for allowing us to establish the LTERS at the arboretum. We thank Markus Schlegel for his help selecting the primers and Matteo Tanadini of the Seminar for Statistics (SfS), ETH Zurich, for his support. Finally, we would like to thank Genome Québec (Montréal, Canada) for conducting 454 pyrosequencing and part of the data analysis.

Literature Cited

Agrawal AA, Kotanen PM, Mitchell CE, Power AG, Godsoe W, Klironomos J. 2005. Enemy release? An experiment with congeneric plant pairs and diverse above- and belowground enemies. *Ecology* **86**:2979–2989.

Anderson MJ, Crist TO, Chase JM, Vellend M, Inouye BD, Freestone AL, Sanders NJ, Cornell HV, Comita LS, Davies KF, Harrison SP, Kraft NJB, Stegen JC, Swenson NG. 2011. Navigating the multiple meanings of β diversity: a roadmap for the practicing ecologist. *Ecology Letters* **14**:19–28.

Bagchi R, Swinfield T, Gallery RE, Lewis OT, Gripenberg S, Narayan L, Freckleton RP. 2010. Testing the Janzen-Connell mechanism: pathogens cause overcompensating density dependence in a tropical tree. *Ecology Letters* **13**:1262–1269.

Bever JD, Westover KM, Antonovics J. 1997. Incorporating the soil community into plant population dynamics: the utility of the feedback approach. *Journal of Ecology* **85**:561–573.

Bever JD, Dickie IA, Facelli E, Facelli JM, Klironomos J, Moora M, Rillig MC, Stock WD, Tibbett M, Zobel M. 2010. Rooting theories of plant community ecology in microbial interactions. *Trends in Ecology and Evolution* **25**:468–478.

Bever JD, Platt TG, Morton ER. 2012. Microbial population and community dynamics on plant roots and their feedbacks on plant communities. *Annual Review of Microbiology* **66**:265–283.

Bonfante P, Anca IA. 2009. Plants, mycorrhizal fungi, and bacteria: a network of interactions. *Annual Review of Microbiology* **63**: 363–383.

Bongard CL, Navaranjan G, Yan W, Fulthorpe RR. 2013. Fungal colonization of the invasive vine *Vincetoxicum rossicum* and native plants. *Plant Ecology and Evolution* **146**:45–52.

Buée M, Reich M, Murat C, Morin E, Nilsson RH, Uroz S, Martin F. 2009. 454 Pyrosequencing analyses of forest soils reveal an unexpectedly high fungal diversity. *New Phytologist* **184**:449–456.

Chave J. 2004. Neutral theory and community ecology. *Ecology Letters* **7**:241–253.

Compant S, van der Heijden MGA, Sessitsch A. 2010. Climate change effects on beneficial plant-microorganism interactions. *FEMS Microbiology Ecology* **73**:197–214.

Costa FO, Antunes PM. 2012. The contribution of the barcode of life initiative to the discovery and monitoring of biodiversity. In: Mendonca A, Cunha A, Chakrabarti R, eds. *Natural resources, sustainability and humanity*. The Netherlands: Springer, 37–68.

Dumbrell AJ, Nelson M, Helgason T, Dytham C, Fitter AH. 2010. Idiosyncrasy and overdominance in the structure of natural communities of arbuscular mycorrhizal fungi: is there a role for stochastic processes? *Journal of Ecology* **98**:419–428.

Dybzinski R, Tilman D. 2007. Resource use patterns predict long-term outcomes of plant competition for nutrients and light. *The American Naturalist* **170**:305–318.

Flory SL, Clay K. 2013. Pathogen accumulation and long-term dynamics of plant invasions. *Journal of Ecology* **101**:607–613.

Gianinazzi-Pearson V, van Tuinen D, Wipf D, Dumas-Gaudot E, Recorbet G, Liu Y, Doidy J, Redecker D, Ferrol N. 2012. Exploring the genome of Glomeromycotan fungi. In: Hock B, ed. *Fungal associations*. Berlin: Springer.

Hubbell SP. 2001. *The unified neutral theory of biodiversity and biogeography*. Princeton: Princeton University Press.

Ihrmark K, Bödeker ITM, Cruz-Martinez K, Friberg H, Kubartova A, Schenck J, Strid Y, Stenlid J, Brandström-Durling M, Clemmensen KE, Lindahl BD. 2012. New primers to amplify the fungal ITS2 region—evaluation by 454-sequencing of artificial and natural communities. *FEMS Microbiology Ecology* **82**: 666–677.

Jeschke JM, Gómez Aparicio L, Haider S, Heger T, Lortie CJ, Pyšek P, Strayer DL. 2012. Support for major hypotheses in invasion biology is uneven and declining. *NeoBiota* **14**:1–20.

Johnson D, Vandenkoornhuyse PJ, Leake JR, Gilbert L, Booth RE, Grime JP, Young JPW, Read DJ. 2004. Plant communities affect arbuscular mycorrhizal fungal diversity and community composition in grassland microcosms. *New Phytologist* **161**:503–515.

Johnson NC, Graham JH, Smith FA. 1997. Functioning of mycorrhizal associations along the mutualism-parasitism continuum. *New Phytologist* **135**:575–586.

Jumpponen A, Trappe JM. 1998. Performance of *Pinus contorta* inoculated with two strains of root endophytic fungus, *Phialocephala fortinii*: effects of synthesis system and glucose concentration. *Canadian Journal of Botany* **76**:1205–1213.

Kauserud H, Kumar S, Brysting AK, Nordén J, Carlsen T. 2012. High consistency between replicate 454 pyrosequencing analyses of ectomycorrhizal plant root samples. *Mycorrhiza* **22**:309–315.

Keane RM, Crawley MJ. 2002. Exotic plant invasions and the enemy release hypothesis. *Trends in Ecology and Evolution* **17**:164–170.

Kiers ET, Duhamel M, Beesetty Y, Mensah JA, Franken O, Verbruggen E, Fellbaum CR, Kowalchuk GA, Hart MM, Bago A, Palmer TM, West SA, Vandenkoornhuyse P, Jansa J, Bücking H. 2011. Reciprocal rewards stabilize cooperation in the mycorrhizal symbiosis. *Science* **333**:880–882.

Kirk PM, Cannon PF, David JC, Stalpers JA. 2001. *Dictionary of the fungi*. Wallingford, UK: CABI Publishing.

Klironomos JN. 2002. Feedback with soil biota contributes to plant rarity and invasiveness in communities. *Nature* **417**:67–70.

Klironomos JN. 2003. Variation in plant response to native and exotic arbuscular mycorrhizal fungi. *Ecology* **84**:2292–2301.

Kõljalg U, Nilsson RH, Abarenkov K, Tedersoo L, Taylor AFS, Bahram M, Bates ST, Bruns TD, Bengtsson-Palme J, Callaghan TM, Douglas B, Drenkhan T, Eberhardt U, Dueñas M, Grebenc T, Griffith GW, Hartmann M, Kirk PM, Kohout P, Larsson E, Lindahl BD, Lücking R, Martín MP, Matheny PB, Nguyen NH, Niskanen T, Oja J, Peay KG, Peintner U, Peterson M, Põldmaa K, Saag L, Saar I, Schüßler A, Scott JA, Senés C, Smith ME, Suija A, Taylor DL, Telleria MT, Weiss M, Larsson K.-H. 2013. Towards a unified paradigm for sequence-based identification of fungi. *Molecular Ecology* **22**:5271–5277.

Kulmatiski A, Beard KH, Stevens JR, Cobbold SM. 2008. Plant-soil feedbacks: a meta-analytical review. *Ecology Letters* **11**: 980–992.

Lekberg Y, Gibbons SM, Rosendahl S, Ramsey PW. 2013. Severe plant invasions can increase mycorrhizal fungal abundance and diversity. *ISME Journal* **7**:1424–1433.

Levine JM. 2000. Species diversity and biological invasions: relating local process to community pattern. *Science* **288**:852–854.

Martínez-García LB, Armas C, Miranda JD, Padilla FM, Pugnaire FI. 2011. Shrubs influence arbuscular mycorrhizal fungi communities in a semi-arid environment. *Soil Biology and Biochemistry* **43**:682–689.

McGill BJ, Etienne RS, Gray JS, Alonso D, Anderson MJ, Benecha HK, Dornelas M, Enquist BJ, Green JL, He F, Hurlbert AH, Magurran AE, Marquet PA, Maurer BA, Ostling A, Soykan CU, Ugland KI, White EP. 2007. Species abundance distributions: moving beyond single prediction theories to integration within an ecological framework. *Ecology Letters* **10**:995–1015.

McGonigle TP, Miller MH, Evans DG, Fairchild GL, Swan JA. 1990. A new method which gives an objective measure of colonization of roots by vesicular-arbuscular mycorrhizal fungi. *New Phytologist* **115**:495–501.

Miller TE, Burns JH, Munguia P, Walters EL, Kneitel JM, Richards PM, Mouquet N, Buckley HL. 2005. A critical review of twenty years' use of the resource-ratio theory. *The American Naturalist* **165**: 439–448.

Mitchell CE. 2003. Trophic control of grassland production and biomass by pathogens. *Ecology Letters* **6**:147–155.

Naeem S, Li S. 1997. Biodiversity enhances ecosystem reliability. *Nature* **390**:507–509.

Nelson EB, Karp MA. 2013. Soil pathogen communities associated with native and non-native *Phragmites australis* populations in freshwater wetlands. *Ecology and Evolution* **3**:5254–5267.

Nürnberger T, Lipka V. 2005. Non-host resistance in plants: new insights into an old phenomenon. *Molecular Plant Pathology* **6**: 335–345.

Oksanen J, Blanchet FG, Kindt R, Legendre P, Minchin PR, O'Hara RB, Simpson GL, Solymos P, Stevens MHH, Wagner H. 2013. *Vegan: community ecology package*. R package version 2.0-10. http://CRAN.R-project.org/package=vegan.

Öpik M, Vanatoa A, Vanatoa E, Moora M, Davison J, Kalwij JM, Reier Ü, Zobel M. 2010. The online database MaarjAM reveals global and ecosystemic distribution patterns in arbuscular mycorrhizal fungi (Glomeromycota). *New Phytologist* **188**:223–241.

Parniske M. 2008. Arbuscular mycorrhiza: the mother of plant root endosymbioses. *Nature Reviews Microbiology* **6**:763–775.

Pendergast TH IV, Burke DJ, Carson WP. 2013. Belowground biotic complexity drives aboveground dynamics: a test of the soil community feedback model. *New Phytologist* **197**:1300–1310.

Pinto AJ, Raskin L. 2012. PCR biases distort bacterial and archaeal community structure in pyrosequencing datasets. *PLoS ONE* **7**: e43093.

Porras-Alfaro A, Bayman P. 2011. Hidden fungi, emergent properties: endophytes and microbiomes. *Annual review of phytopathology* **49**:291–315.

Pringle A, Bever JD, Gardes M, Parrent JL, Rillig MC, Klironomos JN. 2009. Mycorrhizal symbioses and plant invasions. *Annual Review of Ecology, Evolution, and Systematics* **40**:699–715.

R Core Team. 2013. *R: a language and environment for statistical computing*. Vienna, Austria: R Foundation for Statistical Computing. http://www.R-project.org/.

Reinhart KO, Wilson GWT, Rinella MJ. 2012. Predicting plant responses to mycorrhizae: integrating evolutionary history and plant traits. *Ecology Letters* **15**:689–695.

Reininger V, Sieber TN. 2012. Mycorrhiza reduces adverse effects of dark septate endophytes (DSE) on growth of conifers. *PLoS ONE* **7**:e42865.

Reininger V, Sieber TN. 2013. Mitigation of antagonistic effects on plant growth due to root co-colonization by dark septate endophytes and ectomycorrhiza. *Environmental Microbiology Reports* **5**:892–898.

Reininger V, Grünig CR, Sieber TN. 2012. Host species and strain combination determine growth reduction of spruce and birch seedlings colonized by root-associated dark septate endophytes. *Environmental Microbiology* **14**:1064–1076.

Reynolds HL, Haubensak KA. 2009. Soil fertility, heterogeneity, and microbes: towards an integrated understanding of grassland structure and dynamics. *Applied Vegetation Science* **12**:33–44.

Reynolds HL, Packer A, Bever JD, Clay K. 2003. Grassroots ecology: plant-microbe-soil interactions as drivers of plant community structure and dynamics. *Ecology* **84**:2281–2291.

Roy HE, Handley LJL. 2012. Networking: a community approach to invaders and their parasites. *Functional Ecology* **26**:1238–1248.

Schmid B, Hector AMH, Inchausti P, Nijs I, Leadley P, Tilman D. 2002. The design and analysis of biodiversity experiments. In: Loreau M, Naeem S, Inchausti P, eds. *Biodiversity and ecosystem functioning: synthesis and perspectives*. Oxford: Oxford University Press, 61–78.

Schnitzer SA, Klironomos JN, HilleRisLambers J, Kinkel LL, Reich PB, Xiao K, Rillig MC, Sikes BA, Callaway RM, Mangan SA, van Nes EH, Scheffer M. 2011. Soil microbes drive the classic plant diversity-productivity pattern. *Ecology* **92**:296–303.

Schoch CL, Seifert KA, Huhndorf S, Robert V, Spouge JL, Levesque CA, Chen W, Fungal Barcoding Consortium. 2012. Nuclear ribosomal internal transcribed spacer (ITS) region as a universal DNA barcode marker for fungi. *Proceedings of the National Academy of Sciences of the USA* **109**:6241–6246.

Schulz B, Boyle C. 2005. The endophytic continuum. *Mycological Research* **109**:661–686.

Sinclair WA, Howard HL. 2005. *Diseases of trees and shrubs*. Ithaca and London: Comstock Publishing Associates, a Division of Cornell University Press.

Smith SE, Read DJ. 2008. *Mycorrhizal symbiosis*. New York: Elsevier/ Academic Press.

Stockinger H, Krüger M, Schüßler A. 2010. DNA barcoding of arbuscular mycorrhizal fungi. *New Phytologist* **187**:461–474.

Thompson K. 1987. The resource ratio hypothesis and the meaning of competition. *Functional Ecology* **1**:297–303.

Thoms C, Gleixner G. 2013. Seasonal differences in tree species' influence on soil microbial communities. *Soil Biology and Biochemistry* **66**:239–248.

Tilman D. 1985. The resource-ratio hypothesis of plant succession. *The American Naturalist* **125**:827–852.

Treseder KK. 2013. The extent of mycorrhizal colonization of roots and its influence on plant growth and phosphorus content. *Plant and Soil* **371**:1–13.

Van der Heijden MGA, Klironomos JN, Ursic M, Moutoglis P, Streitwolf-Engel R, Boller T, Wiemken A, Sanders IR. 1998. Mycorrhizal fungal diversity determines plant biodiversity, ecosystem variability and productivity. *Nature* **396**:69–72.

Van der Putten WH, Bardgett RD, Bever JD, Bezemer TM, Casper BB, Fukami T, Kardol P, Klironomos JN, Kulmatiski A, Schweitzer JA, Suding KN, Van de Voorde TFJ, Wardle DA. 2013. Plant-soil feedbacks: the past, the present and future challenges. *Journal of Ecology* **101**:265–276.

Voříšková J, Brabcová V, Cajthaml T, Baldrian P. 2014. Seasonal dynamics of fungal communities in a temperate oak forest soil. *New Phytologist* **201**:269–278.

Wang B, Qiu YL. 2006. Phylogenetic distribution and evolution of mycorrhizas in land plants. *Mycorrhiza* **16**:299–363.

Wang Z, Binder M, Schoch CL, Johnston PR, Spatafora JW, Hibbett DS. 2006. Evolution of helotialean fungi (Leotiomycetes, Pezizomycotina): a nuclear rDNA phylogeny. *Molecular Phylogenetics and Evolution* **41**:295–312.

Webster J, Weber RWS. 2007. *Introduction to fungi.* Cambridge, UK: Cambridge University Press.

Wilby A, Brown VK. 2001. Herbivory, litter and soil disturbance as determinants of vegetation dynamics during early old-field succession under set-aside. *Oecologia* **127**:259–265.

Wilson GWT, Hartnett DC. 1998. Interspecific variation in plant responses to mycorrhizal colonization in tallgrass prairie. *American Journal of Botany* **85**:1732–1738.

8

Effects of elevated atmospheric CO_2 concentrations, clipping regimen and differential day/night atmospheric warming on tissue nitrogen concentrations of a perennial pasture grass

Astrid Volder[1*], Roger M. Gifford[2] and John R. Evans[3]

[1] Department of Plant Sciences, University of California – Davis, Davis, CA, USA

[2] CSIRO Agriculture, Canberra, Australian Capital Territory 2601, Australia

[3] Division of Plant Sciences, Research School of Biology, The Australian National University, Linnaeus Building 134, Canberra, Australian Capital Territory 0200, Australia

Associate Editor: Ulo Niinemets

Abstract. Forecasting the effects of climate change on nitrogen (N) cycling in pastures requires an understanding of changes in tissue N. We examined the effects of elevated atmospheric CO_2 concentration, atmospheric warming and simulated grazing (clipping frequency) on aboveground and belowground tissue N concentrations and C : N ratios of a C3 pasture grass. *Phalaris aquatica* L. cv. 'Holdfast' was grown in the field in six transparent temperature gradient tunnels (18 × 1.5 × 1.5 m each), three at ambient atmospheric CO_2 and three at 759 p.p.m. CO_2. Within each tunnel, there were three air temperature treatments: ambient control, +2.2/+4.0 °C above ambient day/night warming and +3.0 °C continuous warming. A frequent and an infrequent clipping treatment were applied to each warming × CO_2 combination. Green leaf N concentrations were decreased by elevated CO_2 and increased by more frequent clipping. Both warming treatments increased leaf N concentrations under ambient CO_2 concentrations, but did not significantly alter leaf N concentrations under elevated CO_2 concentrations. Nitrogen resorption from leaves was decreased under elevated CO_2 conditions as well as by more frequent clipping. Fine root N concentrations decreased strongly with increasing soil depth and were further decreased at the 10–60 cm soil depths by elevated CO_2 concentrations. The interaction between the CO_2 and warming treatments showed that leaf N concentration was affected in a non-additive manner. Changes in leaf C : N ratios were driven by changes in N concentration. Overall, the effects of CO_2, warming and clipping treatments on aboveground tissue N concentrations were much greater than on belowground tissue.

Keywords: C : N ratio; climate change; defoliation; grassland; tissue quality.

* Corresponding author's e-mail address: avolder@ucdavis.edu

Introduction

Long-term responses of plant growth to elevated atmospheric CO_2 and warming will depend in part on the availability of mineral nutrients and the way in which they are utilized by the plant (Stitt and Krapp 1999; Duval *et al.* 2012). For example, the stimulation of biomass accumulation at elevated CO_2 is generally less under nitrogen (N) limited than well-fertilized conditions (Norby and Luo 2004; Finzi *et al.* 2006; Gill *et al.* 2006; Reich *et al.* 2006). Leaf and plant N concentrations generally decrease in response to elevated atmospheric CO_2, and such decreases occur even under conditions of high soil N availability (Cotrufo *et al.* 1998). Thus, decreased tissue N concentration in response to elevated atmospheric CO_2 is not just due to a N limitation in the growth medium (Poorter *et al.* 1997; Lee *et al.* 2011). Decreased leaf N concentrations have been attributed to either lower N demand as plants utilize rubisco protein more efficiently for photosynthesis under elevated atmospheric CO_2 concentrations (Drake *et al.* 1997; Stitt and Krapp 1999), or a N-dilution effect as a result of increased carbohydrate production and storage (Coleman *et al.* 1993; Loladze 2002), or diminished ability of plants to assimilate nitrate under high CO_2 conditions (Rachmilevitch *et al.* 2004; Bloom *et al.* 2010, 2012, 2014). Regarding the negative impact of elevated CO_2 concentrations on nitrate assimilation, in C_3 species, suppression of photorespiration by elevated atmospheric CO_2 concentrations is hypothesized to limit the availability of reductant in the glutamine synthetase and glutamine:2-oxoglutarate amidotransferase cycle. Competition for reductants slows the rate at which nitrate can be converted to glutamate (Bloom 2006), the form of N used in protein synthesis. Since elevated atmospheric CO_2 and higher air temperatures have opposite effects on photorespiration (Ziska and Bunce 1995), higher temperatures may counteract the suppressing effect of elevated atmospheric CO_2 on nitrate assimilation in C_3 plants.

As both warming and elevated atmospheric CO_2 affect N uptake, translocation and assimilation, they can both be expected to affect tissue N concentrations. Warming applied under ambient atmospheric CO_2 conditions led to decreased tissue N concentration in both an unfertilized tallgrass prairie (An *et al.* 2005) and a Mediterranean shrubland (Sardans *et al.* 2008). In an interactive $CO_2 \times$ warming experiment, Lilley *et al.* (2001) found no effect of temperature on herbage N concentration in *Phalaris aquatica*, but elevated atmospheric CO_2 conditions did lower herbage N concentration. Zhou *et al.* (2011) found that elevated atmospheric CO_2 concentrations led to lower leaf N concentration in *P. arundinacea*, yet higher temperatures led to higher leaf N concentrations overall. These findings on the effect of warming contrast with those of An *et al.* (2005) and Sardans *et al.* (2008) who found lower leaf tissue N concentrations when warming was applied under ambient CO_2 conditions. Neither Lilley *et al.* (2001) nor Zhou *et al.* (2011) found an interactive effect of warming and atmospheric CO_2 concentration on leaf N concentration. Anthropogenic global warming is likely greater for night-time minima than for daytime maxima (Karl *et al.* 1991; Vose *et al.* 2005; Gershunov *et al.* 2009), which can also affect plant C balances. For example, Lilley *et al.* (2001) found that *P. aquatica* stores very little carbon (C) as starch, but it maintains high levels of soluble carbohydrates. In this study, soluble carbohydrate concentrations were enhanced by elevated atmospheric CO_2 but decreased by warming under ambient CO_2 conditions. Diurnal patterns of warming with higher night-time temperatures, a trend that has been observed in the global temperature records (Vose *et al.* 2005), may have a different effect on plant tissue quality than continuous warming. Temperate steppe grasses responded to nocturnal warming by depleting more stored carbohydrates at night and compensated for this response by enhancing daytime photosynthesis to such a degree (+19.8 %) that the steppe ecosystem switched from being a minor C source to a C sink (Wan *et al.* 2009). Such strong C balance responses to night-time warming will also likely affect tissue N concentrations and C : N ratios.

Changes in tissue N or C concentrations, or both, will affect tissue C : N ratios, with consequences for growth and decomposition, which can then affect litter decomposition rates and C and N cycling if a change in living tissue C : N ratios translates into altered litter C : N ratios. For example, exposure to increasing atmospheric CO_2 levels along a CO_2 gradient led to increased plant tissue C : N ratios, greater aboveground N storage in plant parts and increased C : N ratios in soil organic matter in a Texas grassland (Gill *et al.* 2006). In this same experiment, N mineralization rates were decreased, while C mineralization increased under elevated CO_2 conditions. One possible explanation was that increased litter C : N ratios may have stimulated microbes to start using the more recalcitrant N-rich C fractions in the soil to meet their N demands.

Few studies have reported on the effect of elevated atmospheric CO_2 on both root C and N concentrations (Nie *et al.* 2013), and data on interactive effects of atmospheric CO_2 concentration and temperature on root C and N concentration are even rarer (Dieleman *et al.* 2012). The response of tissue nutrient concentrations to CO_2 and warming is likely to be dependent on the grazing or clipping regime. Ziter and MacDougall (2013) found that grazing enhanced leaf tissue N concentrations while not affecting root N and sugar concentrations. When heavy grazing was compared with moderate grazing, Biondini *et al.* (1998) found that heavy grazing reduced root N concentrations. As grazing removes substantial amounts of biomass from the system and thus interferes with nutrient cycling, pastures are often fertilized.

The objective of this study was to investigate the interactive effects of both atmospheric warming and elevated atmospheric CO_2 concentration on N concentration and C : N ratio of leaves, roots and litter of *P. aquatica* plantings subjected to two different clipping intensities (infrequent and frequent) to simulate grazing. We reduced the impact of feedbacks due to drought or nutrient limitation by providing ample water and nutrients. Therefore, we report on the direct impacts of both elevated atmospheric CO_2 and warming on tissue quality in a managed pasture system.

Methods

Species description

Phalaris aquatica L., previously also known as *P. tuberosa* L., is a highly productive deep-rooted perennial grass originating from the Mediterranean and Middle East and was first introduced as a pasture grass in Australia in 1877 (Oram *et al.* 2009). The common name for *P. aquatica* is 'phalaris', although sometimes the name 'Hardinggras' is also used (Oram *et al.* 2009). *Phalaris aquatica* is the most widely sown perennial grass in temperate areas of south-eastern Australia. In addition, it is also used in pastures in the USA, South America and New Zealand and to a limited extent in parts of Africa and southern Europe (Oram *et al.* 2009). The cultivar used in this experiment, 'Holdfast', is grazing tolerant, winter active and exhibits low summer dormancy. The cultivar 'Holdfast' was developed by CSIRO, Canberra, Australia, to have reduced aluminium sensitivity and tolerate acid soils (Oram *et al.* 1993; Oram and Culvenor 1994).

Phalaris aquatica is considered an environmental weed in both Australia (Stone 2009) and the Western USA, where it thrives in areas with deep soil and adequate soil moisture (>500 mm rainfall). Due to low seedling vigour, the ability to establish outside cultivation is low, unless bare patches of soil are available (Stone 2009). Once established in natural ecosystems, it can outcompete native species through its ability to form dense clumps with deep root systems that allow it to survive periods of drought (Stone 2009). In general, *P. aquatica* has a high nutrient requirement, especially for N and P, which inhibits its ability to be highly productive outside agricultural situations (Stone 2009). The California Invasive Plant Council rates *P. aquatica* as moderately invasive in California (IPC 2015).

Temperature gradient tunnels and environmental conditions

The experiment site and tunnel controls are described in Volder *et al.* (2004). Briefly, six transparent ventilated temperature gradient tunnels (TGT, 18 × 1.5 × 1.5 m each) were established on a uniform flat fallow field at Ginninderra Experimental Station (Canberra, Australian Capital Territory, Australia, lat. 35.22°S, long. 149.13°E). During the experiment, hourly averaged ambient temperatures ranged from −5.5 °C (24 August 2002) to 43.6 °C (18 January 2003) with a daily mean of 14.3 °C (Volder *et al.* 2004). Three tunnels were kept at ambient atmospheric CO_2 concentrations (i.e. no CO_2 adjustment) and three tunnels were designed to maintain elevated atmospheric CO_2 concentrations (target: 750 p.p.m.) by injecting CO_2 into the airstream at two locations. CO_2 concentrations in the airstream were measured (Model ADC-2000, Analytical Development Co. Ltd, Hoddesdon, UK) at downstream locations every 0.3 s and pulse lengths were adjusted every 1 s accordingly (Volder *et al.* 2004). The daily CO_2 concentration over the whole experiment in the three elevated atmospheric CO_2 tunnels was 759 ± 12.6 p.p.m. In the ambient CO_2 tunnels, the daily average was 403 p.p.m. During the day, the drawdown between the start and the end of the tunnels was ~7 p.p.m. The tunnels were constructed using a thin aluminium framework with Teflon (Nowoflon ET-film 6235, Nowoflon Kunststoffprodukte GmbH and Co., Siegsdorf, Germany) panels. Radiation energy intercepted by the tunnel structure varied between 20 and 40 %, with the highest proportional interception at midday and in winter (Volder *et al.* 2004).

Within each tunnel, three plant sections (3 × 1.5 × 1.5 m each) were established where the air temperature was either kept at ambient, or a constant warming of +3.0 °C above ambient (constant warming), or +2.2 °C warming above ambient during the day and +4.0 °C warming during the night (high night-time warming). Warming was accomplished mostly by passive solar heating and variable fan speed as air moved through the tunnels during the day and by using a combination of air heaters and drawing in cool air at night (Volder *et al.* 2004). Temperatures were controlled using double-shielded, continuously aspirated thermistors, located 1 m above the surface in each plant section, connected to a Microzone II controller that controlled fan speed and activity of the air heaters. Temperatures were measured every 0.1 s, and fan speed and heater outputs were adjusted every 0.5 s based on the measured temperature differential from the target temperature. Thermistors used for control were independent from the thermistors used to log section air temperatures to ensure data integrity. Daily averaged air temperature warming was +2.9 °C in the high night-time warming treatment and +3.0 °C in the constant warming treatment.

The plots were maintained at non-limiting soil water levels, using overhead irrigation with 16 nozzles per section to ensure even distribution. Soil water content was measured using Theta probes (Delta-T devices Ltd., Cambridge, UK) installed at 5 cm depth in the infrequently clipped plots. In Year 1, soil water content was maintained at 0.32–0.38 $m^3\ m^{-3}$ (77–92 % of field capacity). Starting 25 September 2002, soil water content was reduced to between 0.20 and 0.25 $m^3\ m^{-3}$. Soil water levels were checked daily and when a Theta probe read below 0.20 $m^3\ m^{-3}$, water was applied to all plots until the driest probe read 0.25 $m^3\ m^{-3}$. All sections received the same amount of water. There was always a higher soil moisture content (by 0.02–0.05 $m^3\ m^{-3}$) in the elevated CO_2 treatment (Volder *et al.* 2004).

Perennial temperate pasture grass (*P. aquatica* cv. Holdfast) was winter-sown at a density of 250 plants m^{-2} using a row spacing of 8 cm on 3 May 2001. Within each temperature treatment, there were two clipping regimes, both at 7 cm above the ground in 1.5 × 1.5 m areas. One regime was clipped two times as often as the other, using visual criteria to decide the time for clipping. The frequently clipped treatment dates were 25 October and 21 November 2001; 8 January, 19 February, 4 April, 1 May, 28 August, 25 September, 6 November, 11 December 2002 and 5 February, 5 March, 8 April and 20–21 May 2003. Plots for the infrequent clipping were clipped every second date. A border zone of 30 cm width was established around each measured plot to avoid edge effects. Both border and measured plot (90 × 90 cm) zones were clipped at each harvest, but border zone material was discarded. Random subsamples (3–6 % of total dry mass) were taken from the bulk sample at each harvest during the harvesting of each plot. The subsamples were divided into standing dead (i.e. litter), green leaf lamina and remainder (sheath, stem and flowers). Sheath, stems and reproductive parts were placed together as most of the stem growth occurs in support of the reproductive parts. All fractions were dried for a week at 80 °C and weighed. Total C and N were determined using a Europa elemental analyser (Sercon, Cheshire, UK). At the final harvest, stubble remaining below the clipping height was harvested from two 20 × 20 cm quadrats placed randomly within each plot. All stubble material including crowns was cleaned free of soil and any roots were discarded. Total C and N were measured as above.

Soil cores to 30 cm depth were collected with 32-mm diameter push tubes from each CO_2 × temperature × clipping combination on 25 September 2001 (start of treatments), 20 February 2002, 26 September 2002 and 5 March 2003. At the end of the experiment, cores were collected down to 1 m using a hydraulic corer. Cores were cut into 0–10, 10–20 and 20–30 cm segments, with additional segments for the final deeper cores at 50–60 and 80–90 cm. Roots were then washed with distilled water, sieved through a 2-mm sieve, followed by a 0.5-mm sieve (Volder *et al.* 2007) and split into two size classes, fine laterals (first-order roots, less than ~0.3 mm diameter) and lateral-bearing coarse roots (generally >0.3 mm diameter). Fine laterals were removed from the coarse roots and placed in the fine fraction. All fractions were dried for a week at 80 °C, weighed and total C and N was determined using a Europa elemental analyser (Sercon).

All plots were fertilized three times per year with 100 kg N ha^{-1} per occasion using a slow release fertilizer (Osmocote, Scotts Company, Marysville, OH, USA), which also included 26.7 kg P ha^{-1} and 50.6 kg K ha^{-1}. Fertilizer was applied on 28 September 2001; 20 February, 18 June and 29 September 2002 and 6 March 2003.

Statistical analyses

When necessary to improve homogeneity of variances, data were natural log transformed (Zar 1984). Effects of CO_2, temperature and clipping frequency were analysed with ANOVA (residual maximum likelihood procedure) using JMP Pro for windows version 10 (SAS Institute, Cary, NC, USA). The design was a split-split plot with CO_2 level (ambient, 750 p.p.m.) as the main plot factor, warming treatment (+0, +2.2/+4.0 and +3.0 °C) as subplots and clipping frequency (regular, frequent) as randomly assigned sub-subplots within each warming treatment.

Results

Across treatments and harvests, harvested green leaf tissue N concentrations were decreased by 15.6 % under elevated atmospheric CO_2 concentration (Fig. 1A) and increased by 32.0 % by frequent clipping (Fig. 1B). The influence of elevated CO_2 and clipping frequency on leaf N concentration varied with harvest period (Fig. 1, Table 1, **see Supporting Information—Tables S1–S3** for tissue C and N concentrations and C : N ratios on each harvest date). The decline in leaf tissue N concentration due to elevated CO_2 was stronger in spring and summer (November–May and September–March) than during the winter (Fig. 1A, $P_{\text{harvest date} \times CO_2} < 0.001$), while the effect of clipping frequency was stronger in the winter and fall (May–December, March–May) than in the spring and summer (Fig. 1B, $P_{\text{harvest date} \times \text{clipping}} < 0.001$).

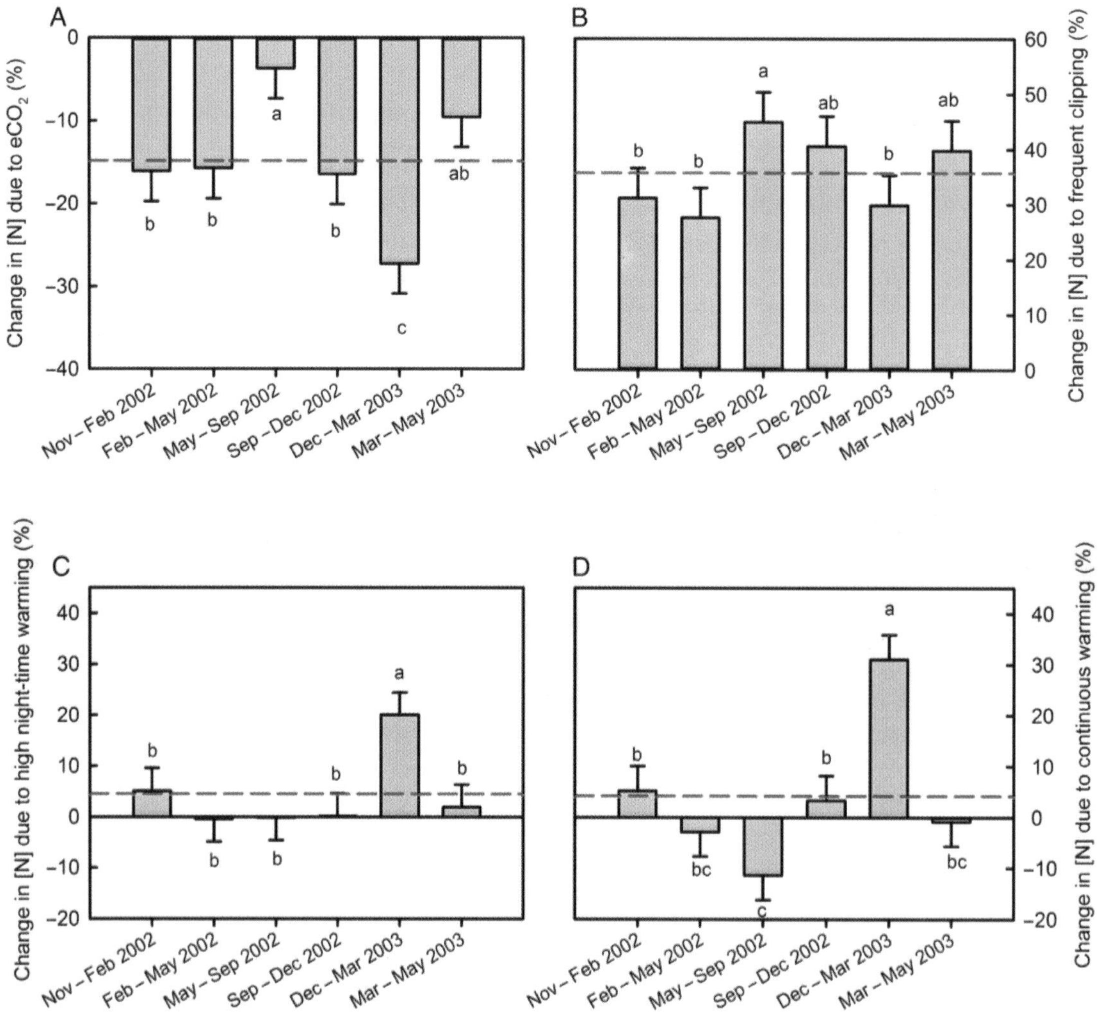

Figure 1. Change in green leaf N concentrations due to (A) increased atmospheric CO_2 concentration (756 p.p.m.) compared with ambient atmospheric CO_2 concentration (405 p.p.m.), (B) increased clipping frequency, (C) higher night-time warming compared with ambient air temperatures (+2.0/+4.4 °C above ambient, day/night) and (D) continuous warming above ambient (+3.0 °C), as affected by harvest period. The dashed lines indicate the average response to elevated atmospheric CO_2 levels (A), increased clipping frequency (B), high night-time warming (C) and continuous warming (D). Data presented are least square means and SEM based on a full model. Different letters indicate statistically significant differences between harvest periods at $P < 0.05$ using Student's *t* LSD test.

Warming was without any overall effect on green leaf N concentration when averaged across all the other treatments and harvest times (Table 1). However, there were strong interactive effects of warming with CO_2 and harvest period. Averaged across CO_2 treatments, both warming treatments greatly increased leaf N concentration in the second summer ($P_{harvest \times warm} < 0.001$, December–March 2003, Fig. 1C and D). In contrast, continuous warming decreased leaf N concentration in comparison with ambient temperatures in the winter (May–September 2002, Fig. 1C). When averaged across all dates and clipping frequencies, warming significantly increased leaf N concentrations under ambient CO_2, but under elevated CO_2, leaf N concentration was not affected by warming (Fig. 2A). The negative effect of increased atmospheric CO_2 concentration on leaf N concentration was substantially enhanced by both warming treatments, from 7.6 % decrease under ambient temperature to 17.5 and 21.4 % decrease under continuous and high night-time warming ($P_{CO_2 \times warm} = 0.036$).

Elevated CO_2 increased leaf C : N ratios, particularly under the high night-time warming scenario ('HN', $P_{CO_2 \times warming} = 0.053$, from 18.0 to 22.9, Fig. 2B). Changes in leaf C : N ratio in response to elevated CO_2 reflect changes in leaf N concentration. There was no relationship between changes in C : N ratio and changes in C concentration in response to elevated CO_2. Increased clipping frequency decreased green leaf C : N ratios by an average of 24 %, from 23.6 to 17.9 (Fig. 3A, $P < 0.001$).

Table 1. *P*-values of the effect of atmospheric CO_2 concentration (CO_2), warming treatment (*W*), clipping frequency (*C*) and harvest period (*H*) on green leaf N concentration, C concentration and C : N ratio. [1]Numerator, denominator. [2]Data were ln transformed prior to analysis. [3]Missing data prevented analysis of the four-way interaction term. *P* values <0.10 but >0.05 are given in italics, while *P* values <0.05 are given in bold.

	df[1]	Green leaf						Litter					
		N		C		C : N ratio		N[2]		C		C : N ratio	
		F	*P*	*F*	*P*	*F*	*P*	*F*	*P*	*F*	*P*	*F*	*P*
CO_2	1,4	130	**<0.001**	0.99	0.375	98.1	**<0.001**	0.90	0.385	6.39	*0.064*	0.22	0.651
Warming (*W*)	2,8	0.83	0.470	5.88	**0.027**	0.38	0.694	4.25	**0.047**	8.61	**0.010**	0.23	0.798
$CO_2 \times W$	2,8	5.17	**0.036**	0.48	0.633	3.48	*0.082*	2.44	0.140	3.37	*0.087*	1.80	0.219
Clipping frequency (*C*)	1,12	196	**<0.001**	0.07	0.799	160	**<0.001**	55.0	**<0.001**	2.10	0.171	29.6	**<0.001**
$C \times CO_2$	1,12	0.16	0.691	0.07	0.797	2.32	0.153	0.00	0.947	0.21	0.652	0.48	0.498
$C \times W$	2,12	0.49	0.615	2.45	0.128	1.24	0.325	0.24	0.788	1.72	0.217	0.18	0.838
$C \times W \times CO_2$	2,12	0.55	0.584	0.41	0.674	0.70	0.518	0.42	0.668	0.68	0.523	0.52	0.607
Harvest (*H*)	5,120	79.5	**<0.001**	61.1	**<0.001**	39.8	**<0.001**	6.25	**<0.001**	15.3	**<0.001**	0.98	0.435
$H \times CO_2$	5,120	6.39	**<0.001**	1.35	0.249	5.00	**<0.001**	1.05	0.392	1.38	0.236	0.33	0.893
$H \times W$	10,120	4.43	**<0.001**	2.89	**0.003**	4.05	**<0.001**	1.15	0.334	1.02	0.434	0.66	0.761
$H \times CO_2 \times W$	10,120	0.75	0.662	1.22	0.286	0.91	0.528	1.17	0.317	1.20	0.296	0.67	0.752
$H \times$ Clip	5,120	5.17	**<0.001**	14.2	**<0.001**	2.02	*0.080*	16.7	**<0.001**	11.9	**<0.001**	13.4	**<0.001**
$H \times$ Clip $\times CO_2$	5,120	*2.05*	*0.075*	0.19	0.964	0.85	0.515	0.47	0.796	1.46	0.207	0.41	0.839
$H \times$ Clip $\times W$	10,120	0.78	0.652	1.48	0.156	0.61	0.806	1.51	0.143	1.47	0.157	1.07	0.392
$H \times$ Clip $\times CO_2 \times W$	10,120	0.42	0.936	1.07	0.393	0.408	0.941		NA[3]		NA[3]		NA[3]
Model r^2			0.898		0.779		0.874		0.649		0.620		0.621
Model *P*			<0.001		<0.001		<0.001		<0.001		<0.001		<0.001

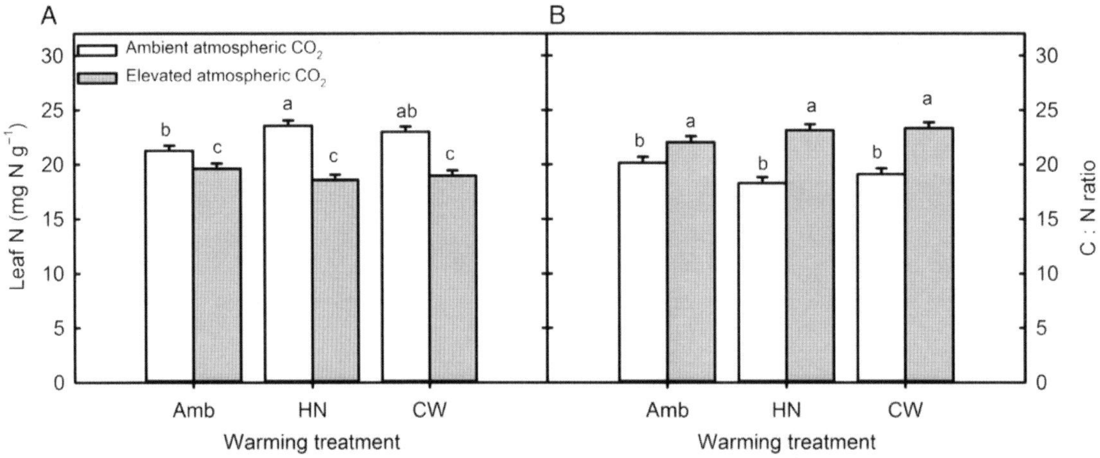

Figure 2. Effect of warming treatment, Amb = ambient temperature, HN = +2.2/+4.0 °C (day/night warming), CW = +3.0 °C continuous warming and atmospheric CO_2 concentration (ambient, 405 p.p.m., and elevated, 756 p.p.m.) on (A) green leaf N concentration and (B) green leaf C : N ratio. Data presented are least square means and SEM averaged across harvests and clipping frequencies. Different letters indicate statistically significant differences at $P < 0.05$ using Student's *t* LSD test.

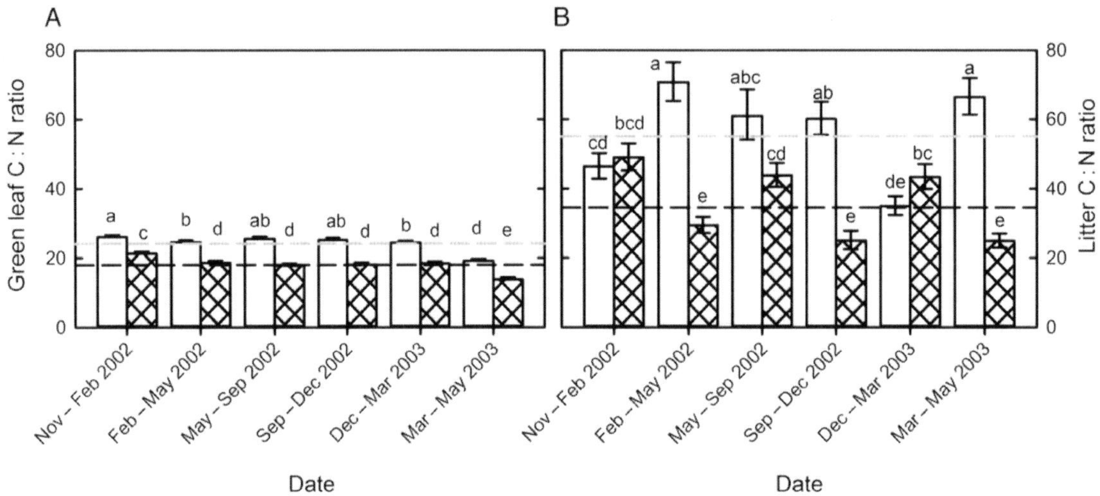

Figure 3. Effect of clipping frequency (open bar = infrequent, hashed bar = frequent) on C : N ratios of (A) green leaf material and (B) litter, through time averaged across CO_2 and warming treatments. Dashed line indicates overall mean across dates, CO_2 and warming treatment for each clipping frequency, infrequent (grey) and frequent (black). Different letters indicate statistically significant differences between C : N ratios at $P < 0.05$ using Student's *t* LSD test.

Litter (i.e. standing dead leaf) N concentrations were unaffected by atmospheric CO_2 concentration (Table 1 and **see Supporting Information—Table S1**). There was a marginally significant warming effect (Table 1, $P_{warming} = 0.047$) where litter N concentration averaged across harvests was higher under a continuous warming scenario (9.1 mg g^{-1}) than under ambient warming (7.6 mg g^{-1}). The average of the two warming treatments increased litter N concentration by 25.3 % (Fig. 4C and D). Litter N concentration was most affected by clipping frequency; increased clipping frequency increased litter N concentrations by an average of 95.9 % (Fig. 4B). Consequently, frequent clipping reduced litter C : N ratio from 55.0 to 34.5 averaged across harvests and CO_2 and warming treatments (Fig. 3B and **see Supporting Information—Table S3**). Nitrogen resorption efficiency, calculated as [100 × (green leaf N concentration − litter N concentration)/green leaf N concentration], was decreased from 56.7 % at ambient atmospheric CO_2 concentration to 50.9 % under elevated CO_2 when averaged across warming and clipping treatments (Fig. 5A). Increased clipping frequency decreased N resorption efficiency from 58.9 % in infrequently clipped plots to 48.7 % in frequently clipped plots when averaged across warming and CO_2 treatments (Fig. 5A). Averaged across clipping and CO_2 treatment, both warming treatments reduced

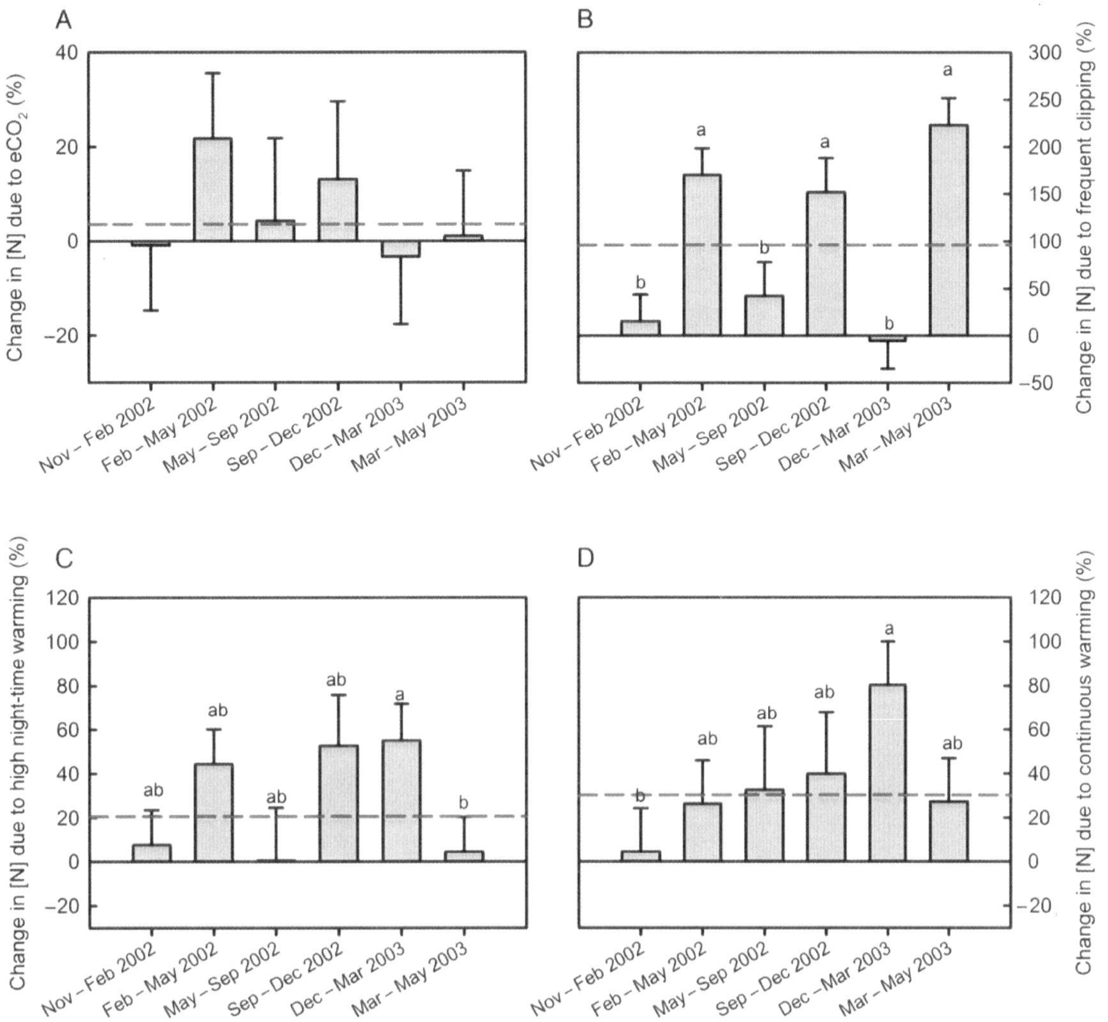

Figure 4. Percent change in litter N concentrations due to (A) increased atmospheric CO_2 concentration (756 p.p.m.) compared with ambient atmospheric CO_2 concentration (405 p.p.m.), (B) increased clipping frequency, (C) higher night-time warming compared with ambient air temperatures (+2.0/+4.4 °C above ambient, day/night) and (D) continuous warming above ambient (+3.0 °C), as affected by harvest period. The dashed lines indicate the average response to elevated atmospheric CO_2 levels (A), increased clipping frequency (B), high night-time warming (C) and continuous warming (D). Data presented are least square means and SEM based on a full model. Different letters indicate statistically significant differences between harvest periods at $P < 0.05$ using Student's *t* LSD test.

N resorption efficiency from 59.1 % in the unwarmed treatment to 52.8 and 49.5 % in the high night-time and continuously warmed treatments, respectively.

Fine root N concentrations decreased strongly with depth in the soil on all three harvest dates (Fig. 6, Table 2). Elevated atmospheric CO_2 had no systematic effect on fine root N concentration, although there was a small significant decline under high CO_2 in March 2003, in the 20–30 cm soil depth (Fig. 6A). Increased clipping frequency had no overall effect on fine root N concentration but did increase fine root N concentration in just the 0–10 cm soil layer in March 2003 (Fig. 6B). Overall, averaged across CO_2, clipping frequency and harvests, warming was without effect on root N concentration. However, small effects of warming varied with depth and date and were limited to the shallower (0–10 and 10–20 cm) soil depths (Fig. 6C). In February 2002, high night-time warming led to slightly increased fine root N concentration compared with continuous warming in the 0–10 and 10–20 cm soil depths, while in March 2003, continuous warming led to higher fine root N concentrations in the continuous warming treatment compared with ambient warming in the 0–10 cm soil depth (Fig. 6C). At the final harvest (March 2003), where we also collected deeper soil cores, fine root N concentration decreased further with soil depth below 50 cm. There was a strong atmospheric CO_2 concentration × depth interaction effect ($P_{CO_2 \times depth} = 0.009$, Table 2) where elevated atmospheric CO_2 levels reduced fine root N concentrations at the 10–20, 20–30 and 50–60 cm core depths,

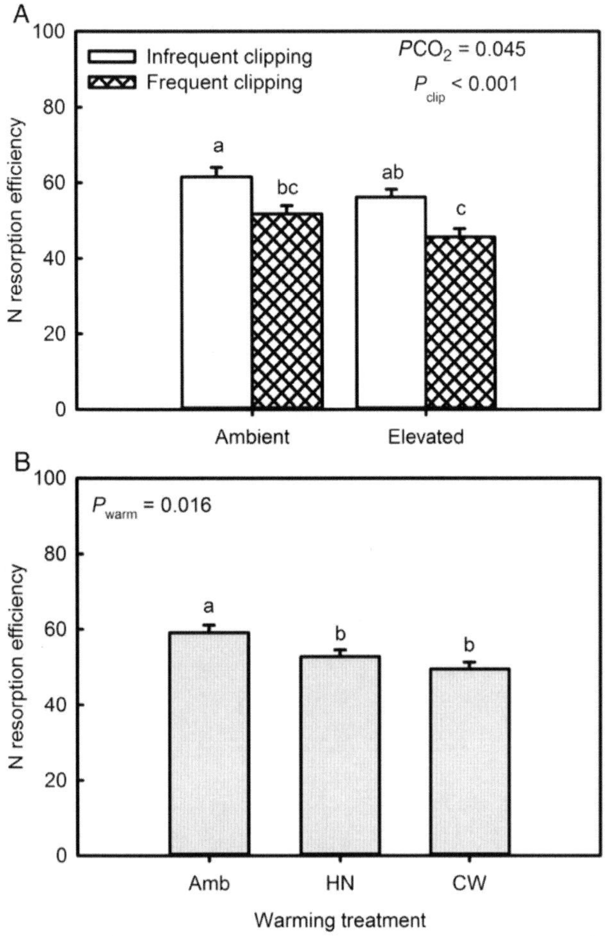

Figure 5. (A) Effect of clipping frequency and atmospheric CO_2 concentration on N resorption efficiency. (B) Effect of warming treatment on N resorption efficiency, Amb = ambient, HN = +2.2 °C/+4.0 °C day/night warming, CW = +3.0 °C continuous warming. Different letters indicate statistically significant differences at $P < 0.05$ using Student's *t* LSD test.

but not in the 0–10 and 80–90 cm core depths (Fig. 7). There were no additional effects of clipping frequency or air warming on fine root N of roots growing below 50 cm soil depth (Fig. 7B and C).

Discussion

Across the three tissues examined, green leaf, litter and fine root, green leaves had the highest tissue N concentration and were the most responsive to elevated atmospheric CO_2 concentration and warming. Elevated atmospheric CO_2 concentration generally reduced tissue N concentrations while the effect of warming depended on the CO_2 concentration. Of the three treatments, frequent clipping had the greatest overall impact on leaf and litter N concentrations, with increased clipping frequency strongly increasing tissue N concentrations. Because treatment impacts varied by tissue type, we will discuss treatment effects by tissue type.

Green leaf

More frequent clipping strongly increased green leaf N concentration by an average of 33 % across the other treatments and harvest dates (Fig. 1B). This was expected because the frequently clipped plants have shoot tissue that is about half the age of shoot tissue of infrequently clipped plants. Younger tissue generally has higher N concentrations (Field 1983). Just why the magnitude of this clipping-frequency effect was greater in the slower growing winter months (May–September) is not clear.

Despite being well fertilized, elevated CO_2 concentration decreased green leaf N concentrations in these field grown C_3 grasses. That result is consistent with the literature (Cotrufo *et al.* 1998). The mean decrease in leaf N concentration in response to elevated CO_2 was 15 % (Fig. 1A), which is similar to the 17 % decrease in N concentration for non-woody C_3 species reported by Cotrufo *et al.* (1998). The magnitude of lower leaf N concentration in response to elevated atmospheric CO_2 varied seasonally with a much smaller response to elevated CO_2 in the slower growing winter season. The biomass production response to elevated atmospheric CO_2 in this same experiment was also smaller during the cold months (Volder *et al.* 2004), consistent with the notion (Coleman *et al.* 1993; Loladze 2002) that dilution of leaf N concentration by enhanced growth in response to elevated CO_2 plays a role at least in part. However, the strong relationship between changes in leaf N concentration and changes in leaf C : N ratio, as well as the lack of a relationship between changes in leaf C concentration and changes in leaf C : N ratio, supports the idea that effects of elevated CO_2 on green leaf tissue C : N ratio are mostly driven by the effects of CO_2 on leaf N concentrations rather than C accumulation in response to elevated CO_2 (Gifford *et al.* 2000).

Atmospheric warming treatments, when averaged across all harvests, increased green leaf N concentrations of plants grown under ambient atmospheric CO_2 concentrations by 11.4 and 8.6 % for the higher night and continuous warming treatments, respectively (Fig. 2A, open bars). However, warming had no significant effect on leaf N concentrations of the swards grown under elevated CO_2 (Fig. 2A, closed bars). The effects of CO_2 concentration and warming on leaf N concentrations were not simply additive; warming of ambient CO_2 plants enhanced green leaf N concentrations by ~10 %, elevated CO_2 decreased leaf N concentrations by 7.6 %, while combined warming and elevated CO_2 decreased green leaf N concentrations by ~11 % (Fig. 2A). Non-additivity of warming and CO_2 responses, with the response to CO_2 dominating, is also consistent with a meta-analysis of leaf tissue N in six temperature × CO_2 manipulation

Table 2. *P*-values of the effect of atmospheric CO_2 concentration (CO_2), warming treatment (*W*), clipping frequency (*C*) and soil depth (*D*) on fine root N and C concentrations, and C : N ratio through time. [1]Depth classes also included 50–60 and 80–90 cm. [2]Numerator, denominator. [3]Data were ln transformed prior to analysis.

Treatment	February 2002 (summer)							September 2002 (early spring)							March 2003 (early fall)[1]						
	df[2]	N		C		C : N[3]		df	N		C		C : N[3]		df	N		C		C : N[3]	
		F ratio	*P*	*F* ratio	*P*	*F* ratio	*P*		*F* ratio	*P*	*F* ratio	df	*F* ratio	*P*		*F* ratio	*P*	*F* ratio	*P*	*F* ratio	*P*
CO_2	1,4	0.01	0.942	1.01	0.370	0.64	0.466	1,4	0.03	0.883	3.63	0.127	0.69	0.453	1,4	6.93	*0.059*	0.62	0.475	7.42	*0.054*
Warming (*W*)	2,8	*3.22*	*0.096*	0.76	0.500	2.45	0.166	2,8	0.10	0.906	0.02	0.976	0.96	0.959	2,8	4.46	**0.049**	1.19	0.352	3.20	*0.095*
CO_2 × *W*	2,8	1.08	0.386	0.01	0.987	0.58	0.589	2,8	0.70	0.527	0.47	0.640	0.60	0.602	2,8	0.30	0.746	0.32	0.737	0.12	0.886
Clipping (*C*)	1,12	1.31	0.275	*3.84*	*0.076*	0.03	0.866	1,12	0.11	0.742	0.45	0.515	0.91	0.915	1,12	2.37	0.150	8.54	**0.013**	22.5	**<0.001**
CO_2 × Clip	1,12	2.39	0.148	0.05	0.825	1.89	0.197	1,12	0.04	0.852	0.21	0.656	0.73	0.734	1,12	1.63	0.226	0.13	0.730	4.50	*0.055*
W × Clip	2,12	0.53	0.602	0.40	0.681	0.10	0.910	2,12	0.41	0.670	0.23	0.799	0.45	0.448	2,12	0.41	0.673	1.96	0.182	2.73	0.105
CO_2 × *W* × Clip	2,12	1.24	0.324	2.08	0.171	0.10	0.910	2,12	0.15	0.859	0.28	0.762	0.48	0.477	2,12	2.27	0.146	0.06	0.940	3.01	*0.087*
Depth (*D*)	2,44	78.1	**<0.001**	3.82	**0.029**	69.5	**<0.001**	2,46	135	**<0.001**	28.9	**<0.001**	47.8	**<0.001**	2,95	514	**<0.001**	16.5	**<0.001**	427	**<0.001**
D × CO_2	2,44	1.07	0.351	0.28	0.758	1.00	0.377	2,46	1.00	0.375	0.37	0.695	1.12	0.335	2,95	3.59	**0.009**	3.62	**0.009**	3.52	**0.010**
D × *W*	4,44	1.40	0.251	1.40	0.249	0.33	0.858	4,46	0.84	0.507	1.64	0.179	0.660	0.665	4,95	1.23	0.291	1.95	*0.062*	1.03	0.420
D × CO_2 × *W*	4,44	0.75	0.566	2.04	0.105	0.66	0.627	4,46	0.79	0.537	1.07	0.381	0.48	0.752	4,95	0.75	0.645	1.41	0.202	1.35	0.228
D × *C*	2,44	0.39	0.676	0.70	0.504	0.38	0.683	2,46	0.56	0.576	0.86	0.432	1.38	0.262	2,95	1.74	0.147	2.36	*0.059*	1.31	0.273
D × CO_2 × *C*	2,44	0.39	0.681	3.53	**0.038**	1.83	0.175	2,46	0.79	0.459	2.74	*0.075*	1.94	0.156	2,95	0.98	0.422	0.73	0.574	0.71	0.584
D × *W* × *C*	4,44	0.95	0.445	0.70	0.599	0.61	0.656	4,46	0.40	0.807	4.84	**0.002**	2.32	*0.071*	4,95	0.54	0.822	0.60	0.774	1.00	0.442
D × CO_2 × *W* × *C*	4,44	0.21	0.932	0.56	0.691	0.84	0.506	4,46	0.64	0.640	1.32	0.277	0.06	0.994	4,95	1.55	0.152	0.63	0.754	1.02	0.425
Model r^2			0.721		0.448		0.741			0.854		0.681		0.716			0.955		0.566		0.948
Model *P*			<0.001		<0.001		<0.001			<0.001		<0.001		<0.001			<0.001		<0.001		<0.001

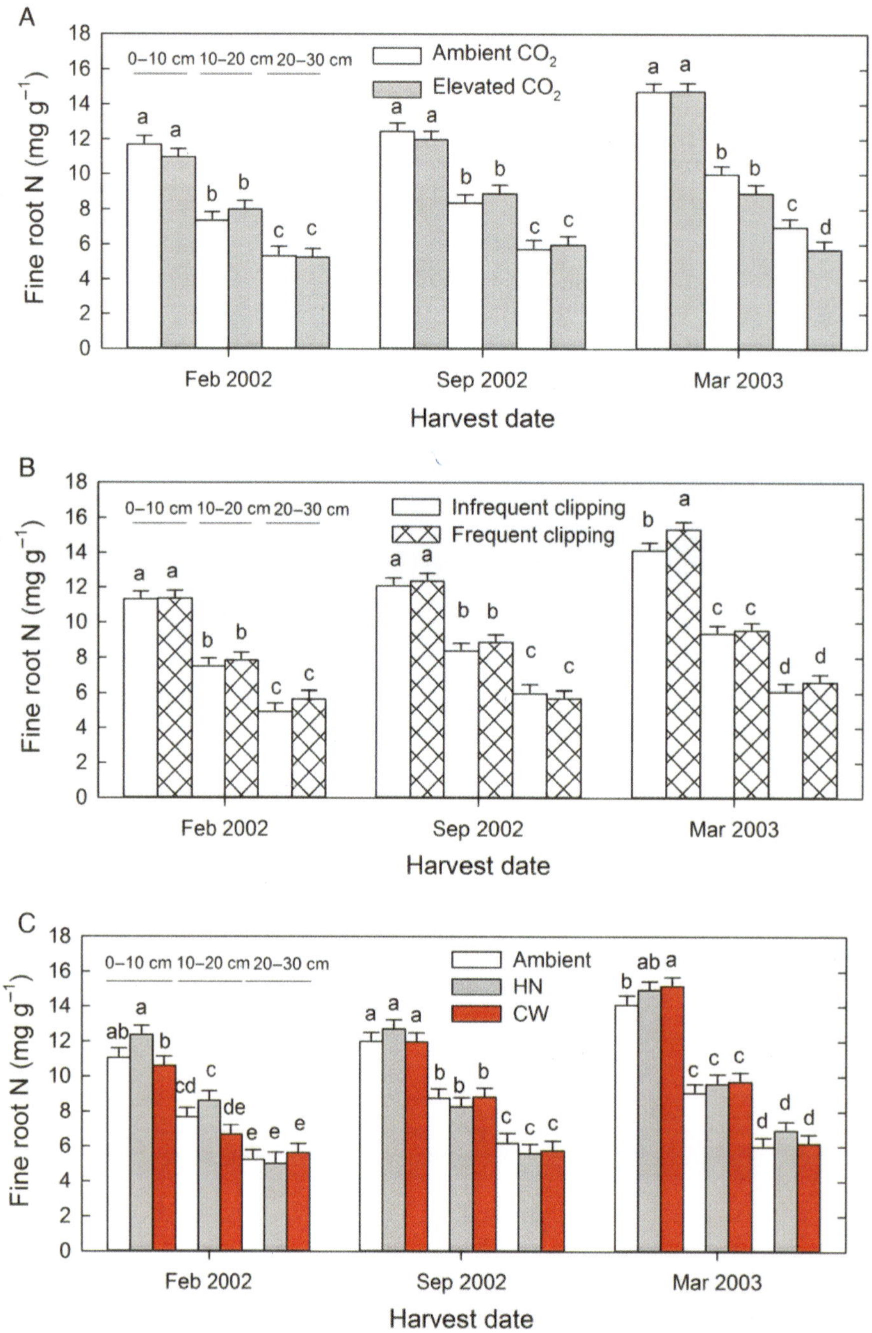

Figure 6. Effect of atmospheric CO_2 concentration (A), clipping frequency (B) and warming treatment (C) on fine root N concentrations at three harvest dates and three depths. HN = +2.2/+4.0 °C day/night, CW = +3.0 °C continuous warming. Different letters indicate statistically significant differences within a harvest date at $P < 0.05$ using Student's *t* LSD test.

studies by Dieleman *et al.* (2012). The increase in tissue N in response to warming at ambient atmospheric CO_2 levels is consistent with the idea that shoot nitrate assimilation depends on photorespiration (Rachmilevitch *et al.* 2004; Bloom *et al.* 2010, 2012). Under conditions where photorespiration is enhanced (i.e. ambient CO_2 and warming), nitrate assimilation is stimulated, while under conditions where photorespiration is repressed (i.e. high atmospheric CO_2 conditions), increasing temperatures will not affect tissue N concentrations. This suggests that the impact of future climate warming on tissue quality and N cycling cannot be predicted based on warming experiments alone.

Our finding that green leaf tissue N concentration increased in response to warming under ambient atmospheric CO_2 concentrations differs from An *et al.* (2005),

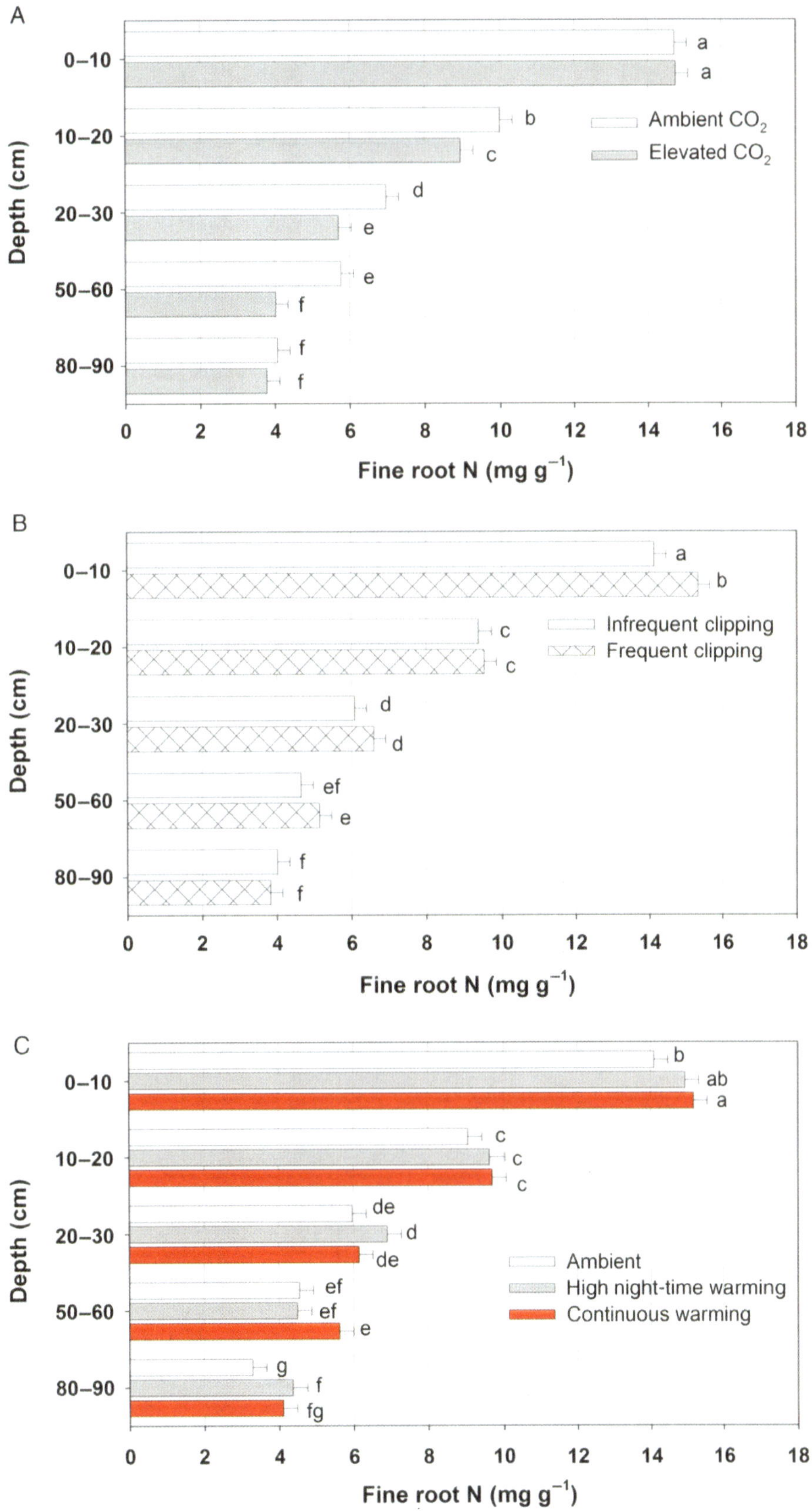

Figure 7. Effect of (A) atmospheric CO_2 concentration, (B) clipping frequency and (C) warming treatment on fine root N concentration at five depths at the end of the experiment (March 2003). Treatments started in September 2001. Data are the least square means + SEM across treatments. Different letters indicate statistically significant differences at $P < 0.05$ using Student's t LSD test.

who found that warming decreased tissue N concentrations in five grassland species. An *et al.* (2005) observed the effects of warming after 1 year of treatment, whereas in our study, warming did not have a statistically significant effect on green leaf N concentrations until well into the second season of treatments. Thus, the effect of warming may not be as immediate and is more subtle than the effects of elevated atmospheric CO_2 and clipping frequency.

In contrast to the effects of combined warming and atmospheric CO_2 on leaf N concentration, there was no interaction between clipping frequency and CO_2 concentration or between clipping frequency and warming treatment. This suggests that the effects of clipping management are additive to the effects of the climate change drivers. Our earlier findings on aboveground biomass production (Volder *et al.* 2004) and new root production (Volder *et al.* 2007) showed non-additivity, where the effect of elevated CO_2 was dependent on clipping frequency. Thus, as expected, whether effects of climate change drivers are additive or not depends on the response variable measured and the management regimen of the ecosystem (Dieleman *et al.* 2012; Xu *et al.* 2013).

Standing dead and leaf litter

The decreases in leaf N concentration due to elevated atmospheric CO_2 concentrations (Figs 1A and 2A) did not translate into reduced standing dead leaf litter N concentration or increased litter C : N ratios (Fig. 4A). Neither litter N concentration nor litter C : N ratio were significantly affected by elevated CO_2 (Table 1). Increased clipping frequency strongly decreased C : N ratios of both leaf (from 24.0 to 18.0) and litter (from 55.0 to 34.5, Fig. 3). Frequent clipping increased green leaf N concentration and also reduced N resorption efficiency from 59.2 to 53.6 %, thus leaving a greater amount of N per gram litter biomass. As more N is left in senescing tissues of frequently clipped vegetation and litter C : N ratios are reduced, it is possible that frequent clipping would speed up the rate of mineralization of litter N from a given amount of litter added to the soil (Booth *et al.* 2005), if soil moisture and soil temperature remain similar. However, whether that would lead to actual increased N availability in soils of frequently cut swards also depends on the total amount of litter returned to the soil. Previous published results (Volder *et al.* 2004) showed that frequent clipping strongly decreased biomass production, and thus the total amount of N returned to the soil from litter may be reduced, even if mineralization rates are increased because of decreased litter C : N ratios.

Roots

Root N concentrations were very strongly affected by soil depth; the concentrations in fine roots at 80–90 cm depth were less than a third of those in the top 10 cm (Fig. 7). Effects of warming, CO_2 and clipping frequency were minor in comparison with the depth effect. These minor effects of elevated CO_2 concentration and warming at depth were most evident at the final harvest. Although most CO_2 responses occurred in roots below 10 cm soil depth in our experiment, others have found negative impacts of elevated CO_2 concentrations on root N concentration in shallow (<10 cm) soil layers in grassland systems (Kitchen *et al.* 2009). The lack of a clipping effect on root N concentration was surprising given that increased clipping frequency increased root turnover rate (Volder *et al.* 2007), which would lead to a younger root system on average. Younger roots have been shown to have higher rates of nitrate uptake and higher tissue N concentrations (Volder *et al.* 2005); however, the proportional change in average root age may not have been large enough to affect the average tissue N concentration in our bulk samples.

Often root research is limited to the top soil layer because that is the zone where generally >50 % of root length occurs (Schenk and Jackson 2002). However, our data suggest that when evaluating the impact of climate on root parameters, some major changes in tissue N concentrations may be occurring deeper in the soil profile. It is important to note that our experiment involved a managed pasture grass system where water and nutrients were supplied at high levels—soil water content was kept at 20 % or higher (Volder *et al.* 2004), and the plots were fertilized three times per year at a rate of 100 kg N ha^{-1} per occasion. Thus, plant responses in our system were mostly decoupled from soil system feedbacks (Type I system, Körner 2006) when compared with other climate change experiments in grasslands, which generally take place in systems where water and/or nutrients are limited (Morgan *et al.* 2004).

Conclusions

While increasing atmospheric CO_2 concentration decreased green leaf N concentrations considerably, this was not propagated to leaf litter, as leaf litter N concentration was unaffected by elevated CO_2. Atmospheric warming increased green leaf N under ambient CO_2 but did not significantly affect leaf N concentration at elevated CO_2 concentration. The increase in leaf N concentration under warming at ambient CO_2 was reflected in increased litter N concentration. For fine roots, elevated CO_2 tended to decrease N concentration ($P = 0.059$) and increase C : N ratio by the end of the experiment with the magnitude of the effect increasing deeper in the soil. The effects of continuous uniform warming were similar to differential day/night warming.

In general, the non-additivity of CO_2, warming and management treatment effects with unexplained time variability of specific interactive effects that are exhibited in this data set presents considerable problems for predicting long-term climate change impacts on pasture ecophysiology by either rules of thumb or simulation modelling.

Sources of Funding

This work was funded by the Cooperative Research Centre for Greenhouse Accounting and internal funding from CSIRO Division of Plant Industry, Australia.

Contributions by the Authors

A.V. collected and processed samples, performed data analysis and wrote the manuscript. R.M.G. and J.R.E. collected samples and co-wrote the manuscript.

Acknowledgements

We thank Bruce Robertson, Philip Measham, Michael Halliday, staff of the CSIRO Plant Industry Workshop and Ginninderra Experiment Station and Richard Blyton for invaluable advice and help with this project.

Supporting Information

The following additional information is available in the online version of this article –

Table S1. Tissue N concentrations (mg N g^{-1}) through time as affected by atmospheric CO_2 concentration, warming treatment, ambient, higher night-time warming and continuous warming, and clipping frequency.

Table S2. Tissue C concentrations (g C g^{-1}) through time as affected by atmospheric CO_2 concentration, warming treatment, ambient, higher night-time warming and continuous warming, and clipping frequency.

Table S3. Tissue C : N ratios through time as affected by atmospheric CO_2 concentration, warming treatment, ambient, higher night-time warming and continuous warming, and clipping frequency.

Literature Cited

An Y, Wan S, Zhou X, Subedar AA, Wallace LL, Luo Y. 2005. Plant nitrogen concentration, use efficiency, and contents in a tallgrass prairie ecosystem under experimental warming. *Global Change Biology* **11**:1733–1744.

Biondini ME, Patton BD, Nyren PE. 1998. Grazing intensity and ecosystem processes in a northern mixed-grass prairie, USA. *Ecological Applications* **8**:469–479.

Bloom AJ. 2006. Rising carbon dioxide concentrations and the future of crop production. *Journal of the Science of Food and Agriculture* **86**:1289–1291.

Bloom AJ, Burger M, Asensio JSR, Cousins AB. 2010. Carbon dioxide enrichment inhibits nitrate assimilation in wheat and *Arabidopsis*. *Science* **328**:899–903.

Bloom AJ, Asensio JSR, Randall L, Rachmilevitch S, Cousins AB, Carlisle EA. 2012. CO_2 enrichment inhibits shoot nitrate assimilation in C_3 but not C_4 plants and slows growth under nitrate in C_3 plants. *Ecology* **93**:355–367.

Bloom AJ, Burger M, Kimball BA, Pinter PJ Jr. 2014. Nitrate assimilation is inhibited by elevated CO_2 in field-grown wheat. *Nature Climate Change* **4**:477–480.

Booth MS, Stark JM, Rastetter E. 2005. Controls on nitrogen cycling in terrestrial ecosystems: a synthetic analysis of literature data. *Ecological Monographs* **75**:139–157.

Coleman JS, McConnaughay KDM, Bazzaz FA. 1993. Elevated CO_2 and plant nitrogen-use: is reduced tissue nitrogen concentration size-dependent? *Oecologia* **93**:195–200.

Cotrufo MF, Ineson P, Scott A. 1998. Elevated CO_2 reduces the nitrogen concentration of plant tissues. *Global Change Biology* **4**: 43–54.

Dieleman WIJ, Vicca S, Dijkstra FA, Hagedorn F, Hovenden MJ, Larsen KS, Morgan JA, Volder A, Beier C, Dukes JS, King J, Leuzinger S, Linder S, Luo Y, Oren R, De Angelis P, Tingey D, Hoosbeek MR, Janssens IA. 2012. Simple additive effects are rare: a quantitative review of plant biomass and soil process responses to combined manipulations of CO_2 and temperature. *Global Change Biology* **18**:2681–2693.

Drake BG, Gonzàlez-Meler MA, Long SP. 1997. More efficient plants: a consequence of rising atmospheric CO_2? *Annual Review of Plant Physiology and Plant Molecular Biology* **48**: 609–639.

Duval BD, Blankinship JC, Dijkstra P, Hungate BA. 2012. CO_2 effects on plant nutrient concentration depend on plant functional group and available nitrogen: a meta-analysis. *Plant Ecology* **213**: 505–521.

Field C. 1983. Allocating leaf nitrogen for the maximization of carbon gain: leaf age as a control on the allocation program. *Oecologia* **56**:341–347.

Finzi AC, Moore DJP, DeLucia EH, Lichter J, Hofmockel KS, Jackson RB, Kim HS, Matamala R, McCarthy HR, Oren R, Pippen JS, Schlesinger WH. 2006. Progressive nitrogen limitation of ecosystem processes under elevated CO_2 in a warm-temperate forest. *Ecology* **87**:15–25.

Gershunov A, Cayan DR, Iacobellis SF. 2009. The great 2006 heat wave over California and Nevada: signal of an increasing trend. *Journal of Climate* **22**:6181–6203.

Gifford RM, Barrett DJ, Lutze JL. 2000. The effects of elevated [CO_2] on the C : N and C : P mass ratios of plant tissues. *Plant and Soil* **224**: 1–14.

Gill RA, Anderson LJ, Polley HW, Johnson HB, Jackson RB. 2006. Potential nitrogen constraints on soil carbon sequestration under low and elevated atmospheric CO_2. *Ecology* **87**:41–52.

IPC California. 2015. http://www.cal-ipc.org/ip/management/plant_profiles/Phalaris_aquatica.php (23 June 2015).

Karl TR, Kukla G, Razuvayev VN, Changery MJ, Quayle RG, Heim RR, Easterling DR, Fu CB. 1991. Global warming: evidence for asymmetric diurnal temperature change. *Geophysical Research Letters* **18**:2253–2256.

Kitchen DJ, Blair JM, Callaham MA Jr. 2009. Annual fire and mowing alter biomass, depth distribution, and C and N content of roots and soil in tallgrass prairie. *Plant and Soil* **323**:235–247.

Körner C. 2006. Plant CO_2 responses: an issue of definition, time and resource supply. *New Phytologist* **172**:393–411.

Lee TD, Barrott SH, Reich PB. 2011. Photosynthetic responses of 13 grassland species across 11 years of free-air CO_2 enrichment is modest, consistent and independent of N supply. *Global Change Biology* **17**:2893–2904.

Lilley JM, Bolger TP, Peoples MB, Gifford RM. 2001. Nutritive value and the nitrogen dynamics of *Trifolium subterraneum* and *Phalaris aquatica* under warmer, high CO_2 conditions. *New Phytologist* **150**:385–395.

Loladze I. 2002. Rising atmospheric CO_2 and human nutrition: toward globally imbalanced plant stoichiometry? *Trends in Ecology and Evolution* **17**:457–461.

Morgan JA, Pataki DE, Körner C, Clark H, Del Grosso SJ, Grünzweig JM, Knapp AK, Mosier AR, Newton PCD, Niklaus PA, Nippert JB, Nowak RS, Parton WJ, Polley HW, Shaw MR. 2004. Water relations in grassland and desert ecosystems exposed to elevated atmospheric CO_2. *Oecologia* **140**:11–25.

Nie M, Lu M, Bell J, Raut S, Pendall E. 2013. Altered root traits due to elevated CO_2: a meta-analysis. *Global Ecology and Biogeography* **22**:1095–1105.

Norby RJ, Luo Y. 2004. Evaluating ecosystem responses to rising atmospheric CO_2 and global warming in a multi-factor world. *New Phytologist* **162**:281–293.

Oram RN, Culvenor RA. 1994. *Phalaris* improvement in Australia. *New Zealand Journal of Agricultural Research* **37**:329–339.

Oram RN, Culvenor RA, Ridley AM. 1993. Breeding the perennial pasture grass *Phalaris aquatica* for acid soils. In: Randall PJ, Delhaize E, Richards RA, Munns R, eds. *Genetic aspects of plant mineral nutrition*. The Netherlands: Springer, 17–22.

Oram RN, Ferreira V, Culvenor RA, Hopkins AA, Stewart A. 2009. The first century of *Phalaris aquatica* L. cultivation and genetic improvement: a review. *Crop and Pasture Science* **60**:1–15.

Poorter H, Van Berkel Y, Baxter R, Den Hertog J, Dijkstra P, Gifford RM, Griffin KL, Roumet C, Roy J, Wong SC. 1997. The effect of elevated CO_2 on the chemical composition and construction costs of leaves of 27 C_3 species. *Plant, Cell and Environment* **20**:472–482.

Rachmilevitch S, Cousins AB, Bloom AJ. 2004. Nitrate assimilation in plant shoots depends on photorespiration. *Proceedings of the National Academy of Sciences of the USA* **101**:11506–11510.

Reich PB, Hobbie SE, Lee T, Ellsworth DS, West JB, Tilman D, Knops JMH, Naeem S, Trost J. 2006. Nitrogen limitation constrains sustainability of ecosystem response to CO_2. *Nature* **440**: 922–925.

Sardans J, Peñuelas J, Estiarte M, Prieto P. 2008. Warming and drought alter C and N concentration, allocation and accumulation in a Mediterranean shrubland. *Global Change Biology* **14**: 2304–2316.

Schenk HJ, Jackson RB. 2002. The global biogeography of roots. *Ecological Monographs* **72**:311–328.

Stitt M, Krapp A. 1999. The interaction between elevated carbon dioxide and nitrogen nutrition: the physiological and molecular background. *Plant, Cell and Environment* **22**:583–621.

Stone L. 2009. *Environmental weed risk analysis: Phalaris aquatica*. (ed. Culvenor RA). Perth, Australia: Future Farm Industries CRC.

Volder A, Edwards EJ, Evans JR, Robertson BC, Schortemeyer M, Gifford RM. 2004. Does greater night-time, rather than constant, warming alter growth of managed pasture under ambient and elevated atmospheric CO_2? *New Phytologist* **162**:397–411.

Volder A, Smart DR, Bloom AJ, Eissenstat DM. 2005. Rapid decline in nitrate uptake and respiration with age in fine lateral roots of grape: implications for root efficiency and competitive effectiveness. *New Phytologist* **165**:493–502.

Volder A, Gifford RM, Evans JR. 2007. Effects of elevated atmospheric CO_2, cutting frequency, and differential day/night atmospheric warming on root growth and turnover of *Phalaris* swards. *Global Change Biology* **13**:1040–1052.

Vose RS, Easterling DR, Gleason B. 2005. Maximum and minimum temperature trends for the globe: an update through 2004. *Geophysical Research Letters*; doi:10.1029/2005GL024379.

Wan S, Xia J, Liu W, Niu S. 2009. Photosynthetic overcompensation under nocturnal warming enhances grassland carbon sequestration. *Ecology* **90**:2700–2710.

Xu X, Sherry RA, Niu S, Li D, Luo Y. 2013. Net primary productivity and rain-use efficiency as affected by warming, altered precipitation, and clipping in a mixed-grass prairie. *Global Change Biology* **19**: 2753–2764.

Zar JH. 1984. *Biostatistical analysis*, 2nd edn. New Jersey: Prentice Hall, 718.

Zhou X, Ge ZM, Kellomäki S, Wang KY, Peltola H, Martikainen P. 2011. Effects of elevated CO_2 and temperature on leaf characteristics, photosynthesis and carbon storage in aboveground biomass of a boreal bioenergy crop (*Phalaris arundinacea* L.) under varying water regimes. *Global Change Biology Bioenergy* **3**:223–234.

Ziska LH, Bunce JA. 1995. Growth and photosynthetic response of three soybean cultivars to simultaneous increases in growth temperature and CO_2. *Physiologia Plantarum* **94**:575–584.

Ziter C, MacDougall AS. 2013. Nutrients and defoliation increase soil carbon inputs in grassland. *Ecology* **94**:106–116.

9

Developmental stages of cultivated strawberry flowers in relation to chilling sensitivity

Maria Teresa Ariza, Carmen Soria and Elsa Martínez-Ferri*
IFAPA, Centro de Churriana, Cortijo de la Cruz s/n, 29140 Churriana, Málaga, Spain

Associate Editor: Michael B. Jackson

Abstract. Environmental factors affecting flower development may limit the yields of fruiting crops worldwide. In temperate regions, chilling temperatures during flower development can compromise fruit production, but their negative effects vary depending on the differing susceptibilities of each developmental stage. The cultivated strawberry (*Fragaria* × *ananassa* Duch.) is widely grown worldwide but financial returns are influenced by sudden shifts to chilling temperatures occurring during the cropping cycle. Despite this important limitation, knowledge of *F.* × *ananassa* flower development is lacking, in contrast to the diploid wild-type strawberry (*F. vesca*). In this study we describe steps in floral development of cultivated strawberry and define their vulnerability to chilling temperatures. To achieve this, flower buds from strawberry plants of cv. 'Camarosa' were labelled and monitored from bud initiation until anthesis. Description of morphological and functional changes during flower development was based on histological sections and scanning electron microscopy. To determine the impact of low temperatures at different developmental stages, plants carrying buds of different sizes were chilled at 2 °C for 24 h. Several parameters related to male and female gametophyte development were later evaluated in flowers as they approached anthesis. *Fragaria* × *ananassa* flower development was divided into 16 stages according to landmark events. These stages were similar to those documented for *F. vesca* but three new additional intermediate stages were described. Timing of developmental processes was achieved by correlating developmental staging with specific bud sizes and days before anthesis. Time to reach anthesis from early bud stages was 17–18 days. During this period, we detected four critical periods vulnerable to low temperatures. These were mostly related to male gametophyte development but also to injury to female organs at late developmental stages. These results provide the essential groundwork on floral biology of cultivated strawberry that is a prerequisite for successful comparative studies of cold tolerance among genotypes during flower formation.

Keywords: Critical periods; flower development; flower differentiation; *Fragaria* × *ananassa*; low temperature; pistil; pollen.

Introduction

Over 95 % of the Earth's surface experiences temperatures below 5 °C each year (Atkin *et al.* 1999). On clear days in winter and early spring, plants of extra-tropical regions are subjected to a combination of chilling temperatures and high solar irradiances. These conditions can often damage flower development of fruiting crops and, consequently, depress fruit production, since the capacity of a flower to become a fruit is established before anthesis (Rodrigo *et al.* 2009). Cool temperatures therefore have

* Corresponding author's e-mail address: elsa.martinez@juntadeandalucia.es

negative effects on fruit set by influencing processes taking place throughout flower development (Mercado *et al.* 1997; Ariza *et al.* 2011, 2012).

These effects can be especially prevalent in annual-fruiting crops such as strawberry that produce flowers throughout the cropping cycle. The commercial strawberry (*Fragaria* × *ananassa* Duch.) is grown from the tropics to the Arctic (Galletta and Bringhurst 1990), with main production areas in the USA (Salamé-Donoso *et al.* 2010), Mexico, Turkey (FAOSTAT 2014) and Spain (López-Medina *et al.* 2001; Dominguez *et al.* 2006). In the current growing areas, chilling temperatures are common during late winter and early spring, detrimentally affecting flower development and yields produced in the early season (from January to March; Ariza *et al.* 2011). Good financial returns are closely related to the production of high-quality yields, good flavour, appropriate fruit size and symmetric shape (Perkins-Veazie 1995), but also with the production of early yields (Rowley *et al.* 2010).

Strawberry flowers that develop in early winter are especially prone to chilling. Injury to the inflorescence, anthers and receptacle occurs at temperatures below 0 °C (Ki and Warmund 1992) and low temperatures 7 weeks before harvest have been related to misshapen fruit (Ariza *et al.* 2012). Low temperatures also affect the integrity of the floral structures (Ariza *et al.* 2011). The negative effects of cold on flower development and fruit production are thought to vary depending on the susceptibility of the processes associated with each floral developmental stage (Ohnishi *et al.* 2010). This is cultivar-dependent (Takeda *et al.* 2002) and depends on the range and duration of the chilling.

The suitability to different environments for cultivating strawberry depends on the cultivar; its genetic complexity giving rise to contrasting acclimation abilities in different environments. *Fragaria* × *ananassa* has therefore been termed a microclimatic crop (Rieger 2005). The genetic complexity of cultivated strawberry is greater than in *F. vesca*, and its octoploid nature gives rise to a multiplicity of flowering responses (Heide *et al.* 2013). It may also result in differences during flower development. Although floral development in the diploid *F. vesca* has been well studied (Hollender *et al.* 2012), there is less information for *F.* × *ananassa* and on the effects of chilling during flower differentiation, morphogenesis and maturation of its reproductive structures.

The aim of this study was to identify the main processes that comprise flower differentiation in *F.* × *ananassa* and to determine whether they are sensitive to chilling. We used cv. 'Camarosa' (University of California 1992; US Plant Patent #08708), a widely cropped early season short day cultivar, suitable for the major cropping areas. It is intended for use as a model for *F.* × *ananassa* flower development. Its adoption will help to detect which processes are potentially vulnerable to cold temperatures in other strawberry cultivars.

Methods

Plant material

Ninety strawberry plants of cv. 'Camarosa' (Larson 2001), from a commercial high-elevation nursery of Castille-Leon (northeastern Spain; lat. 41°30′N, long. 4°55′W, alt. 900 m above sea level), were planted into 6 L pots filled with a commercial substrate (Substrate Projar Professional; Fertilizer N-P-K: 14–16–18 + micronutrients, Projar®, Valencia, Spain) and sand (2 : 1, v/v) and, with an overlayer of Perlite. The plants were grown in a controlled temperature glasshouse (22 ± 2 °C and 70 ± 10 % RH) at the IFAPA Centro de Churriana in Málaga, Spain.

Flower bud size and days to anthesis

To analyse the time to anthesis (days before anthesis), five primary flower buds from 0.1 to 1.4 cm length were labelled, dated and monitored daily until anthesis. The length of flower buds was determined each day using a digital calliper (Kern®) and 14 intervals of 1 mm each were linked to a particular floral stage. To characterize the flower, 30 primary flowers were selected 1 day before anthesis. The total number of pollen grains per flower, the percentage of viable and non-viable pollen, ovule viability and immature stigmas were estimated as described below.

Flower development

To analyse the processes of development, three flower buds from each size category were preserved in FAA (5 : 5 : 90 v/v, formalin : acetic acid : 70 % ethanol; Johansen 1940) for embedding in paraffin (Paraplast, Merk) or preserved into GAA (glutaraldehide 2.5 % in 0.03 M phosphate buffer) for embedding in synthetic resins (Historesin, Technovit 7100, Microm). Samples were then sectioned with a rotary microtome (Jung supercut, 2065, Leica), stained and observed under stereoscope (Leica, MZ 95) or microscope (Leitz Laborlux 12). Photographs were taken with a digital camera (Canon PowerShot S70, 7.1 Megapixels) and analysed using an image processor program (UTHSCA, Imagen Tool 3.0).

Samples embedded in paraffin were sectioned at 10 μm and stained with toluidine blue or PAS (Feder and O'Brien 1968) for observation of the anther and pistil structure and with acetocarmine (40 % acetic acid saturated with carmine; Mercado *et al.* 1994) to examine pollen grains. At different buds sizes, 20 anthers and 10 carpels per region of each bud were analysed for the length of outer and inner anthers, length of the receptacle, diameter of pollen grain and length of apical and basal carpels in the receptacle.

For a functional approach, three floral buds 0.1, 0.4, 0.6, 1.1 and 1.3 cm long were included in synthetic resin and cut 2 μm thick. The tissues were then stained with acridine orange to detect cellular activity (i.e. synthesis of RNA and DNA; Dudley *et al.* 1987), with aniline blue to identify callose (Currier 1957; Linskens and Esser 1957) and with Auramine O to detect lipids (Heslop-Harrison 1977). Observations were made under a light microscope equipped with UV epifluorescence using a band pass 355–425 exciter filter and an LP 460 barrier filter.

Anthers, carpels and pollen were also observed by scanning electron microscopy (SEM; Jeol, JSM 840). Five flower buds of each category were submerged in 100 % ethanol after removing sepals and petals. The samples were dehydrated at critical point in liquid CO_2 in a Balzers CPD-030. Dehydrated material was mounted on stubs and coated with gold in a Jeol, JFC 1100 sputter coater prior to SEM photography.

Effect of chilling on floral development

After flowering initiation, 30 plants were transferred to a phytotron (IBERCEX, ASL S.A.) at 2 °C for 24 h with a photoperiod of 11 h, and then returned to the glasshouse. A batch of 30 plants kept in the glasshouse served as controls. Before chilling, all the flower buds were dated and monitored up to 2 days before anthesis. Five to 10 flowers per treatment at each bud stage were observed under a microscope at the same time.

To evaluate the effect of chilling on male structures, the total number of pollen grains per flower, percentage of germinated pollen grains and percentage of non-viable pollen were recorded in control and chilled plants. Anthers from 5 to 10 flowers per treatment were placed in Petri dishes inside a sealed dry chambers containing silica gel for 24 h to promote release of pollen grains. The pollen was placed in 1.5 mL vials and 150 μL of a germination medium consisting of 10 % glucose and 150 ppm boric acid added (Ariza *et al.* 2006). Two aliquots of the solution were separated to estimate the number of pollen grains and to assess pollen viability. To estimate the number of pollen grains per flower, pollen stained with 1 % acetocarmine was counted using a Neubauer chamber under an optical microscope (Leitz, Laborlux 12). Viability of the pollen at each date was assessed by incubating vials for 4 h at room temperature under light conditions and continuous shaking (Hortynski and Zebrowska 1991) before staining with 1.0 % acetocarmine. Germination *in vitro* was examined for ~200 pollen grains by light microscopy (Leitz, Laborlux 12, Germany; Fig. 2E). Pollen was considered viable and fertile when the length of the tube was greater than or equal to the diameter of the grain (Mercado *et al.* 1994). The percentage of germinated pollen was estimated on the basis of the number of stained pollen grains (undisrupted and potentially viable). The percentage of non-viable pollen was calculated from the number of non-stained pollen grains.

To analyse the effect of the chilling on ovule viability and development of the stigmas, 20 carpels per flower were squashed and stained with aniline blue for observation by light microscopy. Carpels were separated into tubular- (Type I), heart- (Type II) and scalloped-shaped (Type III) (Ariza *et al.* 2011) and the percentage of immature stigmas calculated. Carpels with Type I and II stigmas were considered immature, and Type III stigmas were mature and receptive (Ariza *et al.* 2011). An ovule was considered aborted if it stained with aniline blue (Anvari and Stösser 1978; Ariza *et al.* 2011).

Statistical analysis

Statistical analyses were performed using the analytical software STATISTIX 9.0 (Analytical Software, Tallahassee, FL, USA). To test the relationship between days before anthesis and flower bud sizes, length of anthers, diameter of pollen grain and length of carpels, lineal regression models were used. To evaluate the effect of chilling on the percentage of germinated pollen grains, percentage of non-viable pollen, number of pollen grains, type of stigma and percentage of aborted ovules, data of these variables from flowers at the same bud stage when chilling were subjected to a one-way ANOVA ($n = 5$–10 flowers per date). Prior to ANOVA, percentage data were transformed by arcsine and normality and homogeneity assumptions were tested by Kolmogorov–Smirnov and Cochran's *C* tests. Data were back-transformed to give the percentages of decrease or the fold increase.

Results and Discussion

The morphology and development of the flower in *F.* × *ananassa*

At anthesis, *F.* × *ananassa* flowers have a two whorl calyx (inner and outer) of 12.8 ± 0.2 alternating sepals. A whorl of 5.3 ± 0.1 white petals is present interior to the sepals. Subsequently, two whorls (inner and outer) of 22.8 ± 0.3 stamens form and surround the central receptacle dome. Each flower carries 248 384 ± 12 469 pollen grains; of which 47.9 ± 1.9 % are viable and 13.7 ± 0.7 % are non-viable. The pistil of each flower is the innermost whorl made up of 326 ± 18 carpels. Each carpel consists of a stigma, a style an ovary and one ovule (Hancock 1999). Carpels are embedded in the epidermis of the receptacle in a spiral pattern. The percentage of immature stigmas and aborted ovules was 7.7 ± 1.4 % and 12.3 ± 1.2 %, respectively. These results highlight a degree of functional failure of the reproductive structures in *F.* × *ananassa* under non-stressful conditions. However, these modest levels of failure do not necessarily translate into reduce fruit set or damage to development (Ariza *et al.* 2011, 2012).

The flower of *F.* × *ananassa* has considerably more bracts per whorl and is bigger than *F. vesca* (~1.4 cm versus ~0.7 cm, respectively; Hollender *et al.* 2012). In *F.* × *ananassa*, flower buds increased linearly from 0.1 to 1.5 cm at anthesis over 17–18 days with a growth rate ~1 mm day^{-1} (Fig. 1). Time to reach anthesis may differ between both species but data for *F. vesca* are not available. Changes in bud growth were linked to landmark events, allowing a chronological ranking of various developmental stages.

There was a linear relationship between days before anthesis and length of the anthers (Fig. 2A) and diameter of the pollen grain (Fig. 2B); in contrast, the relationship with carpel development was sigmoidal (Fig. 2C). Anthers from the outer part of the receptacle were larger than those from the inner part (Fig. 2A) indicating a maturity gradient from outside to inside. Similarly, basal carpels were longer than the apical ones (Fig. 2C), suggesting acropetal pistil development (Weberling 1989). These differences in the maturation of the reproductive structures within the same flower could be concomitant with different stages of gametophyte development (i.e. male and female) at a specific bud size. However, the model of floral development presented here for *F.* × *ananassa* integrates the main and significant changes occurring in the overall parts of the flower at a given time.

Male gametophyte development

At early stages of male gametophyte development (<1 cm bud size), a group of cells within the pollen sacs differentiate into archesporial cells. Similarly, primary

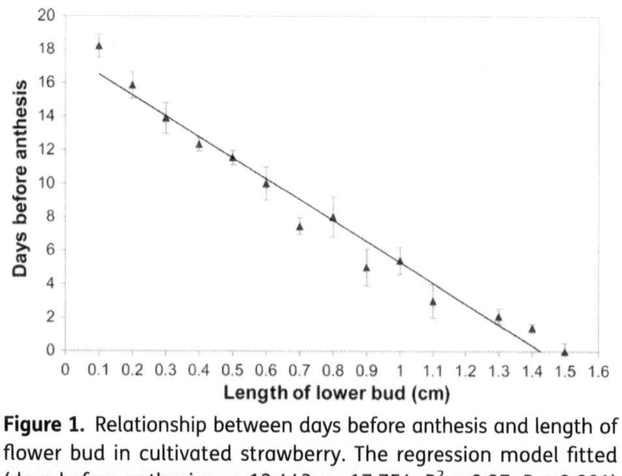

Figure 1. Relationship between days before anthesis and length of flower bud in cultivated strawberry. The regression model fitted (days before anthesis = $-12.443x + 17.754$, $R^2 = 0.97$, $P < 0.001$) is depicted. Each data point is the mean ± SE, $n = 20$.

parietal cells (PPCs) differentiate by mitotic division into sporogenous cells (PSCs; Stanley and Linskens 1974; Strasburger *et al.* 1994). Primary parietal cells give rise to endothecium, the middle layer of the anther and the tapetum (Yang *et al.* 1999), and PSCs differentiate into the pollen mother cells (PMCs) (Lord *et al.* 1989). At stage 8 (0.1 cm bud size), 18–17 days before anthesis, anthers are four-lobed, with walls differentiated into layers and PMCs were observed inside the pollen sacs (Fig. 3A–C). Pollen mother cells were negative to Auramine O, and positive to aniline blue indicating the presence of external callose (Fig. 3B and C, respectively). Also, angular cells were seen in the anthers (Fig. 3E), which might correspond to the tapetum cells aroused from PPCs at previous stages (stage 7, <0.1 cm; Lord *et al.* 1989). At this stage, PMCs began meiosis (Fig. 3D) which lasted until stage 9b (0.3 cm, 14 days before anthesis), when tetrads of four haploid microspores were observed within the locules (Fig. 3F and G). At stage 10a (0.4 cm, 13–12 days before anthesis), single rounded haploid microspores, still with a callose external layer (positive to aniline blue; Fig. 3E), are loose within the locule (Fig. 3H–J), indicating that microsporogenesis and meiotic reduction was completed at the earlier stage (stage 9b; 18–14 days before anthesis). This is in agreement with studies of bell pepper (Erickson and Markhart 2002) although it seems that timing of meiotic division is species dependent (Patterson *et al.* 1987; Porch and Jahn 2001). After meiosis, subsequent microgametogenesis and maturation of microspores takes place (Anger and Weber 2006).

At stage 10b (0.5–0.6 cm, 11–10 days before anthesis), pollen grains were round but only stained with Auramine O (Fig. 3K and L), indicating an outer layer of lipids (i.e. the exine). Secretion of the exine precursors is mediated by tapetum cells (Pacini and Juniper 1984; Scott *et al.* 2004; Yang *et al.* 2007) which also nourish the microspores during microspore maturation (Pacini 1990). These lipids contribute to the ornamentation of the pollen wall. There were also three apertures (i.e. the colpos) along the grain (Fig. 3T and U). Also, at this stage, fibrous bands on the epidermis and endothecium cells of the anther walls become evident (Fig. 3K and L). This may indicate initiation of tapetum degradation (Hollender *et al.* 2012). These cells continue to grow (Fig. 3M, P, R and S).

At stage 11a (0.7–0.9 cm, 9–6 days before anthesis), the final size of pollen grain was reached (Fig. 3Q and T). This is a time when the tapetum becomes fully degraded (Hollender *et al.* 2012) as a consequence of programmed cell death occurring at later stages of pollen development (Kawanabe *et al.* 2006).

At stage 11b (1.0–1.2 cm, 5–3 days before anthesis) pollen mitotic division occurs that creates the vegetative

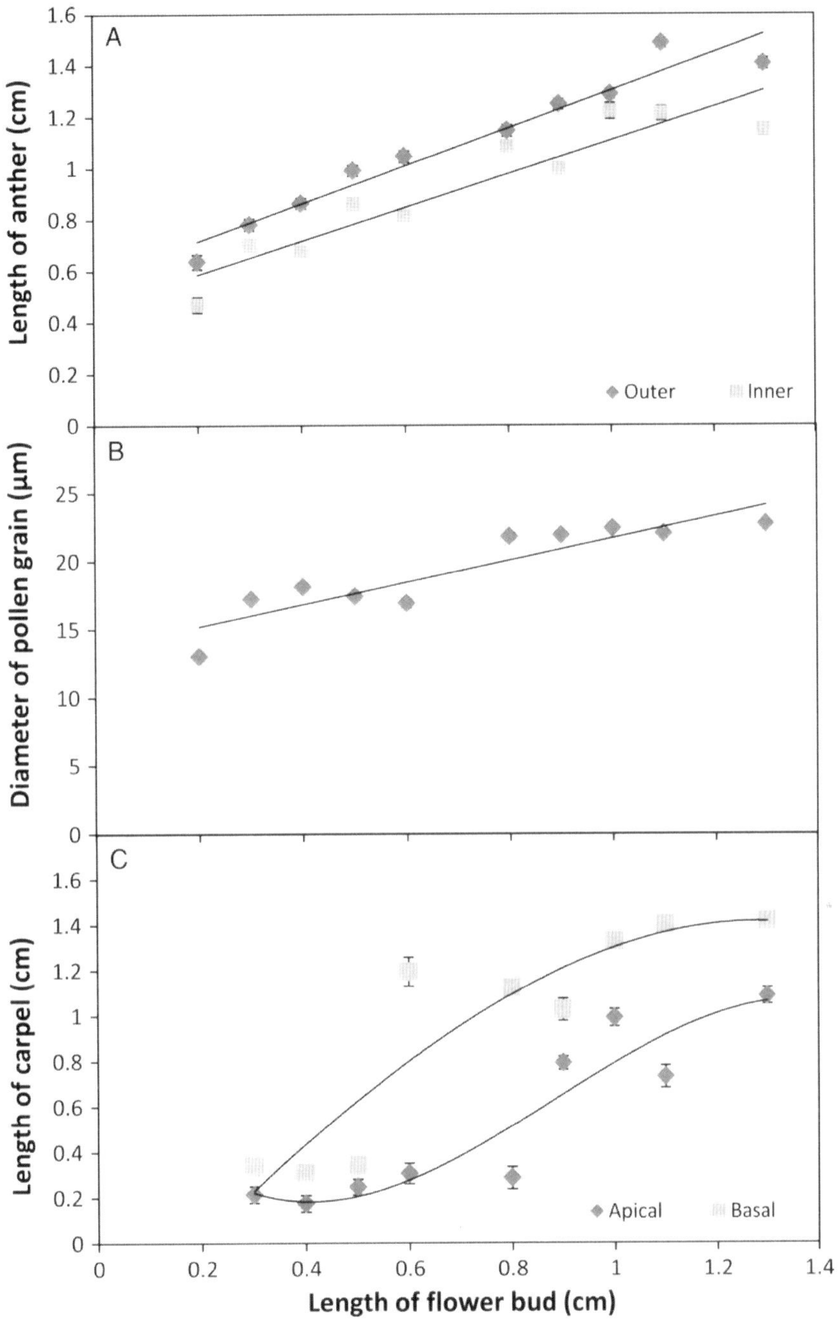

Figure 2. Relationship between the length of flower bud and length of outers and inners anthers (A), diameter of pollen grain (B) and length of carpels in apical and basal region of the receptacle (C). Each data point is the mean ± SE, $n = 9-45$. Fitting models of regression depicted are: length of outer anthers $= 0.734x + 0.5679$, $R^2 = 0.95$; Length of inner anthers $= 0.6513x + 0.456$, $R^2 = 0.87$; Pollen grain diameter $= 8.0943x + 13.593$, $R^2 = 0.82$; Length of apical carpels $= -2.1203x^3 + 5.6x^2 - 3.5177x + 0.8289$, $R^2 = 0.86$; Length of basal carpels $= -0.3122x^3 - 0.3387x^2 + 2.4045x - 0.4533$, $R^2 = 0.84$.

and generative nuclei. This observation agrees with that of Hollender *et al.* (2012) and with our later results on critical periods of susceptibility to low temperature discussed below. Anther dehiscence takes place at stage 13a (1.3 cm) and 1–2 days before anthesis. This occurs when flower buds are 1.4–1.5 cm long (stage 13b).

Pistil and female gametophyte development

At stages 8 and 9 (0.1 and 0.2 cm), the flower receptacle was fully covered by the protruding carpel primordia (Fig. 4A–C). The external lipid layer on the carpel surface and the considerable DNA and RNA synthesis observed in basal and apical carpels from early developmental stages

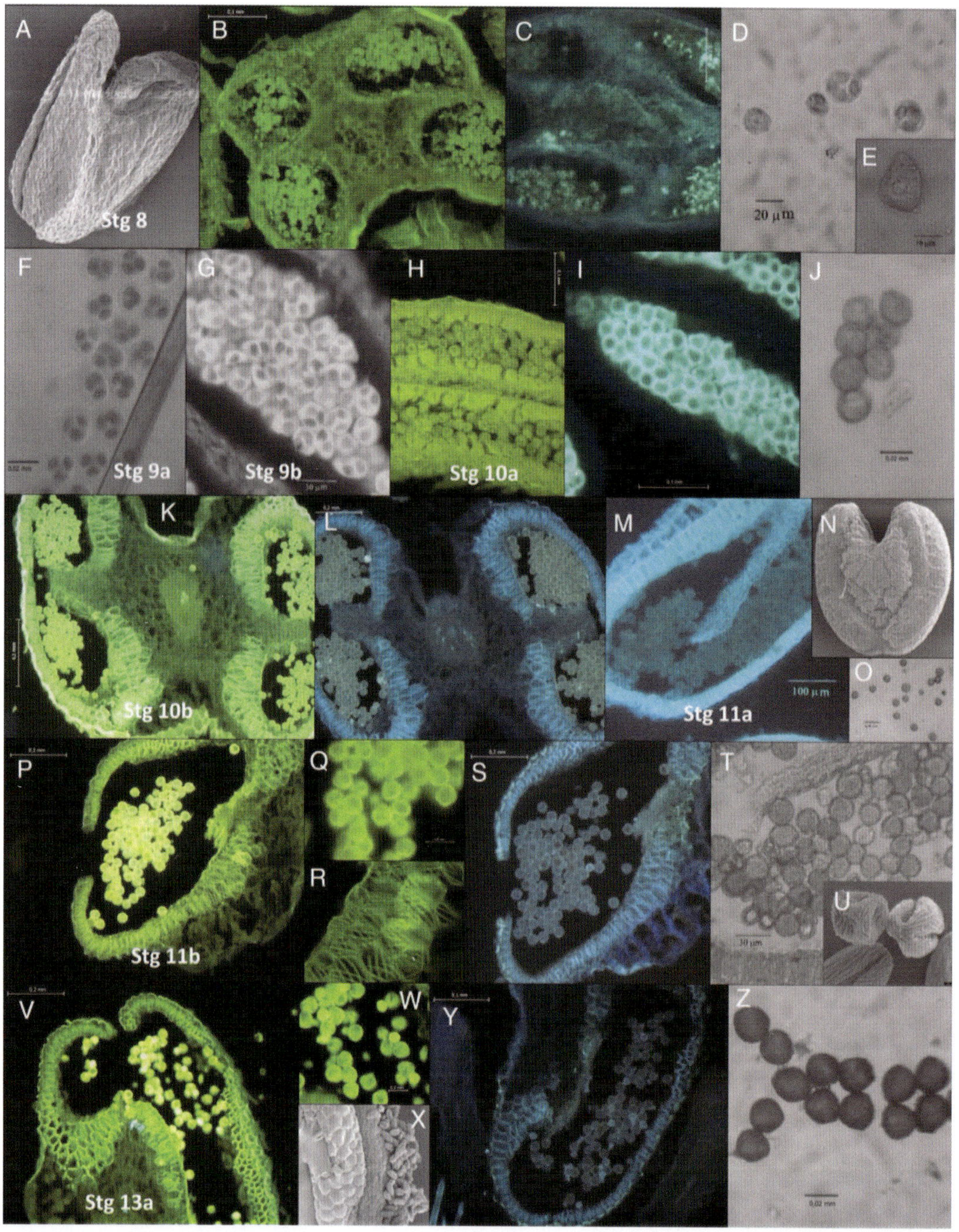

Figure 3. Micrographs of anther and pollen development of *F.* × *ananassa* flower buds at different developmental stages (Stg): stage 8 (A–E), stage 9a (F), stage 9b (G), stage 10a (H–J), stage 10b (K and L), stage 11a (M–O), stage 11b (P–U) and stage 13a (V–Z). Histological sections embedded in historesin are stained with Auramine O (green coloured) and aniline blue (blue coloured). Non-coloured pollen photographs are directly observed after squashing. Three-dimensional images of anthers and pollen were acquired by scanning electron microscopy.

(Fig. 4) point to changes in carpel size and development being closely related to active metabolic processes involved in rapid cell growth and differentiation (Hollender *et al.* 2012). From stages 9 (0.2 cm) to 9b (0.3 cm) carpel primordia undergo rapid morphological changes in 3 days (between 14 and 16 days before anthesis). This results in a transformation from thumb-like protuberances (Fig. 4A–C) to a bowling pin-shaped carpels at the receptacle base (Fig. 4D and E). As described in *F. vesca*, this rapid development involves the differentiation of epidermal carpel cells into several layers and the folding inward of ovary wall margins to envelop the single ovule containing the megasporocyte (megaspore mother cell: MMC) (Hollender *et al.* 2012). This differentiation in *F.* × *ananassa*

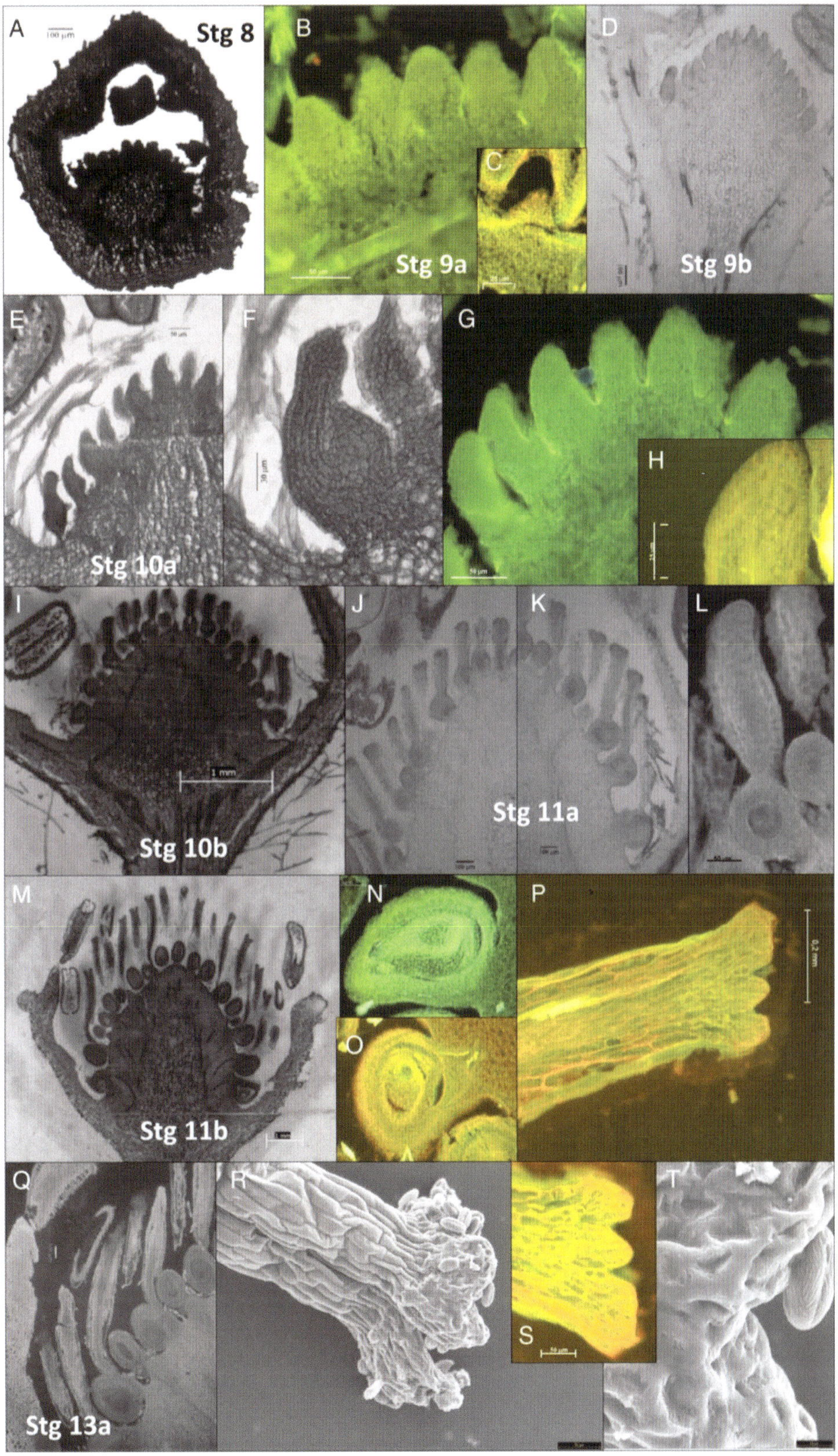

Figure 4. Micrographs of pistil development of *F.* × *ananassa* flower buds at different developmental stages (Stg): stage 8 (A), stage 9a (B and C), stage 9b (D), stage 10a (E–H), stage 10b (I), stage 11a (J–L), stage 11b (M–P) and stage 13a (Q–T). Histological sections embedded in paraffin were stained with PAS or toluidine blue (shown in black and white) for structural examination. Samples embedded in historesin were stained with auramine O (green coloured) and acridine orange (yellow coloured) for a morpho-functional approach. Three-dimensional images of stigmas were acquired by scanning electron microscopy.

was clear at stage 10a (13–12 days before anthesis) when the standard bottle-shape carpel morphology with anacrostily (Weberling 1989) was observed (Fig. 4F). At this stage, meiosis of MMC is probably occurring as reported for *F. vesca* (Hollender *et al.* 2012).

From stage 10b (0.5–0.6 cm, 11–10 days before anthesis) to stage 11a (0.7–0.9 cm, 9–6 days before anthesis), rapid elongation of the style, accompanied by ovary expansion, takes place and carpels shaped like a music-note can be observed at the receptacle base (Fig. 4I–L). This is associated with the prevalence of DNA synthesis.

At stage 11b (1.0–1.2 cm, 5–3 days before anthesis) the 'music-note' carpels reach their maximum size (Fig. 4M), the ovary is completely formed (Fig. 4N and O) and the stigmas become scalloped (Fig. 4P). The three consecutive mitosis of the chalazal megaspore may then generate the embryo sac (Souza *et al.* 2002; Yang 2005). This is in accordance with Hollender *et al.* (2012) and with the observed vulnerability of this stage to low temperatures (see below).

Finally, at stage 13a (1.3 cm, 2–1 days before anthesis), mature and receptive scalloped-shaped stigma were observed (Ariza *et al.* 2009, 2011). These stigmas displayed a high metabolic activity, which is consistent with observed lipid secretion on the stigma surface (Fig. 4R–T) which aids adhesion of the pollen (Avigdori-Avidov 1986). These findings suggest that at 1 or 2 days pre-anthesis flower reproductive structures are fully functional in cultivated strawberry.

Effect of chilling on floral development: critical periods

In flowering plants, the gametophyte phase is known to be sensitive to hot or cold, i.e. freezing and chilling temperatures even for a short period (Zinn *et al.* 2010; Parish *et al.* 2012). In the present study four critically sensitive periods of flower development in *F.* × *ananassa* were identified (Fig. 5) by subjecting flower buds at different stages to chilling.

The earliest stage showing susceptibility to low temperatures is male gametophyte development at stage 8 or before (20–17 days before anthesis). This was revealed as a significant decrease (up to 82 %) in the number of pollen grains, and by a 4-fold increase of the percentage of non-viable pollen (Fig. 5). These results demonstrate that temperature <2 °C negatively affect PMC development and anthers differentiation on endothecium, middle layer and tapetum, respectively.

The second stage showing chilling susceptibility was detected 15–14 days before anthesis (stages 9a and 9b). Here a decrease in the number of pollen grains per flower by up to 57 % was observed. This effect has been previously associated with disruption of PMCs meiosis under high temperatures (Kim *et al.* 2001; Erickson and Markhart 2002). Likewise, our results reveal that meiosis of PMCs in cultivated strawberry is also sensitive to low temperature. This is in agreement with recent studies reporting signs of cytomixis of PMCs at prophase I, contributing to the death of PMCs, as a consequence of chilling (Barton *et al.* 2014).

A third chilling sensitive stage occurs 10–7 days before anthesis (stages 10b and 11a). Dame here decreased the number of pollen grains by 73 %, indicating that microspore development and tapetum degeneration are affected (Parish *et al.* 2012). Pollen sterility induced by cold might be due to a disruption of sugar metabolism in the tapetum, ultimately abolishing starch accumulation (i.e. energy reserves) and exine secretion by the pollen grains (Oliver *et al.* 2005).

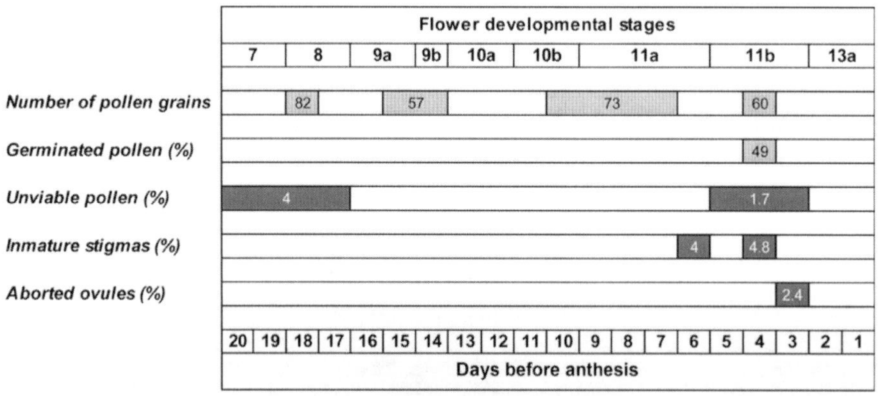

Figure 5. Low-temperature critical periods (grey bands) in days before anthesis of strawberry flower development, detected after subjecting flowers buds at different developmental stages to chilling (24 h at 2 °C). Periods were identified when strawberry flowers at stage 13a displayed significant ($P < 0.05$; $n = 5$–10) differences between the cold and control treatments in any of the following variables: number of pollen grains, percentage of germinated pollen, percentage of unviable pollen, percentage of immature stigmas and percentage of aborted ovules. The *n*-fold increase or the decrease percentages for each variable are depicted in the dark grey bands and light grey bands, respectively.

Table 1. Stages of flower development and key events in pollen and carpel development of cultivated (*F.* × *ananassa*) and wild strawberry (*F. vesca*; Hollender *et al.* 2012). Italics denote those events that are inferred from *F. vesca*. The asterisk indicates those events vulnerable to chilling injury. [1]Described in stage 12 on *F. vesca*.

F. vesca **floral stage**	***F.* × *ananassa*** **floral stage**	***F.* × *ananassa*** **bud length (cm)**	**Days before anthesis**	**Pollen**	**Fig.**	**Carpel**	**Fig.**
7	7	<0.1	>18	*Endothecium, middle layer and tapetum arise from parietal cells. PMCs appear.*		*Round carpel primordia reach the receptacle apex. No cell differentiation of epidermal layer.*	
8	8	0.1	18–17	*PMCs enter in meiosis. Tapetum cells present. Four anther locules are clearly distinct.	3A–E	Receptacle fully covered with carpels.	4A
9	9a	0.2	16–15	*PMCs in meiosis. Tetrads confined and imbibed in callose.	3F	Thumb-like carpel primordia at the receptacle apex.	4B and C
	9b	0.3	14	Meiosis is complete and callose wall holding tetrads together start to disintegrate.	3G	Bowling pin-shaped carpel primordia at the receptacle base. *Indented carpel walls* and *MMC might be visible.*	4D
10	10a	0.4	13–12	Round microspores are loose in the locule surrounded by a callose layer. Complete tetrad disintegration.	3H–J	Central carpel constriction divide organ into two almost equal apical and basal parts. MMC might enter in meiosis.	4E and F
	10b	0.5–0.6	11–10	*Microspore exine wall develops. Tapetum degeneration is initiated.	3K and L	Rapid elongation of style and ovary expansion.	4I
11	11a	0.7–0.9	9–6	*Tapetum is fully degraded and endothecium cells increase in size. Pollen final size reached.	3M–O	Idem. Music-note like carpels at the receptacle base.	4J–L
	11b	1–1.2	5–3	*Pollen mitotic division occurs (not obvious, deduced by chilling at 2 °C)	3P–U	*Fully formed embryo sac. Final size of music-note like carpels.[1] Scalloped stigmas with high metabolic activity.	4M–P
13	13a	1.3	2–1	Anther dehiscence	3V–Z	Scalloped and mature stigmas with moist secretion for pollen adhesion.	4Q–T
	13b	1.4–1.5	0			Anthesis	

Finally, chilling 5–3 days before anthesis (stage 11b) decreased the percentage of germinated pollen grains by 49 % and increased percentage of non-viable pollen by 1.7-fold. This indicates that microspore mitotic processes are being disrupted (Satake and Hayase 1970), an effect probably associated with microtubules disassembly during mitosis. Microtubules are known to be highly thermosensitive (Hepler and Hush 1996). On the other hand, chilling decreased the number of pollen grains by 60 %. This may be a result of physical disruption of pollen grains and of cell death induced by deleterious effect of low temperatures (Zinn *et al.* 2010).

Female reproductive organs were only affected by chilling 3–6 days before anthesis (stages 11a to 11b) when there was up to a 2.4- and 4.8-fold increase in the percentage of aborted ovules or of immature stigmas, respectively (Fig. 5). This suggests that low temperatures damage embryo sac formation and delays stigmas maturation. Both these phases are thus cold sensitive. This agrees with Ebadi *et al.* (1995) who reported a decline in ovule viability by low temperatures before anthesis. In contrast, other reports showed that low temperatures did not affect ovule and stigma receptivity in strawberry (Ariza *et al.* 2011). These discrepancies must be related to differences in the range of chilling temperatures tested by Ariza *et al.* (2011) and in the present study (>4 and 2 °C, respectively). This suggests that female reproductive structures and esporogenesis of cultivated strawberry are fairly tolerant to temperatures >2 °C. These findings underline the lower sensitivity of female reproductive processes to cold temperatures (Dupuis and Dumas 1990; Saini and Lalonde 1997). The chilling sensitivity of other strawberry cultivars now needs to be addressed since responses to environmental stress are often cultivar-related (Ledesma *et al.* 2007).

Conclusions

The present study describes the morpho-functional processes of flower development in cultivated strawberry (*F.* × *ananassa*) and reveals which steps are susceptible to damage from chilling at 2 °C. It identifies a close relationship between flower developmental stages and the size of the flower bud as it develops over ~18 days. Flower developmental stages in *F.* × *ananassa* were defined by macroscopic and microscopic morpho-functional changes in the reproductive structures as flower bud size increased. Most of these changes reflect those described recently for *F. vesca* during flower differentiation (Hollender *et al.* 2012), but additional stages were identified in *F.* × *ananassa* buds of intermediate sizes (Table 1).

In addition to contributing to the knowledge of the floral biology of Rosaceae, and strawberry, the present study implicates particular developmental processes as potentially sensitive to low temperature. The findings will also aid future comparative studies of species or cultivars aimed at identifying differences in tolerance and susceptibility. Such work will benefit breeders and growers working in areas where sudden down-shifts of temperature take place during the cropping cycle, especially during flowering.

Sources of Funding

This research was funded by the Spanish Instituto Nacional de Investigación Agraria (INIA) RTA2004-008 and CC10-009-C02-02 projects, and co-financed by European FEDER funds. M.T.A. was supported by a research grant financed by the Spanish INIA and is currently supported by IFAPA. Junta de Andalucia (20 %) and by the Programa Operativo Fondo Social Europeo (FSE) de Andalucía 2007–2013 (80 %) under the topic *'Andalucía se mueve con Europa'*.

Contributions by the Authors

All authors were involved in designing the experiments of this research, sampling, data handling, statistical analysis and manuscript preparation and submission. Microscopy technical work was mainly done by M.T.A.

Acknowledgements

We thank Dr María Herrero and Dr Araceli Barceló for their help with microscopy, technical advice and valuable comments on the manuscript. We also thank the associate editor Michael Jackson and two anonymous referees for their comments and suggestions on a previous version of the manuscript.

Literature Cited

Anger EM, Weber M. 2006. Pollen-wall formation in *Arum alpinum*. *Annals of Botany* **97**:239–244.

Anvari SF, Stösser R. 1978. Eine neue fluoreszenz-mikroskopiche Methode zur beurteilung der befruchtungsfahigkeit der samenanlegen bei Prunus. *Zeitschrift für Pflanzenzüchtung* **81**: 333–336.

Ariza MT, López-Aranda JM, Soria C. 2006. Optimization of a liquid medium for germination of strawberry pollen. *Acta Horticulturae* **708**:531–534.

Ariza MT, López-Aranda JM, Cid-Atencia MT, Soria C, Medina JJ, Miranda L. 2009. Effect of temperatura in carpel maturation and misshapen fruit in strawberry. *Acta Horticulturae* **842**: 757–760.

Ariza MT, Soria C, Medina JJ, Martínez-Ferri E. 2011. Fruit misshapen in strawberry cultivars (*Fragaria* × *ananassa*) is related to achenes functionality. *Annals of Applied Biology* **158**:130–138.

Ariza MT, Soria C, Medina JJ, Martínez-Ferri E. 2012. Incidence of misshapen fruits in strawberry plants grown under tunnels is affected by cultivar, planting date, pollination and low temperatures. *HortScience* **47**:1–5.

Atkin OK, Hurry VM, Gorsuch P. 1999. Plant responses to cold. In: Atwell BJ, Kriedermann PE, Turnbull CGN, eds. *Plants in action: adaptation in nature, performance in cultivation*, 2nd edn. Melbourne: Macmillan Education Australia.

Avigdori-Avidov H. 1986. Strawberry. In: Monselise SP, ed. *Handbook of fruit set and development*. Boca Raton, FL: CRC Press, 19–448.

Barton DA, Cantrill LC, Law AMK, Phillips CG, Sutton BG, Overall RL. 2014. Chilling to zero degrees disrupts pollen formation but not meiotic microtubule arrays in *Triticum aestivum* L. *Plant, Cell and Environment* **37**:2781–2794.

Currier HB. 1957. Callose substance in plant cells. *American Journal of Botany* **44**:478–488.

Dominguez FJ, Soria C, Medina JJ, Regidor J, Almagro S, Amaya I, Ariza MT, Delgado F, Delgado MD, Domínguez P, Fernández F, Flores F, Gady AL, Gálvez J, Muñoz MJ, Gil D, Gil MI, Gómez E, López-Aranda JM, de Matas J, López J, Miranda L, Palencia P, Peralbo A, Sánchez-Sevilla JF, Sanmiguel JM, Tomás FA, Tudela JA, Villalba R, Vizcaino J, Zorrilla Y. 2006. *RAEA Fresas 2006. Ensayo de variedades adaptables a las condiciones agro-ambientales de Huelva*. Sevilla: Junta de Andalucía.

Dudley ME, Jacobs TW, Long SR. 1987. Microscopic studies of cell divisions induced in alfalfa roots by *Rhizobium meliloti*. *Planta* **171**: 289–301.

Dupuis I, Dumas C. 1990. Influence of temperature stress on in vitro fertilization and heat shock protein synthesis in maize (*Zea mays* L.) reproductive tissues. *Plant Physiology* **94**:665–670.

Ebadi A, May P, Sedgley M, Coombe BG. 1995. Effect of low temperature near flowering time on ovule development and pollen tube growth in the grapevine (*Vitis vinifera* L.), cvs Chardonnay and Shiraz. *Australian Journal of Grape and Wine Research* **1**:11–18.

Erickson AN, Markhart AH. 2002. Flower developmental stage and organ sensitivity of bell pepper (*Capsicum annuum* L.) to elevated temperature. *Plant, Cell and Environment* **25**:123–130.

FAOSTAT Agricultural Data. 2014. http://www.faostat.fao.org (April 2014).

Feder N, O'Brien TP. 1968. Plant microtechnique: some principles and new methods. *American Journal of Botany* **55**:123–142.

Galletta GJ, Bringhurst RS. 1990. Strawberry management. In: Galletta GJ, Himelrick DG, eds. *Small fruit crop management*. Inglewood Cliffs, NJ: Prentice-Hall, 83–156.

Hancock JF. 1999. *Strawberries*. Wallingford: CAB International.

Heide OM, Stavang JA, Sønsteby A. 2013. Physiology and genetics of flowering in cultivated and wild strawberries. *Journal of Horticultural Science and Biotechnology* **88**:1–18.

Hepler PK, Hush JM. 1996. Behaviour of microtubules in living plant cells. *Plant Physiology* **112**:455–461.

Heslop-Harrison Y. 1977. Pollen–stigma interaction—pollen-tube penetration in Crocus. *Annals of Botany* **41**:913–922.

Hollender CA, Geretz AC, Slovin JP, Liu Z. 2012. Flower and early fruit development in a diploid strawberry, *Fragaria vesca*. *Planta* **235**: 1123–1139.

Hortynski JA, Zebrowska J. 1991. The effect of different air temperatures on *in vitro* pollen germination of selected strawberry cultivars (*Fragaria* × *ananassa* Duch.). *Folia Horticulturae* **III**: 107–113.

Johansen DA. 1940. *Plant microtechnique*. New York: McGraw-Hill.

Kawanabe T, Ariizumi T, Kawai-Yamada M, Uchimiya H, Toriyama K. 2006. Abolition of the tapetum suicide program ruins microsporogenesis. *Plant and Cell Physiology* **47**:784–787.

Ki WK, Warmund MR. 1992. Low-temperature injury to strawberry floral organs at several stages of development. *Hortscience* **27**: 1302–1304.

Kim SY, Hong CB, Lee I. 2001. Heat shock stress causes stage-specific male sterility in *Arabidopsis thaliana*. *Journal of Plant Research* **114**:301–307.

Larson KD. 2001. El programa de investigación de fresa de la Universidad de California: Progreso y perspectiva. In: Caja de Ahorros El Monte, ed. *XVIII Jornadas Agrícolas y Comerciales de El Monte*. Ciudad Real: Caja de Ahorros El Monte, 14–17.

Ledesma NA, Nakata M, Sugiyama N. 2007. Effect of high temperature stress on the reproductive growth of strawberry cvs. "Nyoho" and "Toyonoka". *Scientia Horticulturae* **166**:186–193.

Linskens HF, Esser K. 1957. Über eine spezifische Anfärbung der Pollenschläuche im Griffel und die Zahl der Kallosepfropfen nach Selbstung und Fremdung. *Naturwissenschaften* **44**:16.

López-Medina J, Vázquez E, Medina JJ, Domínguez F, López-Aranda JM, Bartual R, Flores F. 2001. Genotype × environment interaction for planting date and plant density effects on yield characters of strawberry. *Journal of Horticultural Science and Biotechnology* **75**:564–568.

Lord EM, Eckard KJ, Crone W. 1989. Development of the dimorphic anthers in *Collomia grandiflora*; evidence for heterochrony in the evolution of the cleistogamous anther. *Journal of Evolutionary Biology* **2**:81–93.

Mercado JA, Fernández-Muñoz R, Quesada MA. 1994. *In vitro* germination of pepper pollen in liquid medium. *Scientia Horticulturae* **57**: 273–281.

Mercado JA, Mar Trigo M, Reid MS, Valpuesta V, Quesada MA. 1997. Effects of low temperature on pepper pollen morphology and fertility: evidence of cold induced exine alterations. *Journal of Horticultural Science* **72**:317–326.

Ohnishi S, Miyoshi T, Shirai S. 2010. Low temperature stress at different flower developmental stages affects pollen development, pollination, and pod set in soybean. *Environmental and Experimental Botany* **69**:56–62.

Oliver SN, Van Dongen JT, Alfred SC, Mamun EA, Zhao X, Saini HS, Fernandes SF, Blanchard CL, Sutton BG, Geigenberger P, Dennis ES, Dolferus R. 2005. Cold-induced repression of the rice anther-specific cell wall invertase gene OSINV4 is correlated with sucrose accumulation and pollen sterility. *Plant, Cell and Environment* **28**:1534–1551.

Pacini E. 1990. Tapetum and microspore function. In: Blackmore S, Knox RB, eds. *Microspores: evolution and ontogeny*. London: Academic Press.

Pacini E, Juniper B. 1984. The ultrastructure of pollen grain development in *Lycopersicum peruvianum*. *Caryologia* **37**:21–50.

Parish RW, Phan HA, Iacuone S, Li SF. 2012. Tapetal development and abiotic stress: a centre of vulnerability. *Functional Plant Biology* **39**:553–559.

Patterson BD, Mutton L, Paull RE, Nguyen VQ. 1987. Tomato pollen development: stages sensitive to chilling and a natural environment for the selection of resistant genotypes. *Plant, Cell and Environment* **10**:363–368.

Perkins-Veazie P. 1995. Growth and ripening of strawberry fruit. *Horticultural Reviews* **17**:267–297.

Porch TG, Jahn M. 2001. Effects of high-temperature stress on microsporogenesis in heat-sensitive and heat-tolerant genotypes of *Phaseolus vulgaris*. *Plant, Cell and Environment* **24**: 723–731.

Rieger M. 2005. Strawberry. In: Rieger M, ed. *Introduction to fruit crops*. New York: Haworth Press, Inc., 383–392.

Rodrigo J, Herrero M, Hormaza JI. 2009. Pistil traits and flower fate in apricot (*Prunus armeniaca*). *Annals of Applied Biology* **154**: 365–375.

Rowley D, Black BL, Drost D, Feuz D. 2010. Early-season extension using june-bearing 'Chandler' strawberry in high-elevation high tunnels. *HortScience* **45**:1464–1469.

Saini HS, Lalonde S. 1997. Injuries to reproductive development under water stress, and their consequences for crop productivity. *Journal of Crop Production* **1**:223–248.

Salamé-Donoso TP, Santos BM, Chandler CK, Sargent SA. 2010. Effect of high tunnels on the growth, yields, and soluble solids of strawberry cultivars in Florida. *International Journal of Fruit Science* **10**: 249–263.

Satake T, Hayase H. 1970. Male sterility caused by cooling treatment at the young microspore stage in rice plants: V. Estimations of pollen developmental stage and the most sensitive stage to coolness. *Japanese Journal of Crop Science* **39**:468–473.

Scott RJ, Spielman M, Dickinson HG. 2004. Stamen structure and function. *Plant Cell* **16**:S46–S60.

Souza MM, Pereira TNS, Hoffmann M, Melo EJT, Louro RP. 2002. Embryo sac development in yellow passion fruit *Passiflora edulis f. flavicarpa* (*Passifloraceae*). *Genetics and Molecular Biology* **25**: 471–475.

Stanley RG, Linskens HF. 1974. *Pollen: biology, biochemistry, management*. Berlin: Springer.

Strasburger E, Noll F, Schenck H, Schimper A. 1994. *Tratado de botánica. 8th edn. castellana*. Barcelona: Omega.

Takeda F, Strik B, Peacock D, Clark JR. 2002. Cultivar differences and the effect of winter temperature on flower bud development in blackberry. *Journal of the American Society of Horticultural Science* **127**:495–501.

Weberling F. 1989. *Morphology of flowers and inflorescences*. Cambridge: Cambridge University Press.

Yang WC. 2005. Female gametophyte development. In: Basra AS, ed. *Handbook of seed science and technology*. New York: The Haworth Press, Inc., 27–62.

Yang C, Vizcay-Barrena G, Conner K, Wilson ZA. 2007. MALE STERILITY1 is required for tapetal development and pollen wall biosynthesis. *The Plant Cell Online* **19**:3530–3548.

Yang WC, Ye D, Xu J, Sundaresan V. 1999. The *SPOROCYTELESS* gene of *Arabidopsis* is required for initiation of sporogenesis and encodes a novel nuclear protein. *Genes and Development* **13**: 2108–2117.

Zinn KE, Tunc-Ozdemir M, Harper JF. 2010. Temperature stress and plant sexual reproduction: uncovering the weakest links. *Journal of Experimental Botany* **61**:1959–1968.

10

An empirical model that uses light attenuation and plant nitrogen status to predict within-canopy nitrogen distribution and upscale photosynthesis from leaf to whole canopy

Gaëtan Louarn[1*], Ela Frak[1], Serge Zaka[1], Jorge Prieto[2] and Eric Lebon[3]

[1] INRA UR4 URP3F, BP6, F86600 Lusignan, France
[2] INTA EEA Mendoza, San Martín 3853, Luján de Cuyo (5507), Mendoza, Argentina
[3] INRA, UMR 759, LEPSE, 2 place Viala, F34060 Montpellier, France

Associate Editor: Ulo Niinemets

Abstract. Modelling the spatial and temporal distribution of leaf nitrogen (N) is central to specify photosynthetic parameters and simulate canopy photosynthesis. Leaf photosynthetic parameters depend on both local light availability and whole-plant N status. The interaction between these two levels of integration has generally been modelled by assuming optimal canopy functioning, which is not supported by experiments. During this study, we examined how a set of empirical relationships with measurable parameters could be used instead to predict photosynthesis at the leaf and whole-canopy levels. The distribution of leaf N per unit area (N_a) within the canopy was related to leaf light irradiance and to the nitrogen nutrition index (NNI), a whole-plant variable accounting for plant N status. N_a was then used to determine the photosynthetic parameters of a leaf gas exchange model. The model was assessed on alfalfa canopies under contrasting N nutrition and with N_2-fixing and non-fixing plants. Three experiments were carried out to parameterize the relationships between N_a, leaf irradiance, NNI and photosynthetic parameters. An additional independent data set was used for model evaluation. The N distribution model showed that it was able to predict leaf N on the set of leaves tested. The N_a at the top of the canopy appeared to be related linearly to the NNI, whereas the coefficient accounting for N allocation remained constant. Photosynthetic parameters were related linearly to N_a irrespective of N nutrition and the N acquisition mode. Daily patterns of gas exchange were simulated accurately at the leaf scale. When integrated at the whole-canopy scale, the model predicted that raising N availability above an NNI of 1 did not result in increased net photosynthesis. Overall, the model proposed offered a solution for a dynamic coupling of leaf photosynthesis and canopy N distribution without requiring any optimal functioning hypothesis.

Keywords: Light; *Medicago sativa*; nitrogen distribution; nitrogen nutrition index; photosynthesis; transpiration; upscaling; within-canopy variability.

* Corresponding author's e-mail address: gaetan.louarn@lusignan.inra.fr

Introduction

A close positive relationship exists between the nitrogen (N) content and photosynthetic capacity of leaves (Field and Mooney 1986; Evans 1989; Hikosaka 2004). Such a relationship is the cornerstone of various approaches proposed to upscale leaf gas exchange models to the whole canopy level (Kull and Jarvis 1995; Leuning 1995; De Pury and Farquhar 1997; Sinoquet *et al.* 2001; Evers *et al.* 2010). Indeed, it has been shown that the within-canopy variability of photosynthetic parameters can be fully specified at a given time by measurements of the spatial distribution of leaf N (Harley *et al.* 1992; Le Roux *et al.* 1999; Braune *et al.* 2009). Changes in leaf N concentration with canopy depth, and the effects of leaf age and leaf light microclimate, have been identified as major sources of spatial variation (Evans 1989). They have been studied extensively in several species and different modelling approaches enable to account for it.

Following the optimization theory, several authors first sought to model N distribution in order to maximize canopy photosynthesis (Charles-Edwards 1981; Field 1983; Hirose and Werger 1987). The outcome was models predicting a leaf N concentration, which paralleled radiation extinction (Kull and Jarvis 1995; Sands 1995). However, there is no *a priori* reason for N distribution to follow such a pattern. On the contrary, empirical observations have consistently indicated that the exponential fall in leaf N with increasing depth into the canopy occurs in most canopies at a slower rate than light extinction (Hirose and Werger 1987; Lemaire *et al.* 1991; Anten *et al.* 1995; Moreau *et al.* 2012). Alternatively, empirical relationships between light extinction and leaf N have been used to mimic a local light acclimation and model leaf N distribution by considering potential departures from the light gradient (Sellers *et al.* 1992; Anten *et al.* 1995; Prieto *et al.* 2012). The approach is usually based on a coefficient of N allocation (k_N) that shapes the N profile with respect to relative light extinction (I/I_0), and a reference leaf N concentration at the top of the canopy (N_{up}):

$$N_a = N_{up}\left(\frac{I}{I_0}\right)^{k_N} \quad (1)$$

Leaf N distribution is not solely a function of light and age, however. It is also dependent on mineral N availability (Hikosaka *et al.* 1994; Lötscher *et al.* 2003), N demand to support plant growth and more generally on the N status of plants (i.e. the relative satisfaction of plant N demand, Lemaire and Gastal 1997). The N demand of a plant at any time in its cycle is generally defined as the amount of N necessary to sustain maximum plant growth. Nitrogen demand is tightly related to the standing crop mass. On a mass increment basis, it decreases as biomass increases, resulting in an apparent dilution of plant N concentration with plant growth (Greenwood *et al.* 1990; Gastal *et al.* 2015). Canopy N content and leaf N distribution, thus, respond not only to changes in the fertilization rate (Bélanger *et al.* 1992; Dreccer *et al.* 2000) but also to all factors that affect the plant growth rate (e.g. temperature and CO_2 concentration; Pettersson and McDonald 1994). Empirical plant N status indices have been developed to account for both aspects and help to diagnose crop N requirements. For instance, the nitrogen nutrition index (NNI) was assessed on plants as different as C3 annual crops (e.g. Justes *et al.* 1994; Colnenne *et al.* 1998), C4 grasses (Plénet and Lemaire 1999) and perennial forage plants (Lemaire *et al.* 1984, 1985). This is based on the concept of critical N dilution that can be applied in dynamic terms and is able to account for temporal changes in the N nutrition of crops (e.g. STICS crop model, Brisson *et al.* 2009).

Modelling the effect of the interaction between light acclimation and N limitations on the distribution of leaf N and photosynthetic characteristics has received comparatively less attention (Thornley 1998). One challenge is that light acclimation is a local process driven by the leaf light microclimate (Evans 1989; Hikosaka *et al.* 1994), whereas N demand, plant N status and N allocation are defined at the whole-plant scale (Givnish 1988; Lemaire and Gastal 1997; Kull 2002; Gastal *et al.* 2015). Empirical relationships between light extinction and leaf N generally refer to static canopies at a given developmental stage, and their parameters need to be adjusted between years, sites or N treatments (Prieto *et al.* 2012). To date, dynamic coupling with plant growth has, thus, mainly been achieved using approaches based on the optimal distribution theory (Johnson *et al.* 2010). Some studies demonstrated a significant relationship between canopy NNI and the k_N and N_{up} parameters (Lötscher *et al.* 2003; Farruggia *et al.* 2004; Gastal *et al.* 2015). These relationships could be tested to make predictions of photosynthetic parameters under contrasting N availabilities without any *a priori* assumptions regarding optimal functioning of the canopy. Such a model would offer a solution to dynamically simulate the interactions between light and N based on parameters that can be directly measured.

During this study, we developed and assessed a model coupling an empirical canopy N distribution model with a leaf gas exchange model derived from Farquhar *et al.* (1980). The distribution of leaf N content per unit area (N_a) was related to leaf light irradiance and to the canopy NNI. The objectives were to determine whether such an empirical approach to leaf N distribution could be used to specify spatial and temporal changes in leaf gas exchange under fluctuating light and N availability. Alfalfa was chosen as a model species because its leaf N distribution has already been described extensively under non-limiting N and because this species presents limited age dependency of leaf characteristics (Lemaire *et al.* 1991, 2005; Evans 1993).

Methods

Model description

Canopy N distribution model. We assumed that spatial and temporal variations in leaf N content per unit area (N_a) within the canopy and in the course of plant growth can be deduced from leaf light exposure and plant N status. The effect of relative leaf irradiance on relative N_a was taken into account using Eq. (1) with the two parameters N_{up} and k_N. The effect of N limitation was assumed to affect whole-canopy N content in leaves by modulating these two parameters. The NNI was considered to account for the effect of canopy N status (integrating the effects of soil mineral N and nodule fixation on internal N availability). At a given time, NNI was defined as:

$$\mathrm{NNI} = \frac{N_m}{N_c} \tag{2}$$

where N_m represents the actual plant N concentration and N_c the critical plant N concentration (g N 100 g^{-1} plant) corresponding to its mass W (given by equation $N_c = 4.8W^{-0.33}$ in alfalfa, Lemaire *et al.* 1985). When NNI is close to 1, the plant N status is considered as near optimum. Departures from 1 indicate deficiency (NNI < 1; the intensity of deficiency is then equal to $1 - \mathrm{NNI}$) or excess N (NNI > 1, the intensity of excess is then equal to $\mathrm{NNI} - 1$). Following Farruggia *et al.* (2004), a linear response of N_{up} to NNI was considered:

$$N_{up} = N_{up}^{opt} + a_2(\mathrm{NNI} - 1) \tag{3}$$

where N_{up}^{opt} represents the N content of leaves exposed to incoming photosynthetically active radiation for a NNI of 1 and a_2 represents the dependency of upper leaf N content on plant N status. Similarly, the coefficient of N distribution relative to the light gradient was assumed to depend on NNI:

$$k_N = k_N^{opt} + a_3(\mathrm{NNI} - 1) \tag{4}$$

where k_N^{opt} represents the allocation coefficient for a NNI of 1 and a_3 represents the dependency of this coefficient on plant N status.

Leaf gas exchange model. The leaf gas exchange model is described in details in Prieto *et al.* (2012) and has originally been assessed on grapevine. It combines the biochemical photosynthetic model developed by Farquhar *et al.* (1980) with a semi-empirical stomatal conductance model that was originally proposed by Ball *et al.* (1987) and then modified by Leuning (1995). All the equations, variables and parameters are presented in Tables A1–A3. The coupling of this leaf gas exchange model with the previously presented canopy N distribution model was performed through the dependency of the principal photosynthesis parameters (value of V_{cmax}, J_{max}, triose phosphate utilization rate (TPU) and R_d at 25 °C) to N_a. A linear relationship was assumed [Eq. (A9)] (Harley *et al.* 1992; Le Roux *et al.* 1999; Braune *et al.* 2009).

Model calibration

Three experiments were carried out at the INRA Lusignan research station, France (46.43N, 0.18E), to calibrate this model and assess the impacts of light, N nutrition and leaf age on the distribution of leaf photosynthetic parameters in alfalfa (*Medicago sativa*). The three experiments were based on the same cultivars (cv. 'Orca' as a regular N fixing material, and cv. 'Agate NF' as a non-N_2-fixing material, Barnes *et al.* 1990).

Experiment 1. The first experiment was performed in a growth chamber between March and June 2010. All plants were grown in 1.5-L pots (10 × 20 cm cylindrical pots) filled with an N-free substrate (fine quartz sand, 0.8–1.4 mm mesh). The pots were arranged in a quincunx and two plants were transplanted into each pot, resulting in a planting density of ~230 plants m^{-2}. Three canopies comprising 81 pots each (i.e. 162 plants each) were grown under contrasting N availabilities at 22 °C/17 °C (day/night) under a 14-h photoperiod. The incident photosynthetic photon flux density (PPFD) was ~400 μmol m^{-2} s^{-1}. Each canopy was surrounded by a row of border plants grown under the same conditions. Two of these canopies were sown using the 'Orca' cultivar and were ferti-irrigated every 4 h (daily amount of 200 mL pot^{-1}) with either a complete nutrient solution (N+, 8 mmol N) or a low N nutrient solution (N−, 0.5 mmol N). The N concentration of the N+ solution was non-limiting for growth and prevented the nodulation of alfalfa roots. With the N− solution however, nodulation and N fixation did occur in the Orca cultivar (with natural strains of rhizobium, since the plants were not inoculated). The third canopy was sown with the 'Agate NF' cultivar grown with the N− nutrient solution, so that N fixation could not compensate for low mineral N availability. In order to induce a size hierarchy into the canopy, and to decorrelate the vertical position of leaves from their age, alternate rows were sown with a 17-day delay in each canopy. The study focussed on the initial growth period (no defoliation). Two samplings were performed in order to characterize the leaf N distribution. The plants were at the

12th visible leaves stage (40 days after the first sowing) and beginning of bloom stage (58 days after the first sowing), based on development of the Orca-N+ canopy. At each date, eight pots (16 plants) were collected from the centre of each canopy.

Experiment 2. The second experiment was performed outdoors between April and August 2009 using the 'Orca' cultivar. The average incident PPFD was ~725 μmol m^{-2} s^{-1}. All plants were grown in individual 1-L pots (5 × 52 cm cylindrical pots), resulting in a plant density of 460 plants m^{-2}. The canopy was made up of 100 study pots surrounded by 3 rows of border plants grown under the same conditions. All pots were filled with a growing medium that comprised sterile potting mix sand and clay-sandy-loam soil from a field in Lusignan (1 : 1 : 1, v/v). They were ferti-irrigated three times a day with the N+ nutrient solution. At the end of the second regrowth (beginning of bloom stage), 20 plants were sampled from the centre of the canopy for the characterization of leaf N distribution. The plants in this canopy had previously been shown to be highly size structured (Baldissera *et al.* 2014).

Experiment 3. The third experiment was carried out in a greenhouse between February and June 2012 using the 'Orca' cultivar. The average incident PPFD was ~540 μmol m^{-2} s^{-1}. All plants were grown in 1.1-L pots (10 × 10 × 11 cm) filled with an N-free substrate (fine quartz sand, 0.8–1.4 mm mesh). A single plant was transplanted into each pot, resulting in a density of 100 plants m^{-2}. Just after transplantation, the seedlings were inoculated with a solid commercial preparation for the coating of alfalfa seeds (*Sinorhizobium meliloti*, strain 2011, Becker Underwood). The pots were automatically ferti-irrigated five times a day with a complete nutrient solution devoid of mineral N (N0, 0 mmol N). The nutrient solution was sampled weekly to determine the absence of NO_3^- and NH_4^+ and ensure that N fixation was the only source of N supplied to the alfalfa plants. The experimental design consisted of 4 contiguous blocks of 49 pots each. At the end of the initial growth period (mid-bloom stage), four plants were sampled from the centre of each block in order to characterize the leaf N distribution.

Measurement of canopy N distribution and NNI. At each sampling date specified in the three experiments, each plant was separated into stems, flowers (when present) and leaves. The leaves were subdivided into 10 cm strata from the bottom to the top of the plant. The leaf area into each strata was determined using an LI-3100 planimeter (LI-COR, Lincoln, NE, USA). Plant samples were dried at 60 °C for 2 days, weighed to determine the dry mass and finally ground in a vibrating ball mill (MM400, Retsch GmbH and Co., Haan, Germany). Each sample was analysed with an elemental analyser to determine the N content (model EA 1108, Carlo Erba Instruments, Milan, Italy). For each canopy, the leaf area index (LAI) was calculated as the average leaf area of the collected plant multiplied by the plant density. The NNI was calculated as the ratio between the canopy N concentration (N_m) and the critical N concentration (N_c, Eq. 2).

Leaf photosynthetic capacity. In Experiments 1–3, gas exchanges were measured using the same procedure with a portable Licor 6400 photosynthesis system (LI-6400, LI-COR). The photosynthetic parameters were determined through the response of A to the internal CO_2 concentration (C_i) at the sub-stomatal level (A–C_i curves). Different levels of C_i were obtained by modifying the ambient CO_2 concentration (C_a) in the leaf measurement chamber. The A–C_i curves were determined as proposed by Long and Bernacchi (2003). Each C_a step was maintained for 5 min in order to record stable values. The three parameters (V_{cmax}, J_{max} and TPU) were estimated simultaneously by fitting the Farquhar model to the whole A–C_i curve according to the procedure proposed by Sharkey *et al.* (2007). All curves were determined at 1500 μmol m^{-2} s^{-1} of PPFD, while the leaf temperature was controlled at 25 °C and the vapour pressure deficit (VPD) between the leaf and the air was kept at 1 ± 0.5 kPa. Night respiration was estimated at the end of the night on a subsample of leaves that had previously been used for photosynthesis measurements (A–C_i curves). A different leaflet from the same leaf was used. Night respiration was considered to be equal to day respiration (R_d).

In each of the canopies studied, leaves were sampled at three to four levels from the bottom to the top of the canopy, just before the plants were collected to assess canopy N content and leaf N distribution. Measurements were made on the central leaflet of primary leaves. A total of 102 A–C_i curves were analysed over the 3 experiments (70, 20 and 12 for Experiments 1, 2 and 3, respectively).

Stomatal conductance. In Experiment 2, the daily evolution of leaf transpiration was recorded on leaves with contrasting positions within the canopy. Measurements were carried out during a series of sunny and cloudy days in summer. The parameters of the stomatal conductance model [Eq. (A10)] were estimated using the data obtained during a period which cumulated ~5 days of measurements.

Leaf traits. For each of the leaves on which A–C_i curves were determined, the leaf age was calculated by the thermal time difference (in °Cd) between the date of measurement and the date of leaf appearance. Thermal time was calculated from the daily integration of air temperatures minus the base temperature (5 °C). Immediately after the gas exchange measurements, the three leaflets were collected and scanned (Konica Minolta C352/C300, Konica Minolta Sensing, Osaka, Japan). The leaf area was determined using image analysis (ImageJ software, http://rsbweb.nih.gov/ij/). The leaves were then dried at 60 °C for 2 days, weighed to determine the dry mass and then ground in a vibrating ball mill (MM400, Retsch GmbH and Co.). Leaf samples were analysed with an elemental analyser (model EA 1108, Carlo Erba Instruments) to determine their N concentration. The specific leaf area (SLA, $m^2\ g^{-1}$), leaf N content per unit dry mass (%) and leaf N content per unit of area (N_a, $g\ N\ m^{-2}$) were then calculated.

Determination of local light conditions. In Experiment 1, the PPFD values at the top of the canopy and at the leaf level (for each leaf used for the $A-C_i$ characterizations) were measured using a portable LI-189 quantum meter (LI-COR). In addition, in all the canopies studied, the vertical distributions of leaf area measured at each sampling date were used to compute light extinction and average PPFD levels corresponding to each leaf stratum, using the RATP model (Sinoquet *et al.* 2001). A leaf angle distribution was derived from measurements of alfalfa architecture in Experiment 3 (see Barillot *et al.* 2011 for details).

Assessment of the leaf gas exchange model at the leaf level

A separate experiment was carried out between March and June 2011 according to the same design as Experiment 1. A data set of 10 leaves was used to assess the ability of the model to predict leaf N from canopy NNI and local leaf irradiance. These leaves were collected from the Orca-N+ and Orca-N− treatments (NNI values of 1.1 and 0.9, respectively) at the beginning of the bloom stage. To evaluate the ability of the model to simulate responses to rapid changes in environmental conditions, the sampled plants were placed outdoors and daily evolutions of the leaf gas exchange were recorded. Measurements were taken during sunny and cloudy days on leaves at different heights within the canopies. On very cloudy days, the plants were placed under a shelter to protect the material from the rain. A total of 14 days were analysed. The incident PPFD, leaf temperature, VPD and C_a were measured.

Assessment of the leaf gas exchange model at the whole-canopy level

The behaviour of the leaf gas exchange model when upscaled to the whole-canopy level was also assessed. The leaf N distribution was simulated for contrasting canopies (i.e. LAI values of 1.5, 3 and 5 $m^2\ m^{-2}$) at NNI values ranging from 0.3 to 1.4. Leaf area was assumed to be distributed homogeneously into eight vertical strata, and the leaf N concentration in each stratum was assumed to be acclimated to the relative light irradiance integrated over the day. A leaf angle distribution was derived from measurements of alfalfa architecture in Experiment 3 (see Barillot *et al.* 2011 for details). The light distribution within the canopy was calculated hourly using the RATP model (Sinoquet *et al.* 2001). Simulations were performed for contrasting days in the series used for validation at the leaf level. Net photosynthesis was calculated within each stratum and then summed to determine aboveground whole-canopy net gas exchanges.

Statistical analyses

Statistical analyses were performed using R software (https://www.r-project.org/). Curve fittings were realized with the nls procedure for Eq. 2 and with the lm procedure for linear regressions (Eqs 3 and 4). Analyses of covariance (ANCOVAs, lm procedure) were used to test for the effects of continuous and categorical variables simultaneously and to compare the slopes and intercepts of linear relationships between N concentration and photosynthetic parameters.

Predicted and measured values of leaf N concentration and net photosynthesis were compared using the root mean square error (RMSE) and bias (Bias) of the model, calculated as follows:

$$\mathrm{RMSE} = \sqrt{\frac{\sum_{i=1}^{n} (s_i - m_i)^2}{n}} \quad (5)$$

$$\mathrm{Bias} = \frac{\sum_{i=1}^{n} (s_i - m_i)}{n} \quad (6)$$

where s_i and m_i are the ith simulated and measured values, respectively, and n is the number of observations.

Results

Impact of NNI on the relationship between leaf irradiance and leaf N

The relationship between N_a and relative leaf irradiance was markedly affected by the N nutrition of the plants. Table 1 summarizes the parameters obtained by fitting Eq. 1 to the different N nutrition situations studied. Parameter N_{up} was the most affected, ranging from ~2.4 $g\ m^{-2}$ under N+ treatments to 0.8 $g\ m^{-2}$ in N− plants reliant on mineral N assimilation alone. It related linearly to the NNI of the plant stand (Fig. 1A; Eq. 3). Variations in N_{up}, thus, reflected variations in N nutrition and internal N availability. By comparison, parameter k_N, which accounted for N allocation with respect to relative leaf irradiance, displayed little

Table 1. Canopy characteristics and N distribution parameters determined during the different experiments and N treatments studied. Parameters were obtained by fitting Eq. 1 to the N content measured in leaves separated in 10-cm strata. Standard errors are indicated in brackets.

Experiment	Nutrient solution	N acquisition mode	LAI	N_{up}	k_N	r^2
1	N+	Assimilation	5.1	2.12 (0.075)	0.23 (0.013)	0.95
1	N+	Assimilation	2.6	2.31 (0.106)	0.24 (0.020)	0.96
2	N+	Assimilation	8.1	2.59 (0.123)	0.21 (0.021)	0.96
1	N−	Assimilation	1	1.09 (0.259)	0.20 (0.088)	0.64
1	N−	Assimilation	0.7	0.76 (0.130)	0.09 (0.312)	0.20
1	N−	Assimilation + fixation	3.3	1.77 (0.134)	0.15 (0.033)	0.87
1	N−	Assimilation + fixation	2.1	1.69 (0.072)	0.29 (0.031)	0.94
3	N0	Fixation	2.0	1.71 (0.181)	0.24 (0.054)	0.75

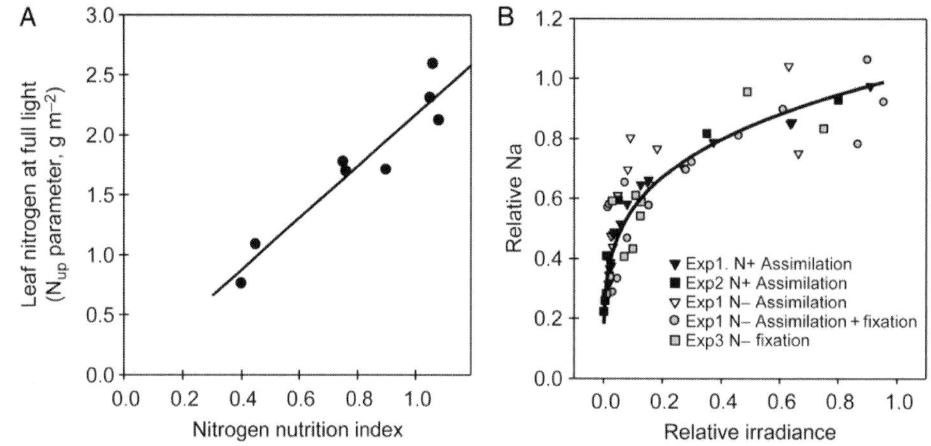

Figure 1. Relationships between (A) canopy NNI and leaf N concentration at the top of the canopy ($N_{up} = 2.15 \times NNI + 0.02$, $r^2 = 0.91$), and (B) relative leaf irradiance and leaf N concentration relative to the leaf N concentration at the top of the canopy ($k_N = 0.247$; $r^2 = 0.73$).

variation. For $k_N = 1$, the N gradient parallels the light gradient within the canopy. All observed values were clearly inferior to unity (<0.5), indicating a more-than-proportional N allocation to leaves with high irradiance and making the N concentration decrease more slowly than relative irradiance. Most k_N values were within the narrow range between 0.2 and 0.3 (except for one 0.09) with no clear relation to NNI. A single k_N parameter (0.25) enabled us to fit the normalized N_a distributions (Fig. 1B).

The effect of leaf age on N_a distributions was also assessed [**see Supporting Information—Figs S1 and S2**]. Due to an upward age gradient in alfalfa canopies, leaf nitrogen per unit area was related to both leaf age and local irradiance when considered separately. A multiple regression analysis confirmed the dominant effect of relative irradiance (t-value = 5.11; $P < 10^{-6}$), but showed a non-significant impact of leaf age *per se* and no interaction with irradiance (t-value = 1.54; $P > 0.12$ for the age term).

The canopy N distribution model was parameterized on the basis of these relationships ($a_2 = 2.15$, $N_{up}^{opt} = 2.17$, $a_3 = 0$ and $k_N^{opt} = 0.25$). **Supporting Information—Figure S3** shows the change in leaf N concentration as a function of relative leaf irradiance and NNI as predicted by this model.

Parameters of the photosynthetic and stomatal conductance model

Parameters V_{cmax}^{25}, J_{max}^{25}, TPU^{25} and R_d^{25} were related linearly to N_a (Fig. 2). The range of values observed for the different leaf parameters varied significantly between experiments, in relation to the minimum and maximum values taken by N_a (i.e. up to 2.8, 2, 1.8 and 1.1 g N m^{-2} in canopies relying on the N+ solution, N− solution and fixation, and fixation only and N− solution only, respectively). The range of variations in N_a values resulted from both the N nutrition of plants and leaf-to-leaf variations in the light microclimate. However, a single relationship

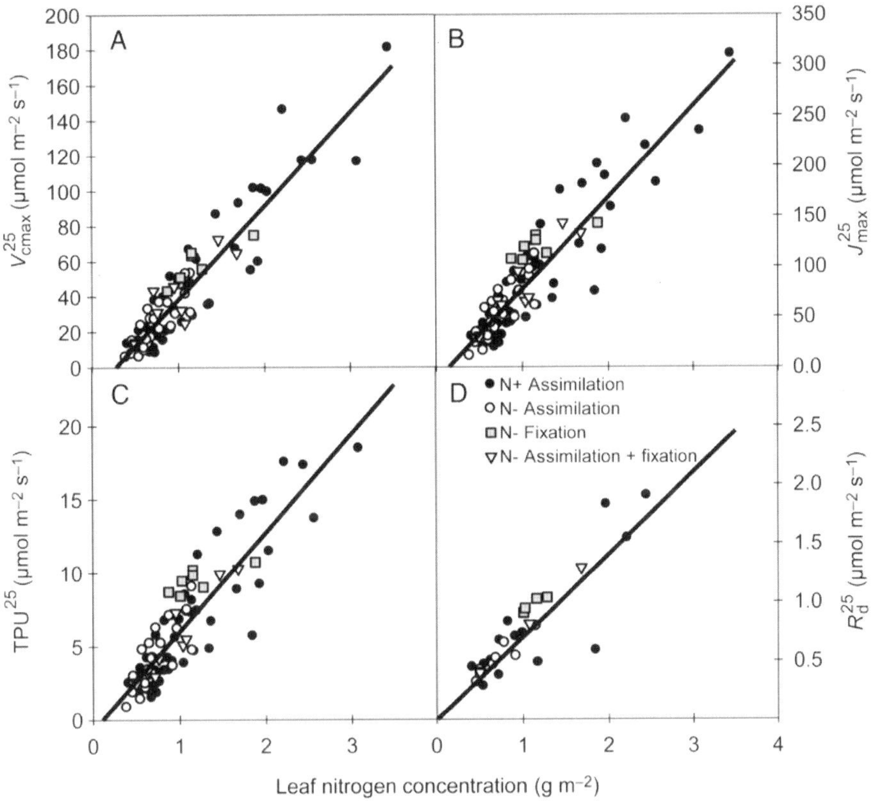

Figure 2. Relationship between the values of photosynthetic parameters at a leaf temperature of 25 °C and leaf N concentration (N_a) across the different experiments and N treatments studied. Linear relationships were found for V_{cmax}^{25} (A, $V_{cmax}^{25} = 53.03N_a - 14.74$, $r^2 = 0.86$), J_{max}^{25} (B, $J_{max}^{25} = 90.91N_a - 13.83$, $r^2 = 0.83$), TPU^{25} (C, $TPU^{25} = 6.72N_a - 0.72$, $r^2 = 0.78$) and R_d^{25} (D, $R_d^{25} = 0.69N_a - 0.005$, $r^2 = 0.77$), respectively.

was found for each parameter between P^{25} and N_a, independently of the N nutrition. No significant difference in the slopes (ANCOVA, *t*-value <0.91; $P > 0.38$ for interaction terms between P^{25} and N_a) and intercepts (ANCOVA, *t*-value <0.31, $P > 0.75$) were found between the N treatments. A larger dispersion of the points within the 'N+-assimilation' data set was observed. This was due to a difference between indoor and outdoor N_a values as shown by slightly higher intercepts in Experiment 2 (ANCOVA, *t*-value < -3.84, $P < 0.001$ for the intercept term).

An unique set of measurements from Experiment 2 was used to determine the stomatal conductance parameters (a_1, D_o). It contained leaves from different positions within the canopy and days with contrasting meteorological conditions. Parameter values are presented in Table A2.

Quantitative assessment of the N distribution model

Figure 3 compares the simulated N_a values (Eqs 1 and 3) with values measured on leaves at various heights within canopies grown under low and high N availability (NNI ranging from 0.45 to 1.1). Most inter-leaf variance in N_a values was explained by the N distribution model ($r^2 = 0.85$).

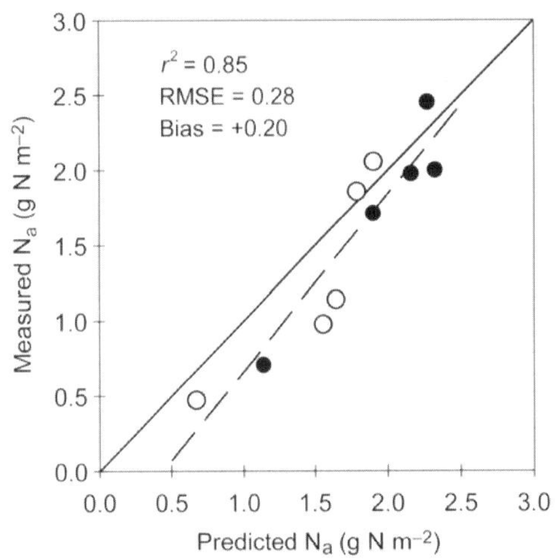

Figure 3. Relationship between leaf N concentrations (N_a) observed at various positions within the canopy and the corresponding values simulated. Open and filled symbols indicate canopies grown with N− and N+ nutrient solutions, respectively.

The model error remained low (RMSE = 0.28 g N m^{-2}), but a significant positive bias was observed. Predicted values of N_a appeared to be slightly higher on average (Bias = +0.20 g N m^{-2}), particularly in leaves at an intermediate height within the canopy.

Quantitative assessment of the leaf gas exchange model

The photosynthesis and transpiration sub-models were further assessed using directly measured leaf N_a. In a first step, the photosynthetic parameters (V_{cmax}^{25}, J_{max}^{25}, TPU^{25} and R_d^{25}) were calculated using measured N_a as an input. The gas exchange model was then run to simulate the daily patterns of *A* and *E* in a range of contrasting leaves (taken from various heights within canopies grown under low and high N availability). The model correctly predicted the diurnal patterns of *A* and *E* in various leaves under contrasting environmental conditions (Fig. 4). Cumulated over a day, the relationship between the observed and simulated values of *A* and *E* did not differ significantly from the 1 : 1 line (Fig. 5; $P < 0.05$). The model accurately estimated the diurnal patterns of *A* and its variation associated with climatic scenarios and leaves under high or low N status (RMSE = 0.04, no significant bias). The predictions also agreed satisfactorily for *E*, but the model errors were greater. Significant discrepancies were observed on *E* predictions for leaves with a high N content on sunny days (e.g. day of the year (DOY) 177 for a leaf at 1.96 g N m^{-2}, Fig. 4H). An underestimation of transpiration of up to 20 % was observed under such conditions. This bias did not result from unpaired temporal predictions at a particular time of the day, but from a general underestimation throughout the day.

Model predictions of the whole-canopy response to N availability

The behaviour of the model integrated at the whole-canopy level was assessed for canopies growing under a range of N availabilities. Examples of daily integrated canopy assimilation are presented in Fig. 6 for three contrasting days (DOY 176, 177 and 157 with an average PPFD decreasing from 709 to 610 and to 263 μmol m^{-2} s^{-1} and average air temperatures of 20.1, 26.3 and 17.9 °C, respectively). All canopies presented a saturating response curve to N availability. As expected, canopy assimilation was lower during cloudy days (Fig. 6A–C). Canopies with a LAI lower than that required for canopy closure (LAI below 3 m^2 m^{-2}) always displayed a lower assimilation rate per unit of soil area. Further increasing the LAI after canopy closure (LAI above 3 m^2 m^{-2}) did not improve canopy assimilation. The threshold at which canopy assimilation ceased to respond to N availability was very close to an NNI value of 1 for closed canopies during sunny and moderately cloudy days (Fig. 6D and E) and for open canopies during very cloudy days (Fig. 6F). Slight shifts of threshold were predicted, depending on the canopy LAI and light availability. Open canopies appeared to be more able to valorize high N availability and displayed delayed thresholds (e.g. at an NNI of ~1.2 on sunny days). In contrast, dense canopies presented anticipated thresholds that were particularly apparent on cloudy days.

Discussion

A simple empirical model to link leaf N distribution with plant N status and light distribution

To date, modelling the interaction between N limitations and light acclimation has been tackled using 'goal seeking' or optimal distribution theory (Chen *et al.* 1993; Thornley 1998; Johnson *et al.* 2010). Our study demonstrated how a combination of empirical relationships might be a promising option for this purpose too. The strategy proposed is based on modulation of the N_{up} and k_N parameters used in the empirical distribution model as a function of plant N status (NNI). A linear relationship was found between N_{up} and NNI over the range of alfalfa canopies studied. Similar results had previously been reported in different grass species, where the relationship was shown to be stable under contrasting growth conditions and canopy structures (Farruggia *et al.* 2004; Gastal *et al.* 2015). In these species, N_{up} has even been used as a routine proxy to facilitate the determination of NNI in the field (Louarn *et al.* 2015; Maamouri *et al.* 2015). The second parameter in the empirical relationship, k_N, was shown to be independent of NNI during the present study. Depending on the species, however, contradictory results have been reported concerning the effect of N limitation on k_N. In some cases, limited effects have been observed (Sinclair and Shiraiwa 1993; Sadras *et al.* 2012), whereas in others, a steeper N gradient has been found in N-stressed plants (Dreccer *et al.* 2000; Milroy *et al.* 2001). Moreau *et al.* (2012) suggested that the size of the canopy (indirectly reduced by N stress), rather than a direct NNI effect, might explain the steeper gradient in N-limited wheat canopies. In line with our results, Lemaire *et al.* (1991) did not show any variation of k_N in alfalfa canopies at contrasting developmental stages. Different types of plant architecture may affect N reallocation strategies and contribute to explaining these differences in the k_N response. Some species, such as alfalfa or sunflower (Archontoulis *et al.* 2011), are made up of leaves distributed in different strata along the vertical light gradient, and may adjust more efficiently than long-leaf species (such as grasses or cereals) in which each leaf may simultaneously experience light conditions from the bottom to the top of the canopy. In those cases, the parameter a_3 representing the dependency of k_N on plant N status (Eq. 3) is likely to take values different from zero.

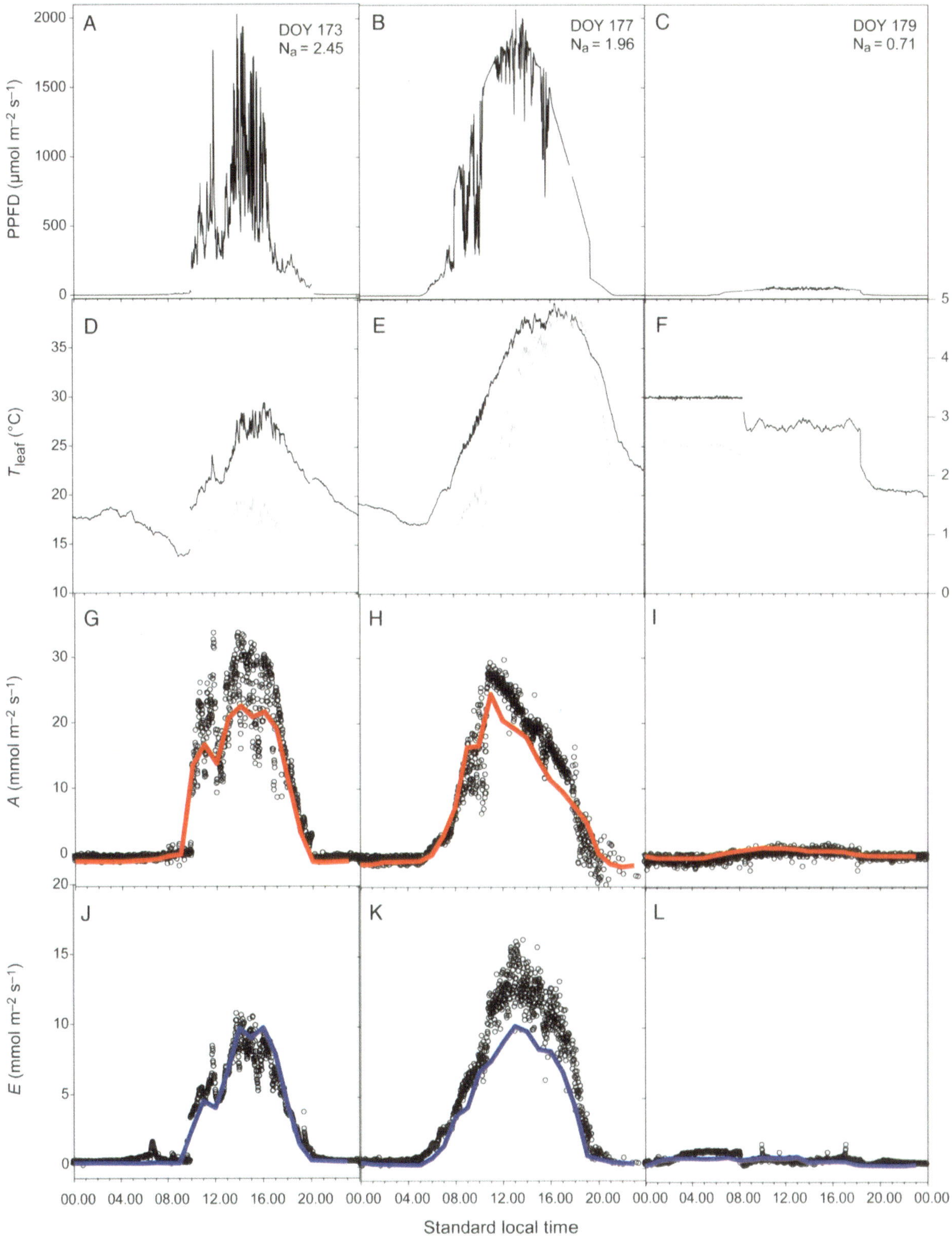

Figure 4. Measured instantaneous PPFD at the leaf level (A–C), leaf temperature (black) and VPD (grey, D–F), and the measured (open circles) and predicted (solid line) net photosynthesis (G–I), and transpiration rates (J–L), for three leaves in Lusignan in 2011. DOY, day of the year.

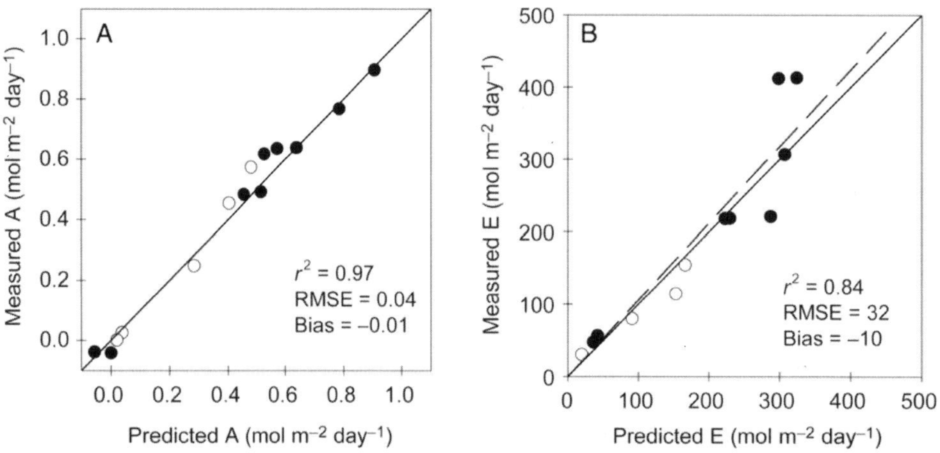

Figure 5. Comparison of measured and predicted values of daily net photosynthesis (A) and transpiration rates (B). Dashed lines: regressions between measured and predicted values; solid lines: 1 : 1 relationships. Open and filled symbols indicate leaves from canopies grown with N− and N+ nutrient solutions, respectively.

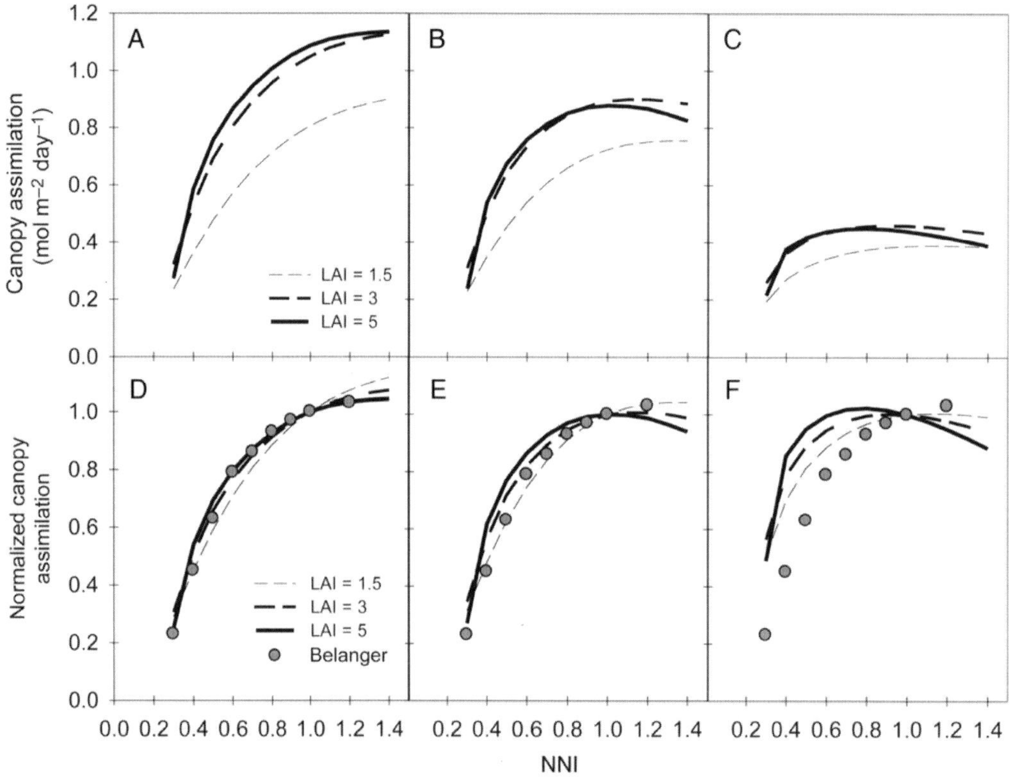

Figure 6. Simulations for three contrasting days of whole-canopy net assimilation in response to changes in the canopy NNI and LAI (A–C) and their corresponding responses normalized by the assimilation rate at an NNI value of 1 (D–F). Grey circles in the lower panels represent the relative reduction in radiation use efficiency measured by Bélanger *et al.* (1992) in response to NNI.

Assumptions and potential limitations of the leaf N distribution model

Species differ in the plasticity of their leaf traits and in the within-canopy variation of photosynthetic characteristics (Niinemets *et al.* 2015). The present model assumes that the distribution of leaf N is mainly driven by two factors: the light gradient within the canopy and the plant N status. No significant age effects were recorded in alfalfa, as previously shown in several other species (Evans 1993; Hikosaka *et al.* 1994). This is not a general feature however, and many plant species display age-dependent leaf traits, such as decreasing SLA in ageing leaves for instance. This can alter the light–N_a relationship and limit the validity of our model (Prieto *et al.* 2012). In their recent review, Niinemets *et al.* (2015) distinguished two main groups of species: a first group with high rates of canopy development and leaf turnover, exhibiting highly dynamic light environments, active change photosynthetic characteristics by N reallocation among leaves, and a second group made up of species with slow leaf turnover exhibiting a passive N_a acclimation response, primarily determined by the acclimation of leaf structure. The proposed model appears clearly best suited to the first group of species because they are less susceptible to leaf ageing effects.

Another limitation of empirical models is their validity out of their domain of calibration. Other environmental factors, such as water stress (Errecart *et al.* 2014), extreme temperatures or extreme light environments (as shown by the indoor/outdoor effect in our data set), can affect leaf growth and leaf traits. In some cases, this is likely to imply a reassessment model parameter values. Finally, the two-parameter model used [Eq. 1, **see Supporting Information—Table S1**] could present a lack of flexibility in some species. This formalism was previously used on other crops (Moreau *et al.* 2012; Sadras *et al.* 2012), but studies comparing a large number of species have generally relied on three-parameter models because they presented an overall better fit (Lötscher *et al.* 2003; Niinemets *et al.* 2015).

The N acquisition mode did not affect the photosynthetic parameters

Our study confirmed in alfalfa a linear relationship between Farquhar photosynthetic parameters (P^{25}) and leaf N per unit leaf area (Field 1983; Evans 1989) and showed that it holds true for leaves in different canopies grown under contrasting mineral N availabilities. Acclimation to light and plant N status both affected the N concentration of leaves, but the N_a–P^{25} relationships remained unchanged, as shown previously (Braune *et al.* 2009). In addition, our study examined the effects of the N acquisition mode in legumes, comparing fixing and non-fixing genotypes of alfalfa under different N nutrition statuses. There was no significant impact of the N acquisition mode on the N_a–P^{25} relationships. We thus showed that, contrary to the whole-plant level (Gosse *et al.* 1986), no extra cost to carbon acquisition was associated with N fixation at the leaf level (Boller and Heichel 1984). No gain, associated with an extra carbon sink, was observed either. Overall, the N_a–P^{25} relationships established during this study on a perennial forage legume displayed slopes (e.g. S_{Na} of the $N_a - V_{cmax}^{25}$ relationship at 53 μmol g N^{-1} s^{-1}) that were intermediate between those of cereals (60 and 63 μmol g N^{-1} s^{-1} for wheat and barley, respectively; Müller *et al.* 2005; Braune *et al.* 2009) and those of C3 trees and vines (e.g. ~30 and 38 μmol g N^{-1} s^{-1} in walnut tree and grapevine, respectively; Le Roux *et al.* 1999; Prieto *et al.* 2012). This positioning was consistent with other productive grassland species (e.g. 36–50 μmol g N^{-1} s^{-1} reported for cocksfoot and red clover; Wohlfahrt *et al.* 1998).

Performance of the leaf gas exchange model at the leaf and whole-canopy levels

The gas exchange model correctly estimated daily cumulated values of net assimilation (A) and transpiration (E) at the leaf level and their diurnal patterns. Model errors, however, were greater with respect to transpiration. This might partly be related to the absence of a direct relationship between the leaf N and stomatal conductance parameters considered in the model. Indeed, the scaling parameter a_1 has been shown to increase for leaves with a very low N concentration (Braune *et al.* 2009). In our case, however, discrepancies in E mainly concerned the top leaves under high N availability. An alternative explanation might be a less robust parameterization of the stomatal conductance model. Leaves from a single experiment were used, covering a more limited range of environmental conditions than that encountered in the validation data set (in terms of VPD in particular). Future work will incorporate the response to water deficit in the model, which should enable the refinement of this parameterization.

When upscaled at the whole-canopy level, the gas exchange model coupled with the N distribution model displayed interesting properties regarding the response to N availability. The relationship between N and the assimilation rate switched from a linear function at the leaf level to a saturating function at the whole-canopy scale. Above a certain threshold, the model predicted that an increase in N did not result in increased canopy assimilation. Such a point had previously been reported during numerous experiments comparing a broad range of N fertilization rates (Justes *et al.* 1994; Lemaire and Gastal 1997), and it defines the critical N concentration on which NNI calculations are based. Remarkably, such behaviour emerged from our canopy-integrated model. Furthermore, the NNI value corresponding to this transition happened to be very close to 1 during sunny and moderately cloudy days. This complied with the theoretical definition of NNI, which states that a value of 1 corresponds to the critical N concentration. The threshold was predicted to be lower during cloudy days. However, as in practice, the critical N concentrations are determined from cumulated values of biomass production, and critical N is likely to be primarily driven by sunny days (accounting for most biomass accumulation) rather than cloudy days.

No direct measurements were carried out to quantitatively assess the gas exchange model on alfalfa canopies. However, the simulated plateau values were consistent with previous studies measuring the daily net carbon exchange in closed canopies of alfalfa under non-limiting N. For instance, Heichel *et al.* (1988) reported net rates of 1.17, 0.81 and 0.45 mol CO_2 m^{-2} day^{-1} in a 2.7 LAI canopy during days with average PPFD values of 1100, 700 and 400 μmol m^{-2} s^{-1}, respectively (corresponding roughly to the light conditions prevailing during the 3 days shown in Fig. 6). Similarly, Woodward and Sheehy (1979) reported rates ranging from 0.27 to 1.39 mol CO_2 m^{-2} day^{-1} after canopy closure during a spring regrowth. As in these two studies, the model outputs concerned the net aboveground carbon exchange of N fertilized alfalfa (with presumably limited N fixation). Allocation to the root system and to the respiration of roots and nodules needs to be implemented in future versions of the model in order to account for a potential cost of N fixation in terms of the carbon balance at the canopy level (Gosse *et al.* 1986). Root growth and respiration indeed represent a substantial share of carbohydrate use in fixing alfalfa (Thomas and Hill 1937; Layzell *et al.* 1981).

Fewer references were available to compare the response induced by N limitation. We, thus, compared the simulation results with those relative to regular non-fixing C3 plants and found a relative reduction in the net assimilation rate measured by Bèlanger *et al.* (1992) in a range of tall fescue canopies (Fig. 6). The simulated response curve of dense canopies on sunny and moderately cloudy days appeared to be particularly close to the measured curve, suggesting a good ability of the upscaled leaf model to capture the N stress response of a whole canopy.

Conclusions

Overall, the set of empirical relationships introduced in this article to distribute leaf N was simple and effective at predicting leaf N concentration in response to light and plant N status. The leaf gas exchange model proved accurate and produced consistent predictions in terms of whole-canopy assimilation under contrasting soil N availability scenarios. Even if the genericity of the coupled model still needs to be challenged in a broader range of species, this work constitutes a further step towards models that can bridge local acclimation to light with N acquisition and global N nutrition status, without presuming an optimal carbon gain or N distribution. Such a model relies on parameters that can all be directly measured and may help us to infer and better understand the differences in N use efficiency observed between species or genotypes.

Sources of Funding

This study received support from the Regional Council for Poitou-Charentes (http://www.poitou-charentes.fr) and INRA's Environment and Agronomy Division (INNI and TransfertN projects).

Contributions by the Authors

G.L., E.L. and J.P. contributed to model development. G.L. and E.F. designed the experiments and conducted measurements. They performed data analyses with the help of S.Z. All of the authors contributed to writing the manuscript.

Acknowledgements

We thank JoAnn Lamb, USDA, St Paul, MN for supplying us with the AgateNF lucerne seeds. We also would like to thank H. Bobille, S. Denis, A. Eprinchard, L. Fernandez, N. Moynet, A. Philiponneau, E. Rivault, N. Romero and J.P. Terrasson for their assistance with the experiments.

Supporting Information

The following additional information is available in the online version of this article –

Figure S1. Impact of leaf ageing and relative leaf irradiance on the specific leaf area and leaf nitrogen concentration per unit area in alfalfa.

Figure S2. Relationship between leaf age and the residuals of the fit of Eq. 1 to leaf nitrogen concentration.

Figure S3. Variations in leaf nitrogen concentrations predicted for alfalfa as a function of the nitrogen nutrition index of the canopy and the relative leaf irradiance.

Table S1. Comparison of two- and three-parameter models to account for distribution of leaf nitrogen concentrations with respect to relative irradiance in alfalfa.

Literature Cited

Anten NPR, Schieving F, Werger MJA. 1995. Patterns of light and nitrogen distribution in relation to whole canopy carbon gain in C_3 and C_4 mono- and dicotyledonous species. *Oecologia* **101**: 504–513.

Archontoulis SV, Vos J, Yin X, Bastiaans L, Danalatos NG, Struik PC. 2011. Temporal dynamics of light and nitrogen vertical distributions in canopies of sunflower, kenaf and cynara. *Field Crops Research* **122**:186–198.

Baldissera TC, Frak E, de Faccio Carvalho PC, Louarn G. 2014. Plant development controls leaf area expansion in alfalfa plants competing for light. *Annals of Botany* **113**:145–157.

Ball MC, Woodrow IE, Berry JA. 1987. A model predicting stomatal conductance and its contribution to the control of photosynthesis under different environmental conditions. In: Biggins J, ed. *Progress in photosynthesis research*. Netherlands: Springer, 221–224.

Barillot R, Louarn G, Escobar-Gutiérrez AJ, Huynh P, Combes D. 2011. How good is the turbid medium-based approach for accounting for light partitioning in contrasted grass-legume intercropping systems? *Annals of Botany* **108**:1013–1024.

Barnes D, Heichel G, Vance C, Peaden R. 1990. Registration of 'ineffective agate' and 'ineffective saranac' non-N_2-fixing alfalfa germplasms. *Crop Science* **30**:752–753.

Bélanger G, Gastal F, Lemaire G. 1992. Growth analysis of a tall fescue sward fertilized with different rates of nitrogen. *Crop Science* **32**: 1371–1376.

Bernacchi CJ, Singsaas EL, Pimentel C, Portis AR Jr, Long SP. 2001. Improved temperature response functions for models of Rubisco-limited photosynthesis. *Plant, Cell and Environment* **24**: 253–259.

Bernacchi CJ, Pimentel C, Long SP. 2003. In vivo temperature response functions of parameters required to model RuBP-limited photosynthesis. *Plant, Cell and Environment* **26**: 1419–1430.

Boller BC, Heichel GH. 1984. Canopy structure and photosynthesis of alfalfa genotypes differing in nodule effectiveness. *Crop Science* **24**:91–96.

Braune H, Müller J, Diepenbrock W. 2009. Integrating effects of leaf nitrogen, age, rank, and growth temperature into the photosynthesis-stomatal conductance model LEAFC3-N parameterised for barley (*Hordeum vulgare* L.). *Ecological Modelling* **220**: 1599–1612.

Brisson N, Launay M, Mary B, Beaudoin N. 2009. *Conceptual basis, formalisations and parameterization of the STICS crop model*. QUAE Editions, Collection "Update Science and Technologies", Versailles, FR. 297 pp.

Charles-Edwards DA. 1981. The mathematics of photosynthesis and productivity. Academic Press, London, 127 pp.

Chen JL, Reynolds JF, Harley PC, Tenhunen JD. 1993. Coordination theory of leaf nitrogen distribution in a canopy. *Oecologia* **93**: 63–69.

Colnenne C, Meynard JM, Reau R, Justes E, Merrien A. 1998. Determination of a critical nitrogen dilution curve for winter oilseed rape. *Annals of Botany* **81**:311–317.

De Pury DGG, Farquhar GD. 1997. Simple scaling of photosynthesis from leaves to canopies without the errors of big-leaf models. *Plant, Cell and Environment* **20**:537–557.

Dreccer MF, Van Oijen M, Schapendonk AHCM, Pot CS, Rabbinge R. 2000. Dynamics of vertical leaf nitrogen distribution in a vegetative wheat canopy. Impact on canopy photosynthesis. *Annals of Botany* **86**:821–831.

Errecart PM, Agnusdei MG, Lattanzi FA, Marino MA, Berone GD. 2014. Critical nitrogen concentration declines with soil water availability in tall fescue. *Crop Science* **54**:318–330.

Evans JR. 1989. Photosynthesis and nitrogen relationships in leaves of C_3 plants. *Oecologia* **78**:9–19.

Evans JR. 1993. Photosynthetic acclimation and nitrogen partitioning within a lucerne canopy. I. Canopy characteristics. *Australian Journal of Plant Physiology* **20**:55–67.

Evers JB, Vos J, Yin X, Romero P, Van der Putten PEL, Struik PC. 2010. Simulation of wheat growth and development based on organ-level photosynthesis and assimilate allocation. *Journal of Experimental Botany* **61**:2203–2216.

Farquhar GD, Von Caemmerer S, Berry JA. 1980. A biochemical model of photosynthetic CO_2 assimilation in leaves of C_3 species. *Planta* **149**:78–90.

Farruggia A, Gastal F, Scholefield D. 2004. Assessment of the nitrogen status of grassland. *Grass and Forage Science* **59**:113–120.

Field C. 1983. Allocating leaf nitrogen for the maximization of carbon gain: leaf age as a control on the allocation program. *Oecologia* **56**:341–347.

Field CB, Mooney HA. 1986. The nitrogen/photosynthesis relationship in wild plants. In: Givnish TJ, ed. *On the economy of plant form and function*. Cambridge, UK: Cambridge University Press, 25–55.

Gastal F, Lemaire G, Durand JL, Louarn G. 2015. Quantifying crop responses to nitrogen and avenues to improve nitrogen-use efficiency. In: Sadras VO, Calderini D, eds. *Crop physiology—applications for genetic improvement and agronomy*, 2nd edn. London, UK: Academic Press, Elsevier, 161–206.

Givnish TJ. 1988. Adaptation to sun and shade: a whole-plant perspective. *Australian Journal of Plant Physiology* **15**:63–92.

Gosse G, Varlet-Grancher C, Bonhomme R, Chartier M, Allirand JM, Lemaire G. 1986. Production maximale de matière sèche et rayonnement solaire intercepté par un couvert végétal. *Agronomie* **6**:47–56.

Greenwood DJ, Lemaire G, Gosse G, Cruz P, Draycott A, Neeteson JJ. 1990. Decline in percentage N of C_3 and C_4 crops with increasing plant mass. *Annals of Botany* **66**:425–436.

Harley PC, Thomas RB, Reynolds JF, Strain BR. 1992. Modelling photosynthesis of cotton grown in elevated CO_2. *Plant, Cell and Environment* **15**:271–282.

Heichel GH, Delaney RH, Cralle HT. 1988. Carbon assimilation, partitioning and utilization. In: Hanson AA, Barnes DK, Hill RR, eds. *Alfalfa and alfalfa improvement*. Madison, WI: American Society of Agronomy, 195–228.

Hikosaka K. 2004. Interspecific difference in the photosynthesis-nitrogen relationship: patterns, physiological causes, and ecological importance. *Journal of Plant Research* **117**:481–494.

Hikosaka K, Terashima I, Katoh S. 1994. Effects of leaf age, nitrogen nutrition and photon flux density on the distribution of nitrogen among leaves of a vine (*Ipomoea tricolor* Cav.) grown horizontally to avoid mutual shading of leaves. *Oecologia* **97**:451–457.

Hirose T, Werger MJA. 1987. Maximizing daily canopy photosynthesis with respect to the leaf nitrogen allocation pattern in the canopy. *Oecologia* **72**:520–526.

Johnson IR, Thornley JHM, Frantz JM, Bugbee B. 2010. A model of canopy photosynthesis incorporating protein distribution through the canopy and its acclimation to light, temperature and CO_2. *Annals of Botany* **106**:735–749.

Justes E, Mary B, Meynard J-M, Machet JM, Thelier-Huché L. 1994. Determination of a critical nitrogen dilution curve for winter wheat crops. *Annals of Botany* **74**:397–407.

Kull O. 2002. Acclimation of photosynthesis in canopies: models and limitations. *Oecologia* **133**:267–279.

Kull O, Jarvis PG. 1995. The role of nitrogen in a simple scheme to scale up photosynthesis from leaf to canopy. *Plant, Cell and Environment* **18**:1174–1182.

Layzell DB, Pate JS, Atkins CA, Canvin DT. 1981. Partitioning of carbon and nitrogen and the nutrition of root and shoot apex in a nodulated legume. *Plant Physiology* **67**:30–36.

Lemaire G, Gastal F. 1997. N uptake and distribution in plant canopies. In: Lemaire G, ed. *Diagnosis of the nitrogen status in crops*. Heidelberg: Springer, 3–43.

Lemaire G, Salette J, Sigogne M, Terrasson J-P. 1984. Relation entre dynamique de croissance et dynamique de prélèvement d'azote pour un peuplement de graminées fourragères. I.—Etude de l'effet du milieu. *Agronomie* **4**:423–430.

Lemaire G, Cruz P, Gosse G, Chartier M. 1985. Etude des relations entre la dynamique de prélèvement d'azote et la dynamique de croissance en matière sèche d'un peuplement de luzerne (*Medicago sativa* L.). *Agronomie* **5**:685–692.

Lemaire G, Onillon B, Gosse G, Chartier M, Allirand JM. 1991. Nitrogen distribution within a lucerne canopy during regrowth: relation with light distribution. *Annals of Botany* **68**:483–488.

Lemaire G, Avice J-C, Kim T-H, Ourry A. 2005. Developmental changes in shoot N dynamics of lucerne (*Medicago sativa* L.) in relation to leaf growth dynamics as a function of plant density and hierarchical position within the canopy. *Journal of Experimental Botany* **56**:935–943.

Le Roux X, Grand S, Dreyer E, Daudet FA. 1999. Parameterization and testing of a biochemically based photosynthesis model for walnut (*Juglans regia*) trees and seedlings. *Tree Physiology* **19**:481–492.

Leuning R. 1995. A critical appraisal of a combined stomatal-photosynthesis model for C_3 plants. *Plant, Cell and Environment* **18**:339–355.

Long SP, Bernacchi CJ. 2003. Gas exchange measurements, what can they tell us about the underlying limitations to photosynthesis? Procedures and sources of error. *Journal of Experimental Botany* **54**:2393–2401.

Lötscher M, Stroh K, Schnyder H. 2003. Vertical leaf nitrogen distribution in relation to nitrogen status in grassland plants. *Annals of Botany* **92**:679–688.

Louarn G, Pereira-Lopès E, Fustec J, Mary B, Voisin A-S, De Faccio Carvalho PC, Gastal F. 2015. The amounts and dynamics of nitrogen transfer to grasses differ in alfalfa and white clover-based grass-legume mixtures as a result of rooting strategies and rhizodeposit quality. *Plant and Soil* **389**:289–305.

Maamouri A, Louarn G, Gastal F, Béguier V, Julier B. 2015. Effects of lucerne genotype on morphology, biomass production and nitrogen content of lucerne and tall fescue in mixed pastures. *Crop and Pasture Science* **66**:192–204.

Milroy SP, Bange MP, Sadras VO. 2001. Profiles of leaf nitrogen and light in reproductive canopies of cotton (*Gossypium hirsutum*). *Annals of Botany* **87**:325–333.

Moreau D, Allard V, Gaju O, Le Gouis J, Foulkes MJ, Martre P. 2012. Acclimation of leaf nitrogen to vertical light gradient at anthesis in wheat is a whole-plant process that scales with the size of the canopy. *Plant Physiology* **160**:1479–1490.

Müller J, Wernecke P, Diepenbrock W. 2005. LEAFC3-N: a nitrogen-sensitive extension of the CO_2 and H_2O gas exchange model LEAFC3 parameterised and tested for winter wheat (*Triticum aestivum* L.). *Ecological Modelling* **183**:183–210.

Niinemets Ü, Keenan TF, Hallik L. 2015. A worldwide analysis of within-canopy variations in leaf structural, chemical and physiological traits across plant functional types. *New Phytologist* **205**: 973–993.

Pettersson R, Mcdonald AJS. 1994. Effects of nitrogen supply on the acclimation of photosynthesis to elevated CO_2. *Photosynthesis Research* **39**:389–400.

Plénet D, Lemaire G. 1999. Relationships between dynamics of nitrogen uptake and dry matter accumulation in maize crops. Determination of critical N concentration. *Plant and Soil* **216**: 65–82.

Prieto JA, Louarn G, Perez Peña J, Ojeda H, Simonneau T, Lebon E. 2012. A leaf gas exchange model that accounts for intra-canopy variability by considering leaf nitrogen content and local acclimation to radiation in grapevine (*Vitis vinifera* L.). *Plant, Cell and Environment* **35**:1313–1328.

Sadras VO, Lawson C, Montoro A. 2012. Photosynthetic traits in Australian wheat varieties released between 1958 and 2007. *Field Crops Research* **134**:19–29.

Sands PJ. 1995. Modelling canopy production. II. From single-leaf photosynthesis parameters to daily canopy photosynthesis. *Australian Journal of Plant Physiology* **22**:603–614.

Schultz HR. 2003. Extension of a Farquhar model for limitations of leaf photosynthesis induced by light environment, phenology and leaf age in grapevines (*Vitis vinifera* L. cvv. White Riesling and Zinfandel). *Functional Plant Biology* **30**:673–687.

Sellers PJ, Berry JA, Collatz GJ, Field CB, Hall FG. 1992. Canopy reflectance, photosynthesis, and transpiration. III. A reanalysis using improved leaf models and a new canopy integration scheme. *Remote Sensing of Environment* **42**:187–216.

Sharkey TD, Bernacchi CJ, Farquhar GD, Singsaas EL. 2007. Fitting photosynthetic carbon dioxide response curves for C_3 leaves. *Plant, Cell and Environment* **30**:1035–1040.

Sinclair TR, Shiraiwa T. 1993. Soybean radiation-use efficiency as influenced by nonuniform specific leaf nitrogen distribution and diffuse radiation. *Crop Science* **33**:808–812.

Sinoquet H, Le Roux X, Adam B, Ameglio T, Daudet FA. 2001. RATP: a model for simulating the spatial distribution of radiation absorption, transpiration and photosynthesis within canopies: application to an isolated tree crown. *Plant, Cell and Environment* **24**: 395–406.

Thomas MD, Hill GR. 1937. The continuous measurement of photosynthesis, respiration, and transpiration of alfalfa and wheat growing under field conditions. *Plant Physiology* **12**:285–307.

Thornley JHM. 1998. Dynamic model of leaf photosynthesis with acclimation to light and nitrogen. *Annals of Botany* **81**: 421–430.

Wohlfahrt G, Bahn M, Horak I, Tappeiner U, Cernusca A. 1998. A nitrogen sensitive model of leaf carbon dioxide and water vapour gas exchange: application to 13 key species from differently managed mountain grassland ecosystems. *Ecological Modelling* **113**: 179–199.

Woodward FI, Sheehy JE. 1979. Microclimate, photosynthesis and growth of lucerne (*Medicago sativa* L.). II. Canopy structure and growth. *Annals of Botany* **44**:709–719.

Appendix

Table A1. Equations for the photosynthesis and stomatal conductance models.

Equation	Description	No.
Photosynthesis model		
$A = V_c - 0.5V_o - R_d = V_c[1-(\Gamma^*/C_i)] - R_d$	Net photosynthetic rate (μmol CO_2 m^{-2} s^{-1})	(A1)
$V_c = \min\{A_c, A_j, A_p\}$	Carboxylation rate (μmol CO_2 m^{-2} s^{-1})	(A2)
$A_c = (V_{cmax}C_i)/[C_i + K_c(1+(O/K_o))]$	RUBISCO-limited photosynthetic rate	(A3)
$A_j = (JC_i)/(4C_i + 8\Gamma^*)$	RuBP regeneration-limited photosynthetic rate	(A4)
$A_p = (3\text{TPU})/[1-(\Gamma^*/C_i)]$	TPU-limited photosynthetic rate	(A5)
$J = (\alpha \cdot \text{PPFD})/\sqrt{1+[(\alpha^2 \cdot \text{PPFD}^2)/J^2_{max}]}$	Electron transport rate, dependence on the radiance level	(A6)
$P = e^{(c-(\Delta H_a/RT_k))}$	Arrhenius function, temperature dependence for K_c, K_o, Γ^* and R_d	(A7)
$P = e^{(c-(\Delta H_a/RT_k))}/(1+e^{(\Delta ST_k-(\Delta H_d/RT_k))})$	Arrhenius function, temperature dependence for V_{cmax}, J_{max} and TPU	(A8)
$P^{25} = S_{Na} \cdot N_a - b$	Nitrogen dependence function for V_{cmax}, J_{max}, TPU and R_d at 25 °C	(A9)
Stomatal conductance model		
$g_s = g_o + (a_1A)/([1+(\text{VPD}/D_o)](C_s - \Gamma))$	Stomatal conductance	(A10)
$C_s = C_a\text{-} A \cdot (1.37/g_b)$	CO_2 partial pressure at the leaf surface	(A11)
$C_i = C_a\text{-} A \cdot [(1.6/g_s) + (1.37/g_b)]$	C_i value by coupling A and g_s	(A12)

Table A2. Symbols, values and units of different parameters, variables and constants used in the photosynthetic and stomatal conductance models. [1]Values taken from Schultz (2003). [2]Values taken from Harley *et al.* (1992). [3]Constant used for measurement in the leaf chamber of the LcPro (ADC Lcpro, BioScientific Ltd, Hoddesdon, Hertfordshire, UK).

Symbol	Value	Unit	Description
Photosynthesis model			
α	0.20[1]	μmol CO_2 μmol photon^{-1}	Photochemical efficiency or initial quantum yield
Γ^*	–	Pa	Compensation point for CO_2 in the absence of mitochondrial respiration
A	–	μmol CO_2 m^{-2} s^{-1}	Net photosynthetic rate
A_c	–	μmol CO_2 m^{-2} s^{-1}	RUBISCO-limited photosynthetic rate
A_j	–	μmol CO_2 m^{-2} s^{-1}	Electron transport rate-limited photosynthetic rate
A_p	–	μmol CO_2 m^{-2} s^{-1}	Triose phosphate utilization-limited photosynthetic rate
c	–	–	Scaling constant
C_a	–	Pa	Ambient CO_2 partial pressure
C_i	–	Pa	Intercellular CO_2 partial pressure
ΔH_a	–	kJ mol^{-1}	Enthalpy of activation
ΔH_d	200[2]	kJ mol^{-1}	Enthalpy of deactivation
K_c	–	Pa	Michaelis–Menten constant of RUBISCO for CO_2
K_o	–	kPa	Michaelis–Menten constant of RUBISCO for O_2
J	–	μmol electron m^{-2} s^{-1}	Electron transport rate

Continued

Table A2. *Continued*

Symbol	Value	Unit	Description
J_{max}	–	$\mu mol\ m^{-2}\ s^{-1}$	Maximum electron transport rate
N_a	–	$g\ m^{-2}$	Area based N content
N_{amin}	–	$g\ m^{-2}$	Minimum value of N_a at which $P^{25} \rightarrow 0$
O	21	kPa	Oxygen partial pressure
P^{25}	–	$\mu mol\ m^{-2}\ s^{-1}$	Value of V_{cmax}, J_{max}, TPU or R_d at 25 °C
PPFD	–	$\mu mol\ m^{-2}\ s^{-1}$	Photosynthetic photon flux density
R	0.00831	$kJ\ mol^{-1}\ K^{-1}$	Universal gas constant for perfect gases
R_d	–	$\mu mol\ m^{-2}\ s^{-1}$	Mitochondrial respiration in light
ΔS	0.635[2]	$kJ\ mol^{-1}$	Entropy term
S_{Na}	–	$\mu mol\ g^{-1}\ s^{-1}$	Slope of the relationship between N_a and V_{cmax}, J_{max}, TPU or R_d
T_{leaf}	–	°C	Leaf temperature in degrees Celsius
T_k	–	Kelvin degrees	Leaf temperature in Kelvin
TPU	–	$\mu mol\ m^{-2}\ s^{-1}$	Triose phosphate utilization rate
V_c	–	$\mu mol\ m^{-2}\ s^{-1}$	Carboxylation rate
V_o	–	$\mu mol\ m^{-2}\ s^{-1}$	Oxygenation rate
V_{cmax}	–	$\mu mol\ m^{-2}\ s^{-1}$	Maximum rate of RUBISCO carboxylation
Stomatal conductance model			
C_s	–	Pa	CO_2 partial pressure at the leaf surface
g_b	2.357[3]	$mol\ m^{-2}\ s^{-1}$	Boundary layer conductance
g_s	–	$mmol\ m^{-2}\ s^{-1}$	Stomatal conductance
g_o	0.020	$mmol\ m^{-2}\ s^{-1}$	Residual stomatal conductance when $A \rightarrow 0$
VPD	–	kPa	Water VPD
D_o	2.86	kPa	Empirical factor assessing stomata sensitivity to VPD
a_1	12.5	–	Empirical stomatal conductance factor

Table A3. Values of c (scaling constant), enthalpies of activation (ΔH_a) describing the temperature response for parameters of the photosynthesis model. [1]Values taken from Bernacchi *et al.* (2001). [2]Values taken from Bernacchi *et al.* (2003). [3]Values taken from Harley *et al.* (1992).

Parameter	Value at 25 °C	c	ΔH_a ($kJ\ mol^{-1}$)
V_{cmax}	–	26.35[1]	65.33[1]
J_{max}	–	17.7[2]	43.9[2]
TPU	–	21.46[3]	53.1[3]
R_d	–	18.72[1]	46.39[1]
Γ^*	42.75[1]	19.02[1]	37.83[1]
K_c	404.9[1]	38.05[1]	79.43[1]
K_o	278.4[1]	20.30[1]	36.38[1]

11

Root hydraulic conductivity and adjustments in stomatal conductance: hydraulic strategy in response to salt stress in a halotolerant species

Victoria Vitali[1], Jorge Bellati[1], Gabriela Soto[2], Nicolás D. Ayub[2] and Gabriela Amodeo[1*]

[1] Departamento de Biodiversidad y Biología Experimental, Facultad de Ciencias Exactas y Naturales, Instituto de Biodiversidad y Biología Experimental, Universidad de Buenos Aires and Consejo Nacional de Investigaciones Científicas y Técnicas, C1428EGA Buenos Aires, Argentina

[2] Instituto de Genética "Ewald A. Favret", CICVyA, INTA-Castelar and Consejo Nacional de Investigaciones Científicas y Técnicas, 1686 Buenos Aires, Argentina

Associate Editor: Tim J. Brodribb

Abstract. Recent advances at the molecular level are introducing a new scenario that needs to be integrated into the analysis of plant hydraulic properties. Although it is not yet clear to what extent this scenario alters the current proposal for the hydraulic circuit models, it introduces new insights when studying plants that are able to easily overcome water restrictions. In this context, our aim was to explore water adjustments in a halotolerant model (*Beta vulgaris*) by studying the coordination between the root in terms of root hydraulic conductivity (L_{pr}) and the shoot as reflected in the stomatal conductance (g_s). The root water pathways were also analysed in terms of root suberization (apoplastic barrier) and aquaporin transcript levels (cell-to-cell pathway). *Beta vulgaris* showed the ability to rapidly lose (4 h) and gain (24 h) turgor when submitted to salt stress (200 mM). The reduction profile observed in L_{pr} and g_s was consistent with a coupled process. The tuning of the root water flow involved small variations in the studied aquaporin's transcripts before anatomical modifications occurred. Exploring L_{pr} enhancement after halting the stress contributed to show not only a different profile in restoring L_{pr} but also the capacity to uncouple L_{pr} from g_s. *Beta vulgaris* root plays a key role and can anticipate water loss before the aerial water status is affected.

Keywords: Aquaporins; *Beta vulgaris*; root hydraulic conductivity; salt stress; soil–plant–atmosphere continuum; stomatal conductance; suberization; water relations.

Introduction

Water flow through plants has been described as a passive mechanism (diffusion and bulk flow) based on the analogy with Ohm's law (Van den Honert 1948). The movement of water along a hydraulic circuit with resistances (R, m^{-3} s MPa) to the water flow at the root, shoot and canopy levels is known as the soil–plant–atmosphere continuum or SPAC (Tardieu and Davies 1993; Suku *et al.* 2014). Given a water potential gradient ($\Delta\Psi$, MPa), an increase or decrease in the water flow (J_v, $m^{-3}\ s^{-1}$) will reflect a change in the hydraulic conductance (L_o, $m^3\ s^{-1}\ MPa^{-1}$) along the plant's hydraulic circuit. In this model, the daytime transpiration demand of the aerial part of the plant—modulated by stomatal

* Corresponding author's e-mail address: amodeo@bg.fcen.uba.ar

conductance (g_s, mmol m^{-2} s^{-1})—is the main contributor to the driving force that ensures water entry through the roots (Sack and Holbrook 2006).

Thus, hydraulic integration can be considered to be a trait with important implications for plant structure and function (Schenk *et al.* 2008). In the last decade, considerable attention has been given to discover how root hydraulic properties affect the overall water uptake. Despite analysing changes in the absorbing surface area or modifications in the driving force, a new approach is provided that considers the intrinsic water uptake properties of the root (hydraulic conductivity, L_{pr}) as a key component of the capacity to transport water per unit surface and per driving force (Steudle and Peterson 1998; Steudle 2000; Tyree 2003). The discovery of aquaporins has contributed to a reconsideration of the paradigm of the membrane transport capacity in terms of water and/or certain solutes or gases (Maurel 1997; Javot and Maurel 2002; Tyerman *et al.* 2002; Hachez and Chaumont 2010; Alleva *et al.* 2012; Chaumont and Tyerman 2014). According to the 'composite transport model' (Steudle 2000), the magnitude of the osmotic and hydrostatic forces will determine which path is the primary contributor to water flow: the apoplastic pathway (with low resistance) and/or the cell-to-cell pathway (i.e. symplastic plus transcellular, with high resistance) (Steudle and Peterson 1998; Suku *et al.* 2014). However, it is not only a question of how limiting the radial water flow could be but also to what extent these two pathways can be modified to rapidly adjust the L_{pr}. Recent findings emphasize that aquaporins might substantially contribute to water uptake (e.g. barley: Knipfer and Fricke 2011; soybean: Vandeleur *et al.* 2014). Evidence for the contribution of the radial water flow has been identified by applying hydrostatic pressure to the root medium (Boursiac *et al.* 2005; Hachez *et al.* 2012; Vandeleur *et al.* 2014) or by dissecting the hydrostatic and osmotic gradients in the entire plant (Fritz *et al.* 2010; Fricke *et al.* 2013; Gambetta *et al.* 2013).

The impact of the radial water flow on the hydraulic circuit could be analysed by studying the response of plants in conditions where the hydraulic driving force limits water absorption. For instance, salt stress is a condition in which both the excessive Na^+ in the soil environment and the water deficit act as linked factors that severely affect the plant growth rate. High salt concentration reduces soil water potential and not only makes water absorption harder for the roots but also introduces toxicity through a gradual accumulation of ions in the plant tissues (Munns and Tester 2008). Thus, the fine regulation between the ion redistribution and the water flow pathways is crucial in the tolerance response. The relevance of membrane pathways involved in ion redistribution—particularly between Na^+ and K^+—has been well described (Niu *et al.* 1995; Peng *et al.* 2004; Karley and White 2009; Shabala *et al.* 2010; Gajdanowicz *et al.* 2011). It is still necessary to understand how water pathway resistances (or conductances) contribute to improve plant salt tolerance.

Beta vulgaris—a member of the Chenopodiaceaea family—is considered a halotolerant (Clarke *et al.* 1993) or moderately salt-tolerant glycophyte (Bartels and Sunkar 2005; Bartels and Dinakar 2013). This behaviour among beet subspecies is related to their versatile ability to accomplish a rapid osmotic adjustment by regulating their ion and water uptake (Daoud *et al.* 2008). In these plants, the decrease in the water potential imposed by salinity is overcome by osmotic regulatory mechanisms, and the plants gain the capacity to take up water from the saline medium and maintain their turgor. An isolated enriched fraction of *B. vulgaris* plasma membrane shows very high water permeability ($P_f = 542\ \mu m\ s^{-1}$; Alleva *et al.* 2006) that favours a highly permeable cell-to-cell pathway. To date, three *B. vulgaris* plasma membrane intrinsic proteins (*Bv*PIP1;1, *Bv*PIP2;1 and *Bv*PIP2;2) have been described (Qi *et al.* 1995; Barone *et al.* 1997, 1998) and characterized in a heterologous system (Bellati *et al.* 2010; Jozefkowicz *et al.* 2013). Because the *B. vulgaris* genome was very recently announced (Dohm *et al.* 2014), transcriptome global sequencing (Mutasa-Göttgens *et al.* 2012) as well as expressed sequence tag libraries provide excellent sources for open reading frame identification for tissue and/or different growth conditions (http://compbio.dfci.harvard.edu). The latter sources are precise enough to provide confidence that, to date, the three identified *Bv*PIPs described in this work remain the consistently abundant and highly expressed ones (Skorupa-Kłaput *et al.* 2015).

In an environmental condition with low water availability in the soil, the root water pathways can combine anatomical/architectural changes with the adjustment of aquaporin contribution, which might finally be reflected in the L_{pr} (Hachez *et al.* 2006; Maurel *et al.* 2010; Chaumont and Tyerman 2014). In particular, plants under salt stress might decrease L_{pr} by means of different strategies, including (i) the modulation of aquaporin by post-transductional mechanisms (Boursiac *et al.* 2005, 2008; Luu *et al.* 2012) or by transcriptional changes (Jang *et al.* 2004; Mahdieh *et al.* 2008; Horie *et al.* 2011; Muries *et al.* 2011; Liu *et al.* 2012) and (ii) changes in the root architectural arrangement (Galvan-Ampudia and Testerink 2011; Horie *et al.* 2012) and/or anatomical changes (Bramley *et al.* 2009), including suberin deposition (Krishnamurthy *et al.* 2011; Sutka *et al.* 2011). Thus far, the above-mentioned mechanisms described in (i) are associated with faster and reversible responses (hours–days), while those described in (ii) are related to long and irreversible acclimation triggered days after the onset of the stress (Munns and Tester 2008; Horie *et al.* 2012).

The aim of this work was to explore how hydraulic adjustments improve the tolerance response in a halotolerant species by analysing L_{pr} and g_s changes. The dynamics of root water adjustment (including water pathways) was explored under two salt treatments (200 mM NaCl and 200 mM KCl). Sodium ion was replaced with K^+ to provide a source of a different monovalent cation as an inorganic osmolyte (Rahnama *et al.* 2010). This experimental design (NaCl versus KCl) was introduced because the ion redistribution is different, i.e. Na^+ linked to the apoplast versus K^+ linked to the transcellular pathway (Tester and Davenport 2003; Shabala and Cuin 2008). These redistributions will affect not only the water fluxes but also the water pathways involved. Our working hypothesis is that changes in resistances (or conductances) should also be accomplished to rapidly adjust the plant hydraulics. Although ABA and signalling crosstalk have been extensively addressed in the literature (Finkelstein 2013; Geng *et al.* 2013; Mittler and Blumwald 2015), the contribution of our work is to analyse in detail the hydraulic continuum associated with tolerance by performing a biophysical study to quantify the water adjustments.

To achieve this goal, our experimental design (NaCl versus KCl salt treatment) included (i) exploration of the plant hydraulic dynamics analysing two conditions that reflect different root–shoot water status in *B. vulgaris* (loss of turgor and gain of turgor) after being submitted to salt treatments and (ii) exploration of the hydraulic adjustment capacity to recover after the salt treatment is halted, thus assessing the contribution of the water pathways. We analysed physiological parameters linked to the water adjustment capacity at the whole-plant level: water potential, g_s and L_{pr}, together with the amount of BvPIPs aquaporin's transcripts and root anatomical modifications. Our hypothesis is that the tolerance of *B. vulgaris* to salt stress may be explained in terms of a high capacity to perform hydraulic adjustments and that this capacity might quantitatively reflect root plasticity that functions as a rheostat in the SPAC (Maurel *et al.* 2010).

Methods

Characterization of a new state for *B. vulgaris* under salt stress

Plant growth and experimental design. *Beta vulgaris* was grown under controlled environmental conditions with a 16/8 h light/dark cycle in a 21 °C conditioned growth chamber (light intensity conditions were 148 ± 10 μmol m^{-2} s^{-1}). Red beet seeds were germinated in plastic containers filled with sterilized sand and moistened with hydroponic culture: 1.25 mM KNO_3, 0.75 mM $MgSO_4$, 1.5 mM $Ca(NO_3)_2$, 0.5 mM KH_2PO_4, 50 μM FeEDTA, 50 μM H_3BO_3, 12 μM $MnSO_4$, 0.70 μM $CuSO_4$, 1 μM $ZnSO_4$, 0.24 μM Na_2MoO_4 and 100 μM Na_2SiO_3 (Javot *et al.* 2003). Ten days after germination, the healthy seedlings were transplanted into aerated hydroponic culture containers. Distilled water was added on the 10th day to compensate for the losses by evapotranspiration. For all of the studied parameters, a nutrient solution was complemented or not with NaCl or KCl (200 mM) at 21 days after planting, i.e. when the first true leaf was completely mature. The treatments were always started at the beginning of the light cycle (9:00 AM), which was considered to be time 0 h. The subsequent harvest time(s), where any parameter was measured and/or samples taken, are in reference to this initial (t= 0 h) time. All treatments were applied in a completely randomized design. At least three to four independent biological replicates were used in each experiment. Data are expressed as the mean of three performed independent experiments. The final salt concentration was selected by analysing the plant's response to different NaCl treatments (50, 100, 250 and 500 mM) **[see Supporting Information—Fig. S1]**. Our strategy was to find a physiological condition where hydroponically grown plants were able to rapidly show a clear change in their phenotype (loss of turgor), followed by a gain of turgor after the onset of salt stress. This phenotype change was remarkable at 200 mM NaCl ($\Psi_{medium} = -0.90$ MPa).

Relative water content. The first true leaf was collected from different plants at different time intervals after treatment and employed to determine relative water content (RWC), as described by Turner (1981). The turgid weight was measured on the same leaves after immersing them for 24 h (until the final weight value was constant) in distilled water, and the oven-dry weight (DW) was obtained after drying them at 70 °C for 24 h (until the final weight value was constant).

Transpiration rate. The volume of water transpired per plant was measured gravimetrically. The plants were grown as follows: 1 day before the treatment was applied, each plant was moved to an individual plastic container, which was sealed to prevent evaporation. Every plant was weighed every hour between 9:00 AM and 5:00 PM during four consecutive days. In each plant, the slope of mass = f(time) was employed to calculate the average mass lost per hour per leaf area per day for all treatments (6–9 plants). In all cases, we determined the leaf area only on the fourth day, and this value was used to calculate the transpiration rate.

Relative growth rate of leaf area. The leaf blades (first true leaf) of the plants were photographed with a digital camera, and the leaf area was measured with image analysis software (Image J ver. 1.37; http://rsb.info.

nih.gov/ij). The relative growth rate (RGR) was calculated with respect to the ratio of A_i (leaf area in a given time) and A_o (leaf area at the beginning of experiment), and the results were expressed as the natural logarithm of the relative leaf area (A_i/A_o) as a function of time (from $t = 0$ h—onset of the salt treatment—up to 48 h). The slope of the curve estimates RGR.

Shoot–root ratio. To analyse the biomass distribution, the shoot–root ratio was determined from the fresh weight in each experimental condition (control, 200 mM NaCl, 200 mM KCl at 0, 4, 8, 24 and 48 h of the imposed treatment).

Apparent leaf water potential. The leaves of the treated or control plants were placed in a plastic bag covered with Parafilm® foil prior to measurement in a Scholander pressure chamber (BioControl, Model 4, Argentina) to determine Ψ_{leaf} (Scholander *et al.* 1965). The measured leaf water potential in this work is referred to as the apparent leaf water potential (Ψ'_{leaf}) because in species such as *B. vulgaris*—which shows halotolerant features—the osmotic potential (Ψ_{osm}) of the xylem is not negligible (Boyer 1969; Kaplan and Gale 1974). It is, therefore, considered as an estimator of the water potential (Turner 1981).

Apparent turgor-pressure component. The pressure component of the water potential (Ψ'_p) in the leaf was calculated as $\Psi'_p = \Psi'_{leaf} - \Psi_{osm}$. In another set of leaves, we determined the Ψ_{osm} following a freezing protocol as previously described (Mahdieh *et al.* 2008). The osmolality of each sample was measured in a vapour pressure osmometer (Vapro 5520, Wescor, USA) **[see Supporting Information—Table S1]**.

Linking root hydraulic response to the overall SPAC

Stomatal conductance measurements. Stomatal conductance was measured with a steady-state porometer (SC-1, Decagon Devices, Pullman, WA, USA) on the first true leaf in each plant, a completely expanded mature one. To avoid time-consuming measurements, we first demonstrated that the measurement of one leaf was sufficient per plant, i.e. the g_s profile was similar between leaves in each plant during the day (data not shown).

Root hydraulic conductivity measurement. Measurements were performed as previously described (Javot *et al.* 2003; Boursiac *et al.* 2005). In these experiments, the entire root system of a freshly detopped plant was inserted into a 50-mL tube filled with the same nutrient solution bathing the intact plant, and the root was then placed inside the pressure chamber (BioControl, Model 2, Argentina). The hypocotyl was carefully connected to a glass capillary tube using a low-viscosity dental paste (A+ Silicone, Densell) and was then threaded through the metal lid of the chamber. We determined the exudated flow (J_v) induced by the pressure. Briefly, the excised roots were subjected to three pressures in a stepwise manner: 0.3, 0.4 and 0.2 MPa. The exudated flow was constant in all time periods of measurement (5–10 min in each pressure). After measurements, the DW of the root was obtained. The L_{pr} of each individual root system (in mL mg^{-1} h^{-1} MPa^{-1}) was calculated from the slope of a plot of flow (J_v) versus pressure divided by the DW of the root system **[see Supporting Information—Graph S1]**. Diurnal effects were discarded measuring both properties during the day in control plants. The change in treated plants was statistically significant and independent of the time of the day.

Exploring root adjustments in terms of water pathways

Root anatomy. As described by Sharp *et al.* (2004), roots were cut in an equivalent position with respect to both root meristem and whole root length to warranty identical ontogenetic state for all the treatments. Fresh roots were cut into pieces 10 mm in length and incubated in 0.3 % w/v Sudan IV (Sigma-Aldrich) (in ethanol 70 %, v/v) for 1 h (Sutka *et al.* 2011). The root fragments were then rinsed in distilled water and finely chopped using a razor blade. The samples were mounted on slides in glycerol and observed with a microscope (Zeiss Axioskop 2, Japan). We found a better pattern for the Sudan IV red-stained root with respect to autofluorescence in the free-hand cross-sections, and measurements of L_{pr} can be made in the same sample without fixing the material.

Quantitative real-time polymerase chain reaction for aquaporin gene expression

The roots were carefully and quickly harvested, frozen in liquid nitrogen and stored at −70 °C. The total RNA was isolated from 70 to 80 mg of tissue using 'RNeasy Plant Extraction kit' with 'Plant RNA Isolation Aid' (Ambion, Austin, TX, USA) according to the manufacturer's recommendation, ending the isolation with a digestion with DNaseI. For each sample, 500 ng of total RNA were converted into cDNA using oligo(dT) and M-MLV reverse transcriptase (Promega, Madison, WI, USA) according to the manufacturer's recommendation.

The transcript expression of *BvPIP2;1*, *BvPIP2;2*, *BvPIP1;1*, *BvUBIep* and *BvGAPDH* genes was studied by real-time polymerase chain reaction (RT-PCR). The primers were designed based on published sequences of the aquaporins found in *B. vulgaris* **[see Supporting Information—Table S2]**. The selection of *BvUBlep* and *BvGAPDH* as the housekeeping genes was based on genes reported in *B. vulgaris* and information available in the literature (Reid *et al.* 2006; Wan *et al.* 2010). The mRNA abundance of *Bv*GAPDH and *Bv*UBIep was not significantly different between the treatments (data not shown).

Quantitative PCR (qPCR) was performed with a MyiQ cycler (Bio-Rad) in a reaction volume of 25 μL containing 12.5 μL of IQ Sybr Green Super Mix (Bio-Rad), 320 nM primers and 5 μL of a 1/500 dilution of cDNA. The RT qPCR conditions comprised 1 cycle at 95 °C for 5 min and 34 cycles at 95 °C for 45 s, 60 °C for 30 s and 72 °C for 1 min. Amplification data were collected during the extension step (72 °C). The efficiency of the primer binding was determined by linear regression by plotting the cycle threshold value versus the log of the cDNA dilution (Soto *et al.* 2010). The absolute RNA amount for each gene was determined in every qPCR experiment. The relative gene expression was calculated as the ratio of the initial gene quantity to the initial mean quantity of the housekeeping genes (Soto *et al.* 2011). Quantitative PCR experiments were independently performed three times with comparable results. The three PCR reactions were carried out in duplicate. The transcript levels of the three studied aquaporins under salt treatments were compared with an osmotic treatment imposed by a non-charge and non-permeable solute [polyethylene glycol (PEG) 6000] at a concentration of 23 % (*p*/*v*), which induces a Ψ_{medium} of −0.90 MPa. The purpose was to contrast aquaporin transcripts between ion signals (NaCl and KCl) versus non-charged non-permeable osmolyte (data not shown).

Restoring salt-treated plants to control medium: L_{pr} and g_s recovery profiles

To measure the L_{pr} and g_s recovery profiles, the plants were first submitted to a salt treatment (200 mM NaCl or KCl) for 4 or 24 h and then transferred to a control solution (considered now $t = 0$ h). Root hydraulic conductivity and g_s were then measured at different time intervals (0, 1 and 24 h) to characterize the plant's capacity to restore L_{pr} and g_s when the salt treatment was halted. As a control, we first determined the L_{pr} values immediately after changing the detopped roots to a control medium. This protocol was crucial to discard the flows that could be artefacts due to injury exacerbated with the salinity treatment. Root hydraulic conductivity values in control medium were not significantly different from those measured in the saline treatment **[see Supporting Information—Graph S2]**. All values shown are the average of three independent experiments. In each experiment, g_s was measured in one leaf per plant in three different plants and two or three roots were detopped for measuring L_{pr} under each condition.

Statistical analysis

Statistical analysis was performed using software GraphPad Prism 5.00 for Windows, GraphPad Software, San Diego, CA, USA, www.graphpad.com. Differences were accepted as significant with at least $P < 0.05$ employing analysis of variance (ANOVA) and Bonferroni tests as indicated in the figure legends.

Results

Characterization of a new state for *B. vulgaris* under salt stress

Salt stress was achieved by the addition of 200 mM NaCl ($\Psi_{medium} = -0.90$ MPa). Under this condition, plants were able to rapidly show a clear change in their phenotype (loss of turgor) followed by a gain in turgor in <24 h (Fig. 1). The phenotype observed in plants exposed to 200 mM KCl was indistinguishable from the NaCl treatment (Fig. 1A), and no chlorosis symptoms were observed in the leaf. As expected, the tolerant phenotype shows a reduction in transpiration rate and in the leaf RGR in both salt treatments, although growth was not arrested **[see Supporting Information—Graph S3]**. Moreover, during the whole experiment, the shoot-root ratio was not significantly modified, so the growth rate changes in the leaf were also translated to the root growth rate (Fig. 1B).

The water status of the hydroponically grown *B. vulgaris* plants was characterized. The RWC was reduced according to the phenotype observed (Fig. 1C). The Ψ'_{leaf} was analysed at different time intervals (0, 4, 8, 24 and 48 h) for the control and treated plants. The mean Ψ'_{leaf} in the control plants was -0.17 ± 0.02 MPa and remained constant during the whole experiment (Fig. 1D). When Ψ'_{leaf} was measured after 4 h of the onset of the stress treatment, its mean value was significantly reduced in the NaCl condition (-0.20 ± 0.03 to -0.74 ± 0.01 MPa) and in the KCl condition (-0.19 ± 0.02 to -0.79 ± 0.01 MPa). The initial drop in Ψ'_{leaf} is well correlated with plant turgor loss in both salt treatments (Fig. 1A). The Ψ'_{leaf} remained at these low values up to 48 h although the turgid phenotype changed (Fig. 1). The leaf Ψ_{osm} remained constant in the control plants and showed a reduction in the plants submitted to stress after 24 h **[see Supporting Information—Table S1]**. The patterns of Ψ_p versus time were well correlated with the observed phenotype of loss and gain in turgor (Fig. 1A and E).

All of these parameters allowed us to define two distinguishable time intervals in terms of water adjustment during salt stress response, 4 h, where there is loss of turgor and 24 h, where there is gain of turgor. Our next step was to analyse these two conditions in terms of overall hydraulic adjustments.

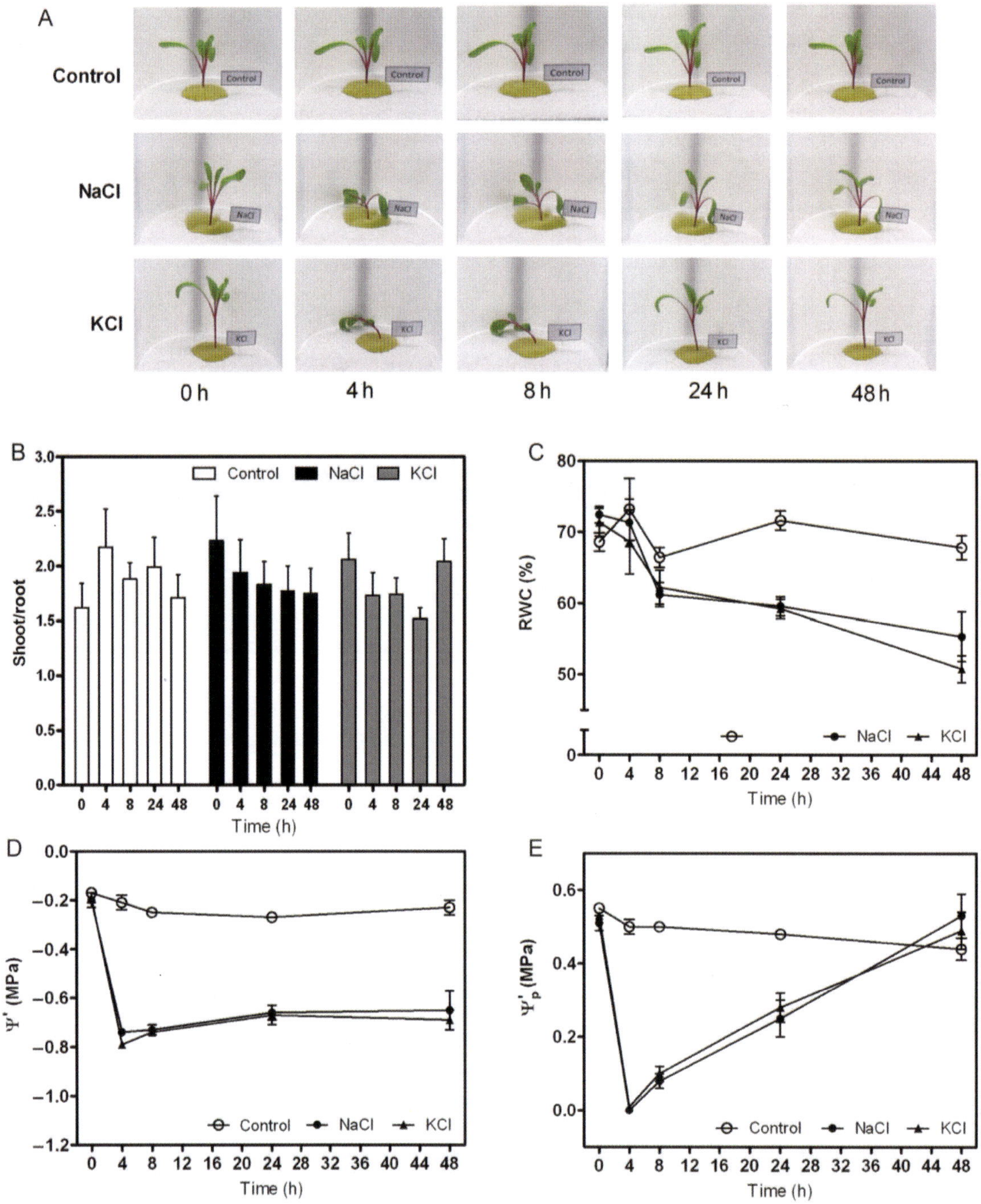

Figure 1. Effect of salinity treatments on *B. vulgaris* plants: control ($\Psi_{medium} = -0.04$ MPa); 200 mM NaCl ($\Psi_{medium} = -0.90$ MPa) or 200 mM KCl ($\Psi_{medium} = -0.90$ MPa). (A) Images of the same hydroponically grown plant taken in each condition at the indicated times after the onset of the treatment. (B) Shoot–root ratio values are given as bars representing mean $\pm$ SE of three independent experiments ($n = 3$). The biological replicates were five to six plants per treatment in each experiment. No differences were observed between treatments ($F_{(2,68)} = 0.16, P = 0.8513$). (C) Relative water content values are expressed as mean $\pm$ SE of three independent experiments ($n = 3$). The biological replicates were three plants per treatment in each experiment, and two leaves per plant were analysed as a duplicate. (D) The Ψ'_{leaf} values (in MPa) measured at each indicated time are expressed as mean $\pm$ SE of three independent experiments. (E) The calculated apparent leaf pressure potential (Ψ'_p in MPa) is expressed as mean values $\pm$ SE.

Linking root hydraulic response to the overall SPAC

As expected, the salt added to the medium triggered a rapid decrease in g_s, which remained low even up to 24 h (Fig. 2A). In the control conditions, the plants showed mean L_{pr} values of 72.3 ± 21.1 mL g^{-1} h^{-1} MPa^{-1} (Fig. 2B). Both salt treatments induced a rapid and indistinguishable decrease

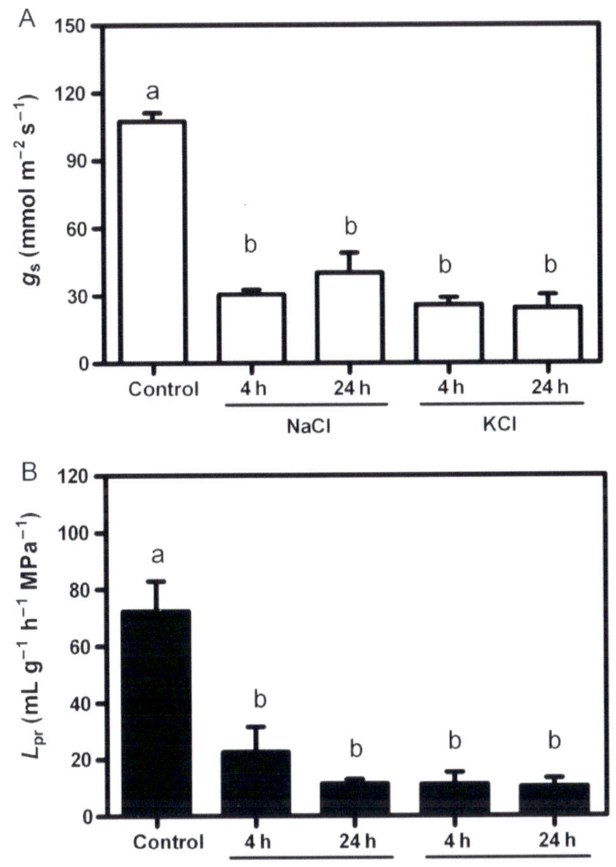

Figure 2. Integrating SPAC key points: g_s and L_{pr}. (A) Stomatal conductance values (g_s, in mmol m^{-2} s^{-1}) are given as bars representing mean ± SE of three independent experiments (at least three plants per treatment). Different letters indicate statistical differences between treatments ($P < 0.001$; Bonferroni test). (B) Hydraulic conductivity values (L_{pr}; in mL mg^{-1} h^{-1} MPa^{-1}) are given as bars representing mean ± SE of three independent experiments (in each one, two to three individual root systems were measured). Different letters indicate statistical differences between treatments ($P < 0.001$; Bonferroni test).

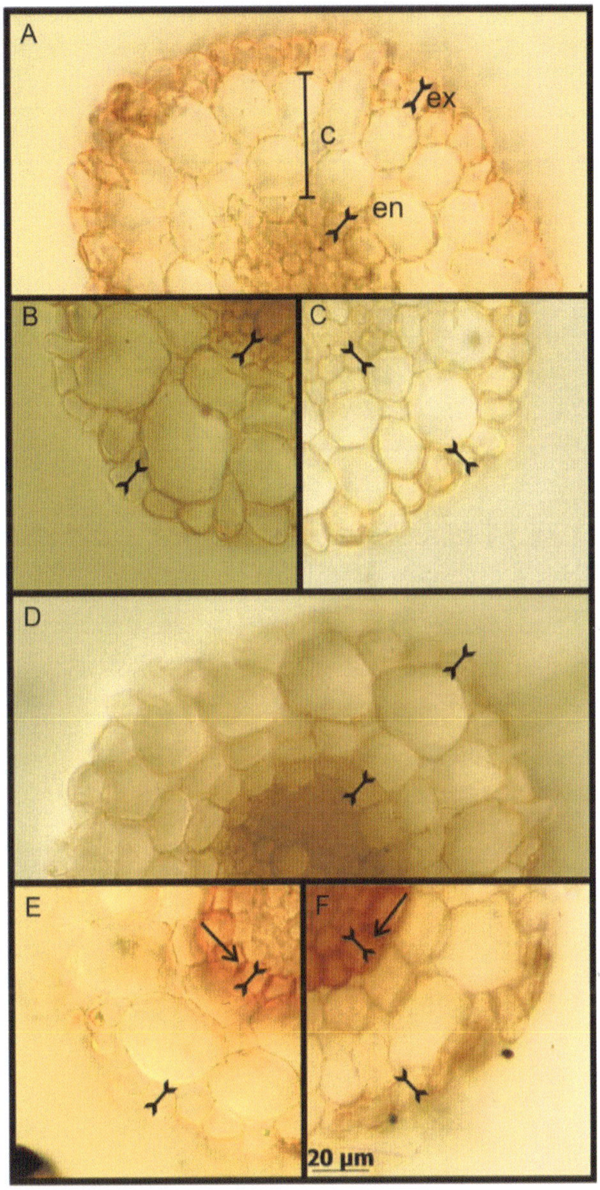

Figure 3. Photographs of *B. vulgaris* fresh root cross-sections stained with Sudan IV, the bar represents 20 μm. (A and D) Control root cuts indicating cortex (C), endodermis (en) and exodermis (ex). (B and C) Representative cuts of roots from plants treated for 4 h with 200 mM NaCl and 200 mM KCl, respectively. (E and F) Representative cuts of roots from plants treated for 24 h with 200 mM NaCl and 200 mM KCl, respectively. The arrows show suberization of endodermis. The images are one sample per condition of 10 independent experiments ($n = 10$).

in L_{pr}. The L_{pr} inhibition was 80 % compared with the control condition after 4 h of treatment, and this low L_{pr} value was maintained up to 24 h of treatment. The g_s modifications are similar to the profile shown by L_{pr}, suggesting that the change in the root water flow is coupled to g_s.

Exploring root adjustments in terms of water pathways

The anatomical changes and the presence of aquaporins provide some insight into the putative involvement of the different water pathways at the root level for the two selected time intervals (4 and 24 h of treatment in 200 mM NaCl or 200 mM KCl). We, therefore, incubated root sections in the presence of Sudan IV in order to check suberization (Fig. 3). The plant roots challenged by either NaCl or KCl for a period of 4 h showed undetectable suberization changes of the endodermis and/or exodermis, as in the control plants. For longer exposures (24 h), the suberization of the endodermis increased independently of the ion treatment (Fig. 3E and F), whereas the control roots do not present enhanced intensity for Sudan IV. Similar results were observed when the autofluorescence of the cell wall was analysed (data not shown). These clear changes observed in the endodermis suberization were not observed in the exodermis. The exodermis suberization was very low and random, mostly attributable to higher thickness in the fresh cuts (Fig. 3A).

Quantitative RT-PCR analysis was performed to accurately determine the transcript levels of the PIP genes *Bv*PIP2;1, *Bv*PIP2;2 and *Bv*PIP1;1 in whole roots under salt stress (NaCl or KCl, Fig. 4) at different time intervals. *Bv*PIP2;1 might have a circadian behaviour as described in other PIPs (Takase *et al.* 2011; Caldeira *et al.* 2014).

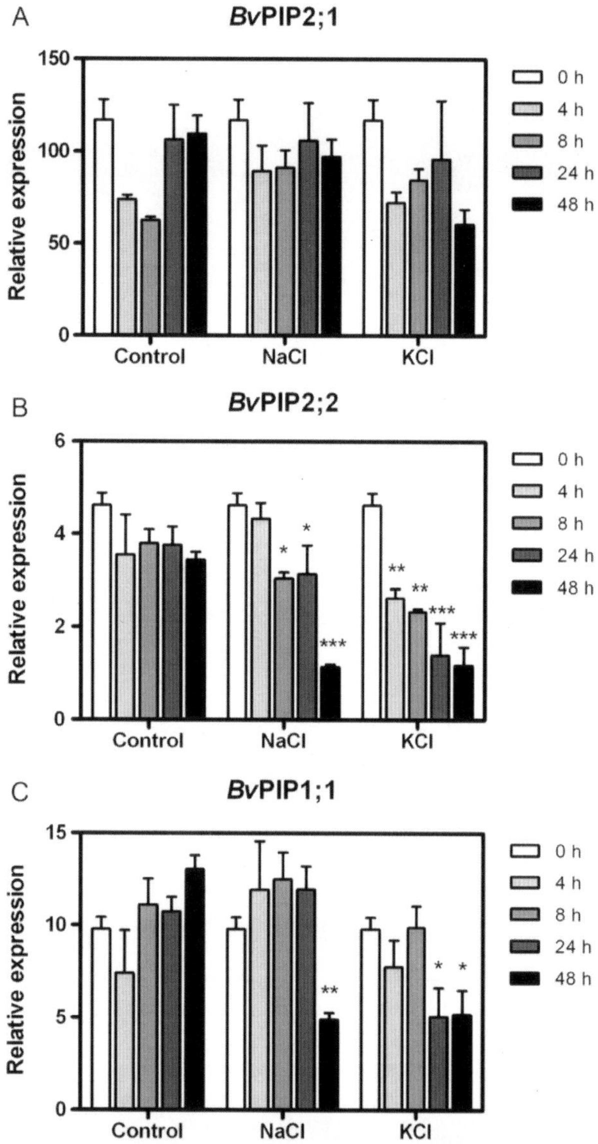

Figure 4. Relative gene expression of *Bv*PIP2;1 (A), *Bv*PIP2;2 (B) and *Bv*PIP1;1 (C) in roots from control, 200 mM NaCl or 200 mM KCl treatments. Data are given as bars representing mean values ± SE of three independent experiments ($n = 3$), and asterisks indicate statistical differences from initial condition ($t = 0$ h) for each treatment (*$P < 0.05$; **$P < 0.01$; ***$P < 0.001$; Bonferroni test) **[see Supporting Information—Tables S3–S5** for statistical analysis].

The studied aquaporins showed a subtle down-regulation profile, except the relative expression level of *Bv*PIP2;1, which did not decrease at all, independently of the treatment (Fig. 4A, **see Supporting Information—Tables S3–S5** for the statistical analyses). Interestingly, the profile of the *Bv*PIP2;2 and *Bv*PIP1;1 expressions for the NaCl stress condition did not show down-regulation at the same pace as observed in the KCl stress (Fig. 4B and C). The differences of the ion treatments became more evident for *Bv*PIP1;1, which showed down-regulation at 24 h when the plants were exposed to KCl, while the decrease became significant at 48 h under NaCl stress (Fig. 4C and **see Supporting Information—Tables S3–S5** for the statistical analyses).

To determine whether the lack of a strong down-regulation of these three aquaporins is only observed under salt treatment, the salt was replaced by a non-permeable and non-charged molecule as PEG. The plants submitted to $\Psi_{medium} = -0.90$ MPa generated with PEG triggered an expression decrease of 56–67 % in these aquaporins after 4 h of treatment, a strong transcript down-regulation compared with both salt treatment (NaCl versus PEG: $P < 0.001$, KCl versus PEG: $P < 0.05$, Bonferroni post-test).

Restoring salt-treated plants to control medium: L_{pr} and g_s recovery

Our results confirm that the root water pathways are different in the two selected intervals (4 and 24 h) and that the three studied aquaporins are relatively stable upon salt treatments and only strongly down-regulated when a non-charge solute is imposed. However, the results do not allow us to completely dissociate the water pathways and ion redistribution (NaCl versus KCl). We decided to explore whether halting the salt treatment allows us to describe the shoot–root water dynamics through the analysis of g_s and L_{pr} recovery.

As shown in Fig. 5, the g_s recovery profile of the plants returned to the control medium reflected dependence of the time of the preceding salt treatment and dependence of the ion involved in the salt treatment. Thus, g_s recovered faster in the 4-h salt-treated plants than in the 24-h salt-treated plants (Fig. 5A and B). The 4-h salt-treated plants restored to the control solution for 1 h were able to increase $g_s \cong 50$ % with respect to the g_s values before halting the treatment (Fig. 5A). In this analysed point (4 h of salt treatment before the halting), the recovery trend is independent of the involved cation (NaCl versus KCl; $F_{(1,22)} = 3.06$, $P = 0.0940$, two-way ANOVA). The plants subjected to NaCl or KCl for 24 h differed in their kinetic to increase g_s when they were restored to the control medium. In this condition (24 h of salt treatment), the ion involved in the salt stress significantly affects the recovery profile of g_s (ion accounts for

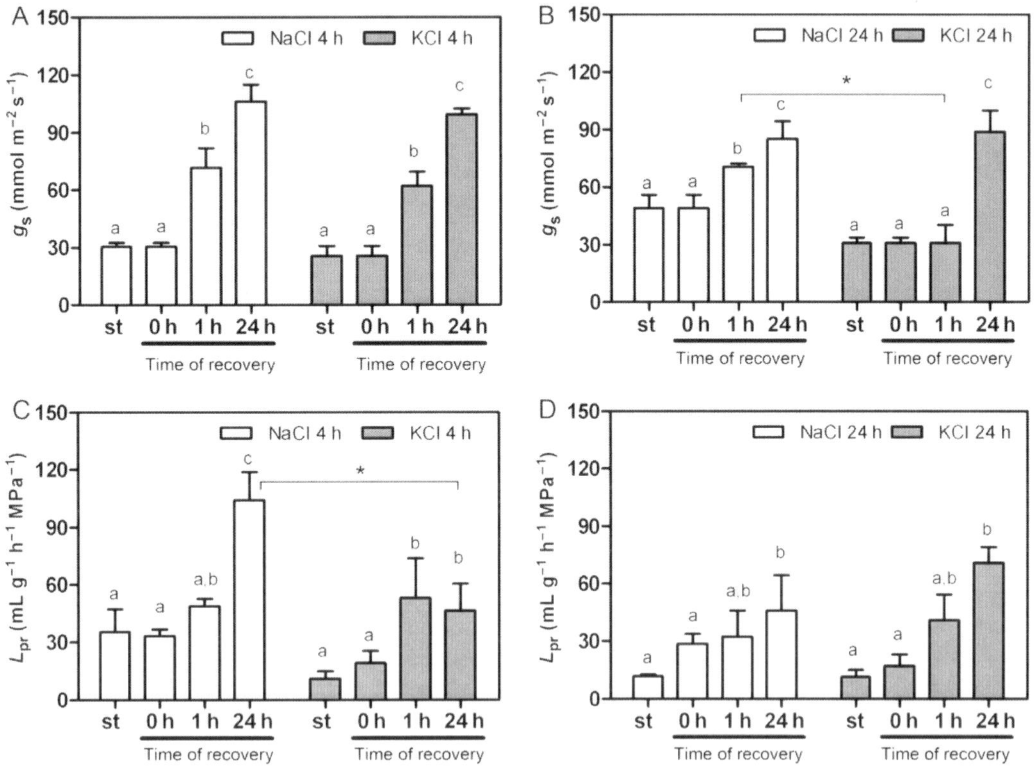

Figure 5. Recovery profile of g_s (A and B) and L_{pr} (C and D) after halting the salt treatment (st). The recovery profiles are separated for plants initially submitted to short salt treatments ($t = 4$ h, A and C) versus a longer interval period ($t = 24$ h, B and D). Data are expressed as mean values $\pm$ SE of three to four independent experiments (in each one, three to four plants were measured). Different letters indicate statistical differences between bars ($P < 0.05$; Bonferroni test), and an asterisk indicates differences between bars from different salt treatments ($P < 0.001$, Bonferroni test).

13.25 % of total variance, $F_{(1,16)} = 12.99$, $P = 0.0024$, two-way ANOVA). Even though g_s reached the same final value after 24 h of recovery in control solution, the recovery trend of g_s is much faster in NaCl-treated plants than in KCl-treated plants (Fig. 5B).

Conversely, the L_{pr} recovery profile of the plants restored to the control medium reflected a completely different strategy in terms of time dependence and ion dependence with respect to g_s. Moreover, the recovery patterns of both hydraulic parameters (g_s and L_{pr}) seem to be uncoupled although both presented a coupled reduction in salt treatment (Figs 5 and 6). In the first analysed condition (4 h of salt treatment), the recovery trend of L_{pr} is significantly affected by the ion involved before halting stress (Fig. 5C, $F_{(1,17)} = 8.62$, $P = 0.0092$). The 4-h KCl-treated plants turned into control solution showed a quick L_{pr} increment that remained unchanged for 24 h. In 4-h NaCl-treated plants, L_{pr} gradually rose to higher values (*$P < 0.001$; Bonferroni test). For the 24-h-treated plants, the trend of L_{pr} increment was independent of the salt treatment ($F_{(1,16)} = 0.58$, $P = 0.4588$, Fig. 5D). In the case of the NaCl-treated plants, L_{pr} recovery was affected by the extension of treatment (4 versus 24 h; Fig. 5C and D). The 4-h salt-treated plants presented a significantly higher L_{pr} value ($P < 0.05$; Bonferroni test) after 24 h of restoring the plants to the control solution. On the contrary, in the case of the KCl-treated plants, the L_{pr} recovery profile is independent of the extension of the treatment, i.e. 4 or 24 h (Fig. 5C and D). The 4-h-treated and 24-h-treated plants presented a similar L_{pr} value after 1 h of restoring the plants to the control solution, which was significantly different from the L_{pr} value observed under salt stress. Figure 6 illustrates g_s and L_{pr} recovery profile observed for the four conditions (NaCl or KCl; and/or the selected time points, 4 and 24 h).

Discussion

Water homeostasis is linked to ion redistribution in plants as an important defence strategy against salt stress (Tester and Davenport 2003; Shabala and Cuin 2008; Shabala 2013; Flowers and Colmer 2015). *Beta vulgaris* showed great plasticity reflecting its ability to rapidly gain turgor due to osmotic adjustment, consistent with the maintenance of a low Ψ'_{leaf} during the entire treatment (Fig. 1). Under our experimental conditions, low

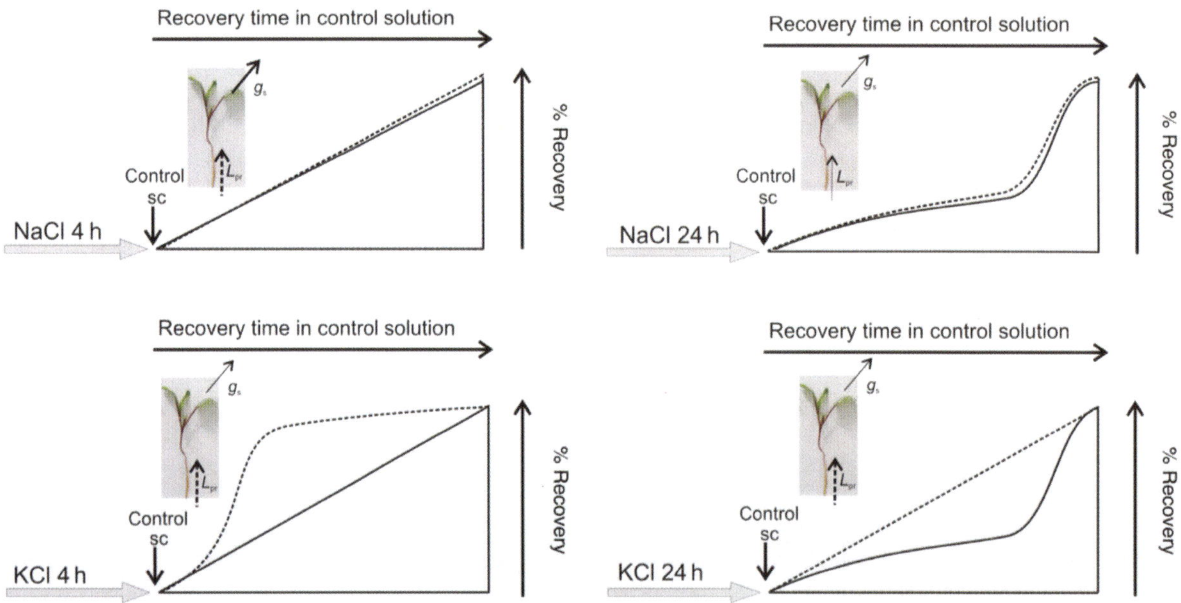

Figure 6. Schematic representation of recovery profiles of L_{pr} (dotted line) and g_s (continuous line) after stress treatments.

RWC values—even for plants gaining turgor after 24 h of salt treatment—might be associated with an underestimation of the RWC as a result of an osmotic adjustment **[see Supporting Information—Table S1]**. The gain of turgor under salt treatment requires solute synthesis and/or recirculation of cations, and this should also be reflected in the obtained RWC values (Weatherley 1950; Boyer *et al.* 2008). The aerial parts only modified 1 % of the water content (data not shown), even in the phenotype that lost turgor (4 h of salt treatment; Fig. 1), a trait consistent with an isohydric-like behaviour (Sade and Moshelion 2014). Thus, the overall strategy is successful for the adjustment of the water content. These data are supported by other studies performed using members of the Chenopodiaceae family under salt stress (Lindhauer *et al.* 1990; Ghoulam *et al.* 2002; Pakniyat and Armion 2007; Abbas *et al.* 2012), where Ψ_{osm} is the key element in turgor recovery. In our experimental design, the analyses were performed at 4 h (loss of turgor) and 24 h (gain of turgor) of salt treatment because these are two distinguishable transition states before a new water plant status is achieved.

The transpiration rate and L_{pr} have not always been reported as a coupled process. For instance, changes in shoot transpiration are not reflected by changes in L_{pr} in *Lotus japonicus* (Henzler *et al.* 1999), while in wheat, it was reported an important correlation between increasing L_{pr}, the cortex cell hydraulic conductivity, transpiration and the root expression of aquaporins—*Ta*PIP1;2 and *Ta*PIP2;5 (Wang *et al.* 2013). Our results clearly showed that under salt treatment, there is a correlated decrease in g_s and root hydraulic properties (L_{pr}) (Fig. 2), as both parameters presented an 80 % reduction compared with the control condition. The decrease in g_s (Fig. 2A) remained low even up to 24 h, which is consistent with the decrease in the leaf water potential values (Fig. 1D). This occurs for both NaCl and KCl treatments and is in agreement with observations performed in other species, such as wheat, that similarly decreased their g_s when exposed to either NaCl or KCl (Rahnama *et al.* 2010). The transition of the phenotypes—loss (4 h) and gain (24 h) of turgor—is not reflected in the two key water balance modulators (L_{pr} and g_s) that remained coupled and similarly low. The hydraulic parameters only reflect a centred strategy of water loss avoidance.

Thus, it is necessary to explore how the root copes with water loss not only in terms of hydraulic properties but also in the analysis of the water pathways. The L_{pr} decrease (Fig. 2B) in our experimental set-up was consistent with other observations for different species (Martínez-Ballesta *et al.* 2003; Boursiac *et al.* 2005; Postaire *et al.* 2010; Muries *et al.* 2011). The roots showed a marked ability to adjust their L_{pr} during the first 4 h of treatment (our first hydraulic transition point) even before plants display any anatomical or morphological change (Fig. 3). It is consistent with faster responses that are usually present in the initial time lapse response to tolerance (Horie *et al.* 2012). In both 4 h salt treatments (NaCl and KCl), the suberization is indistinguishable from the control plants. The root apoplastic pathways were not modified, so the cell-to-cell pathway could be limiting (or maximizing) root resistance to the water flows both in favour of (or restricting) water entry and/or exit. Thus, membrane permeability not only to ions but also to water can contribute to plasticity together with the change in xylem tension as a consequence of the decrease in leaf water potential.

In our second hydraulic transition point (24 h treatment), the low L_{pr} values involved also an anatomical restriction enhancing the hydraulic resistance to water flows along the roots, suggesting an increment in the water flows through the cell-to-cell pathways (a more resistive pathway). The increase in suberization observed after 24-h treatment can be attributed to a completely different strategy. This is consistent with recent reports demonstrating that the cell-to-cell pathway might contribute significantly to the radial water uptake particularly during development (Knipfer and Fricke 2010; Knipfer *et al.* 2011; Gambetta *et al.* 2012, 2013; Caldeira *et al.* 2014; Suku *et al.* 2014). In wheat plants, a non-membranous pathway (apoplast) contributes to increase radial water uptake in the control but not in the NaCl-stressed plants (Fricke *et al.* 2013).

It is possible that the effectiveness of *B. vulgaris* to tolerate the saline stress could be associated with its capacity to maintain the expression level of the AQPs in the salt treatments (Fig. 4), as reported for other specific proteins strictly involved in salt tolerance (Chinnusamy *et al.* 2004, 2006). This statement cannot be made with certainty because aquaporin activity and protein expression were not tested here. The root strategy to maintain water flow is based on water and ion redistribution and adjusting the cell-to-cell pathway by means of its selected membrane permeability (Steudle 2000). A solely osmotic stress (PEG solute) shuts down the transcripts of the *BvPIP* characterized in $<$4 h of treatment, which might contribute by increasing the root cell resistance to the water pathway. On the contrary, the cell-to-cell pathway in salt-treated *B. vulgaris* plants might contribute by increasing the capability to regulate water transfers because water permeability can be tuned to limiting (or maximizing) the resistance in concert with ion redistribution.

We could experimentally dissociate L_{pr} from g_s employing two strategies: (i) different cations—Na^+ versus K^+—to promote the stress and (ii) analysing L_{pr} enhancement when the salt treatment is interrupted (Fig. 5). Most of the studies in the literature are based on the analysis of L_{pr} decrease by means of an imposed stress condition or the presence of aquaporin inhibitors (e.g. Ehlert *et al.* 2009; Vandeleur *et al.* 2009). To our knowledge, this is the first work that explored altogether L_{pr} decrease and enhancement, as most of the works show L_{pr} inhibition and not its recovery. Whereas K^+ and Na^+ have distinct redistribution profiles, different L_{pr} recovery pathways for water are expected to be involved even in the presence of an equivalent change in the driving force along the SPAC for both situations. After stress treatments, the recovery of both hydraulic parameters (L_{pr} and g_s) denoted two different strategies (Fig. 6). The enhancement in L_{pr} shows a profile (Figs 5C and 6) that is linked to ion redistribution (Na^+ versus K^+) and this is part of the root plasticity to prevent water loss. In the first transition point (4 h of salt treatment), the cation dependence of the L_{pr} profile highlights the participation of membrane permeability in root plasticity together with the change in xylem tension. On the other hand, L_{pr} recovery profiles observed after 24 h of salt treatment suggest that under our experimental conditions, root resistance to water flow does not differ between the ion source of the stress (Na^+ or K^+). This is consistent with an increase in the total root resistance and the observed strong endodermis suberization in both salt treatments (Fig. 3E and F). The recovery profile of NaCl treatments shows a coupled temporal dependence strategy where g_s and L_{pr} increase at the same rate. Both parameters increase at a slower pace when the plants were treated for 24 h compared with 4 h. Conversely, g_s and L_{pr} enhancement are clearly uncoupled in the KCl treatments (Fig. 6). The root shows the capacity to restore the water transport capacity before the water is transpired through stomata. It is well described in the literature that under salt stress, Na^+ is redistributed to avoid toxicity, while K^+ functions as an interchangeable ion all along the vasculature (particularly phloem) (Peng *et al.* 2004; Munns and Tester 2008; Karley and White 2009; Shabala *et al.* 2010; Flowers and Colmer 2015). In this context, the potassium gradient might be crucial in the root–shoot hydraulic signalling (Gajdanowicz *et al.* 2011). The profiles observed in Fig. 6 are consistent with sustaining a 'hydraulic' adjustment in the presence of NaCl compared with a 'tuned' adjustment caused by the redistribution in the case of KCl, which is clearly reflected in g_s and L_{pr} changes.

The proposed initial two set points—4 and 24 h extension in the imposed salt treatment—were selected because of the triggered distinguishable phenotypes in *B. vulgaris*. At 4 h of an imposed 200 mM salt stress, plants have lost turgor and osmotic adjustment has not been completed. In this situation, the L_{pr} recovery profile suggests a much higher root tuning capacity to modulate the water dynamics that affects the whole-plant water loss avoidance strategy. At 24 h of 200 mM salt stress, plants are gaining turgor, and the L_{pr} recovery profile suggests that the root versatility is more restricted as tolerance has already been triggered.

Conclusions

Tolerance involves limiting water movement by increasing the total plant hydraulic resistance. *Beta vulgaris* osmotic adjustment is sustained by tuning L_{pr} and g_s.

Our work presents a quantitative analysis of the coordinated link between L_{pr} and g_s when the ion and water redistribution strategy takes place. Even when the xylem tension and apoplast pathway mediate plant water flows, the cell-to-cell pathway contributes as a key component to the capacity to transport water per unit surface and driving force in the SPAC (nicely demonstrated in the enhancement of L_{pr} after halting KCl treatment). Future research should explore the molecular basis for the different strategies that plants use to regulate their water balance and identify the imposed threshold of the cell-to-cell pathways in terms of hydraulic resistance.

Sources of Funding

This work was supported by the Agencia Nacional para la Promoción Científica y Técnica [Préstamo BID PICT11-2239 and PICT14-0744], Consejo de Investigaciones Científicas y Técnicas (CONICET) Proyecto de Investigación Plurianual (PIP12-14) and Universidad de Buenos Aires UBACyT14-17, all grants to G.A.

Contributions by the Authors

V.V., J.B. and G.A. were involved in the study conception and design. V.V. and J.B. planned and performed experiments and analysed data. J.B., G.S. and N.D.A. additionally participated in the design and data acquisition of the qRT-PCR experiments. V.V. and G.A. were involved in the analysis and interpretation of the data, discussion and writing the manuscript. All authors had intellectual input into the project.

Acknowledgements

We thank Milena Manzur, Moira Sutka, Agustín Yaneff, Esteban Tubert, Karina Alleva and Cintia Jozefkowicz for critical comments in the manuscript. We also thank the anonymous reviewers and all the editors for their helpful comments and suggestions on this paper.

Supporting Information

The following additional information is available in the online version of this article –

Figure S1. Different concentrations of salt treatments.

Table S1. Osmotic potential measured for the leaf sap.

Graph S1. Root hydraulic conductivity determination.

Table S2. Accession number of genes and sequences of primer pairs used for qRT-PCR.

Graph S2. Root hydraulic conductivity determination after halting salt treatment.

Graph S3. Plant transpiration rate and RGR of leaf area.

Tables S3–S5. Quantitative real-time polymerase chain reaction statistical analysis.

Literature Cited

Abbas F, Mohanna A, Al-Lahham Gh, Al-Jbawi E. 2012. Osmotic adjustment in sugar beet plant under salinity stress. *Journal of Sugar Beet* **28**:37–43.

Alleva K, Niemietz CM, Sutka M, Maurel C, Parisi M, Tyerman SD, Amodeo G. 2006. Plasma membrane of *Beta vulgaris* storage root shows high water channel activity regulated by cytoplasmic pH and a dual range of calcium concentrations. *Journal of Experimental Botany* **57**:609–621.

Alleva K, Chara O, Amodeo G. 2012. Aquaporins: another piece in the osmotic puzzle. *FEBS Letters* **586**:2991–2999.

Barone LM, Shih C, Wasserman BP. 1997. Mercury-induced conformational changes and identification of conserved surface loops in plasma membrane aquaporins from higher plants. Topology of PMIP31 from *Beta vulgaris* L. *Journal of Biological Chemistry* **272**:30672–30677.

Barone LM, Mu HH, Shih CJ, Kashlan KB, Wasserman BP. 1998. Distinct biochemical and topological properties of the 31- and 27-kilodalton plasma membrane intrinsic protein subgroups from red beet. *Plant Physiology* **118**:315–322.

Bartels D, Dinakar C. 2013. Balancing salinity stress responses in halophytes and non-halophytes: a comparison between *Thellungiella* and *Arabidopsis thaliana*. *Functional Plant Biology* **40**:819–831.

Bartels D, Sunkar R. 2005. Drought and salt tolerance in plants. *Critical Reviews in Plant Science* **24**:23–58.

Bellati J, Alleva K, Soto G, Vitali V, Jozefkowicz C, Amodeo G. 2010. Intracellular pH sensing is altered by plasma membrane PIP aquaporin co-expression. *Plant Molecular Biology* **74**:105–118.

Boursiac Y, Chen S, Luu DT, Sorieul M, Van den Dries N, Maurel C. 2005. Early effects of salinity on water transport in *Arabidopsis* roots. Molecular and cellular features of aquaporin expression. *Plant Physiology* **139**:790–805.

Boursiac Y, Boudet J, Postaire O, Luu DT, Tournaire-Roux C, Maurel C. 2008. Stimulus-induced downregulation of root water transport involves reactive oxygen species-activated cell signalling and plasma membrane intrinsic protein internalization. *The Plant Journal* **56**:207–218.

Boyer JS. 1969. Free-energy transfer in plants. *Science* **163**:1219–1220.

Boyer JS, James RA, Munns R, Condon AG, Passioura JB. 2008. Osmotic adjustment leads to anomalously low estimates of relative water content in wheat and barley. *Functional Plant Biology* **35**:1172–1182.

Bramley H, Turner NC, Turner DW, Tyerman SD. 2009. Roles of morphology, anatomy, and aquaporins in determining contrasting hydraulic behavior of roots. *Plant Physiology* **150**:348–364.

Caldeira CF, Bosio M, Parent B, Jeanguenin L, Chaumont F, Tardieu F. 2014. A hydraulic model is compatible with rapid changes in leaf elongation under fluctuating evaporative demand and soil water status. *Plant Physiology* **164**:1718–1730.

Chaumont F, Tyerman SD. 2014. Aquaporins: highly regulated channels controlling plant water relations. *Plant Physiology* **164**:1600–1618.

Chinnusamy V, Schumaker K, Zhu JK. 2004. Molecular genetic perspectives on cross-talk and specificity in abiotic stress signalling in plants. *Journal of Experimental Botany* **55**:225–236.

Chinnusamy V, Zhu J, Zhu JK. 2006. Salt stress signaling and mechanisms of plant salt tolerance. *Genetic Engineering: Principles and Methods* **27**:141–177.

Clarke NA, Hetschkun H, Jones C, Boswell E, Marfaing H. 1993. Identification of stress tolerance traits in sugar beet. In: Jackson MB, Black CR, eds. *Interaction stresses on plants in a changing climate.* Berlin: Springer, 511–529.

Daoud S, Harrouni C, Huchzermeyer B, Koyro HW. 2008. Comparison of salinity tolerance of two related subspecies of *Beta vulgaris*: the sea beet (*Beta vulgaris* ssp. *maritima*) and the sugar beet (*Beta vulgaris* ssp. vulgaris). In: Abdelly C, Öztürk M, Ashraf M, Grignon C, eds. *Biosaline agriculture and high salinity tolerance*, Part Section I. Basel: Birkhäuser Publisher, 115–129.

Dohm JC, Minoche AE, Holtgräwe D, Capella-Gutiérrez S, Zakrzewski F, Tafer H, Rupp O, Sörensen TR, Stracke R, Reinhardt R, Goesmann A, Kraft T, Schulz B, Stadler PF, Schmidt T, Gabaldón T, Lehrach H, Weisshaar B, Himmelbauer H. 2014. The genome of the recently domesticated crop plant sugar beet (*Beta vulgaris*). *Nature* **505**: 546–549.

Ehlert C, Maurel C, Tardieu F, Simonneau T. 2009. Aquaporin-mediated reduction in maize root hydraulic conductivity impacts cell turgor and leaf elongation even without changing transpiration. *Plant Physiology* **150**:1093–1104.

Finkelstein R. 2013. Abscisic acid synthesis and response. *The Arabidopsis Book* **11**:e0166.

Flowers TJ, Colmer TD. 2015. Plant salt tolerance: adaptations in halophytes. *Annals of Botany* **115**:327–331.

Fricke W, Bijanzadeh E, Emam Y, Knipfer T. 2013. Root hydraulics in salt-stressed wheat. *Functional Plant Biology* **41**:366–378.

Fritz M, Lorenzen S, Popova M, Ehwald R. 2010. Transient and permanent changes of xylem sap exudation by root systems of *Zea mays* after application of hydrostatic and osmotic forces. *Functional Plant Biology* **37**:813–827.

Gajdanowicz P, Michard E, Sandmann M, Rocha M, Corrêa LG, Ramírez-Aguilar SJ, Gomez-Porras JL, González W, Thibaud JB, Van Dongen JT, Dreyer I. 2011. Potassium (K^+) gradients serve as a mobile energy source in plant vascular tissues. *Proceedings of the National Academy of Sciences of the USA* **108**:864–869.

Galvan-Ampudia CS, Testerink C. 2011. Salt stress signals shape the plant root. *Current Opinion in Plant Biology* **14**:296–302.

Gambetta GA, Manuck CM, Drucker ST, Shaghasi T, Fort K, Matthews MA, Walker MA, Mcelrone AJ. 2012. The relationship between root hydraulics and scion vigour across Vitis rootstocks: what role do root aquaporins play? *Journal of Experimental Botany* **63**:6445–6455.

Gambetta GA, Fei J, Rost TL, Knipfer T, Matthews MA, Shackel KA, Walker MA, Mcelrone AJ. 2013. Water uptake along the length of grapevine fine roots: developmental anatomy, tissue-specific aquaporin expression, and pathways of water transport. *Plant Physiology* **163**:1254–1265.

Geng Y, Wu R, Wee CW, Xie F, Wei X, Chan PMY, Tham C, Duan L, Dinneny JR. 2013. A spatio-temporal understanding of growth regulation during the salt stress response in *Arabidopsis*. *The Plant Cell* **25**:2132–2154.

Ghoulam C, Foursy A, Fares K. 2002. Effects of salt stress on growth, inorganic ions and proline accumulation in relation to osmotic adjustment in five sugar beet cultivars. *Environmental and Experimental Botany* **47**:39–50.

Hachez C, Chaumont F. 2010. Aquaporins: a family of highly regulated multifunctional channels. *Advances in Experimental Medicine and Biology* **679**:1–17.

Hachez C, Zelazny E, Chaumont F. 2006. Modulating the expression of aquaporin genes in planta: a key to understand their physiological functions? *Biochimica et Biophysica Acta (BBA) - Biomembranes* **1758**:1142–1156.

Hachez C, Veselov D, Ye Q, Reinhardt H, Knipfer T, Fricke W, Chaumont F. 2012. Short-term control of maize cell and root water permeability through plasma membrane aquaporin isoforms. *Plant, Cell and Environment* **35**:185–198.

Henzler T, Waterhouse RN, Smyth AJ, Carvajal M, Cooke DT, Schäffner AR, Steudle E, Clarkson DT. 1999. Diurnal variations in hydraulic conductivity and root pressure can be correlated with the expression of putative aquaporins in the roots of *Lotus japonicus*. *Planta* **210**:50–60.

Horie T, Kaneko T, Sugimoto G, Sasano S, Panda SK, Shibasaka M, Katsuhara M. 2011. Mechanisms of water transport mediated by PIP aquaporins and their regulation via phosphorylation events under salinity stress in barley roots. *Plant and Cell Physiology* **52**:663–675.

Horie T, Karahara I, Katsuhara M. 2012. Salinity tolerance mechanisms in glycophytes: an overview with the central focus on rice plants. *Rice* **5**:11.

Jang JY, Kim DG, Kim YO, Kim JS, Kang H. 2004. An expression analysis of a gene family encoding plasma membrane aquaporins in response to abiotic stresses in *Arabidopsis thaliana*. *Plant Molecular Biology* **54**:713–725.

Javot H, Maurel C. 2002. The role of aquaporins in root water uptake. *Annals of Botany* **90**:301–313.

Javot H, Lauvergeat V, Santoni V, Martin-Laurent F, Güçlü J, Vinh J, Heyes J, Franck KI, Schäffner AR, Bouchez D, Maurel C. 2003. Role of a single aquaporin isoform in root water uptake. *The Plant Cell* **15**:509–522.

Jozefkowicz C, Rosi P, Sigaut L, Soto G, Pietrasanta LI, Amodeo G, Alleva K. 2013. Loop A is critical for the functional interaction of two *Beta vulgaris* PIP aquaporins. *PLoS ONE* **8**:e57993.

Kaplan A, Gale J. 1974. Modification of the pressure-bomb technique for measurement of osmotic potential in halophytes. *Journal of Experimental Botany* **25**:663–668.

Karley AJ, White PJ. 2009. Moving cationic minerals to edible tissues: potassium, magnesium, calcium. *Current Opinion in Plant Biology* **12**:291–298.

Knipfer T, Fricke W. 2010. Root pressure and a solute reflection coefficient close to unity exclude a purely apoplastic pathway of radial water transport in barley (*Hordeum vulgare*). *New Phytologist* **187**:159–170.

Knipfer T, Fricke W. 2011. Water uptake by seminal and adventitious roots in relation to whole-plant water flow in barley (*Hordeum vulgare* L.). *Journal of Experimental Botany* **62**:717–733.

Knipfer T, Besse M, Verdeil JL, Fricke W. 2011. Aquaporin-facilitated water uptake in barley (*Hordeum vulgare* L.) roots. *Journal of Experimental Botany* **62**:4115–4126.

Krishnamurthy P, Ranathunge K, Nayak S, Schreiber L, Mathew MK. 2011. Root apoplastic barriers block Na^+ transport to shoots in rice (*Oryza sativa* L.). *Journal of Experimental Botany* **62**: 4215–4228.

Lindhauer MG, Haeder HE, Beringer H. 1990. Osmotic potentials and solute concentrations in sugar beet plants cultivated with varying potassium/sodium ratios. *Journal of Plant Nutrition and Soil Science* **153**:25–32.

Liu C, Li C, Liang D, Wei Z, Zhou S, Wang R, Ma F. 2012. Differential expression of ion transporters and aquaporins in leaves may contribute to different salt tolerance in *Malus* species. *Plant Physiology and Biochemistry* **58**:159–165.

Luu DT, Martinière A, Sorieul M, Runions J, Maurel C. 2012. Fluorescence recovery after photobleaching reveals high cycling dynamics of plasma membrane aquaporins in *Arabidopsis* roots under salt stress. *The Plant Journal* **69**:894–905.

Mahdieh M, Mostajeran A, Horie T, Katsuhara M. 2008. Drought stress alters water relations and expression of PIP-type aquaporin genes in *Nicotiana tabacum* plants. *Plant and Cell Physiology* **49**:801–813.

Martínez-Ballesta MC, Aparicio F, Pallás V, Martínez V, Carvajal M. 2003. Influence of saline stress on root hydraulic conductance and PIP expression in *Arabidopsis*. *Journal of Plant Physiology* **160**:689–697.

Maurel C. 1997. Aquaporins and water permeability of plant membranes. *Annual Review of Plant Physiology and Plant Molecular Biology* **48**:399–429.

Maurel C, Simonneau T, Sutka M. 2010. The significance of roots as hydraulic rheostats. *Journal of Experimental Botany* **61**:3191–3198.

Mittler R, Blumwald E. 2015. The roles of ROS and ABA in systemic acquired acclimation. *The Plant Cell* **27**:64–70.

Munns R, Tester M. 2008. Mechanisms of salinity tolerance. *Annual Review of Plant Biology* **59**:651–681.

Muries B, Faize M, Carvajal M, Martínez-Ballesta MC. 2011. Identification and differential induction of the expression of aquaporins by salinity in broccoli plants. *Molecular BioSystems* **7**:1322–1335.

Mutasa-Göttgens ES, Joshi A, Holmes HF, Hedden P, Göttgens B. 2012. A new RNASeq-based reference transcriptome for sugar beet and its application in transcriptome-scale analysis of vernalization and gibberellin responses. *BMC Genomics* **13**:99.

Niu X, Bressan RA, Hasegawa PM, Pardo JM. 1995. Ion homeostasis in NaCl stress environments. *Plant Physiology* **109**:735–742.

Pakniyat H, Armion M. 2007. Sodium and proline accumulation as osmoregulators in tolerance of sugar beet genotypes to salinity. *Pakistan Journal of Biological Sciences* **10**:4081–4086.

Peng YH, Zhu YF, Mao YQ, Wang SM, Su WA, Tang ZC. 2004. Alkali grass resists salt stress through high [K^+] and an endodermis barrier to Na^+. *Journal of Experimental Botany* **55**:939–949.

Postaire O, Tournaire-Roux C, Grondin A, Boursiac Y, Morillon R, Schäffner AR, Maurel C. 2010. A PIP1 aquaporin contributes to hydrostatic pressure-induced water transport in both the root and rosette of *Arabidopsis*. *Plant Physiology* **152**:1418–1430.

Qi X, Tai CY, Wasserman BP. 1995. Plasma membrane intrinsic proteins of *Beta vulgaris* L. *Plant Physiology* **108**:387–392.

Rahnama A, James RA, Poustini K, Munns R. 2010. Stomatal conductance as a screen for osmotic stress tolerance in durum wheat growing in saline soil. *Functional Plant Biology* **37**:255–263.

Reid KE, Olsson N, Schlosser J, Peng F, Lund ST. 2006. An optimized grapevine RNA isolation procedure and statistical determination of reference genes for real-time RT-PCR during berry development. *BMC Plant Biology* **6**:27.

Sack L, Holbrook NM. 2006. Leaf hydraulics. *Annual Review of Plant Biology* **57**:361–381.

Sade N, Moshelion M. 2014. The dynamic isohydric–anisohydric behavior of plants upon fruit development: taking a risk for the next generation. *Tree Physiology* **34**:1199–1202.

Schenk HJ, Espino S, Goedhart CM, Nordenstahl M, Cabrera HIM, Jones CS. 2008. Hydraulic integration and shrub growth form linked across continental aridity gradients. *Proceedings of the National Academy of Sciences of the USA* **105**:11248–11253.

Scholander PF, Hammel HT, Bradstreet ED, Hemmingsen EA. 1965. Sap pressure in vascular plants: negative hydrostatic pressure can be measured in plants. *Science* **148**:339–346.

Shabala S. 2013. Learning from halophytes: physiological basis and strategies to improve abiotic stress tolerance in crops. *Annals of Botany* **112**:1209–1221.

Shabala S, Cuin TA. 2008. Potassium transport and plant salt tolerance. *Physiologia Plantarum* **133**:651–669.

Shabala S, Shabala S, Cuin TA, Pang J, Percey W, Chen Z, Conn S, Eing C, Wegner LH. 2010. Xylem ionic relations and salinity tolerance in barley. *The Plant Journal* **61**:839–853.

Sharp RE, Poroyko V, Hejlek LG, Spollen WG, Springer GK, Bohnert HJ, Nguyen HT. 2004. Root growth maintenance during water deficits: physiology to functional genomics. *Journal of Experimental Botany* **55**:2343–2351.

Skorupa-Kłaput M, Szczepanek J, Kurnik K, Tretyn A, Tyburski J. 2015. The expression patterns of plasma membrane aquaporins in leaves of sugar beet and its halophyte relative, *Beta vulgaris* ssp. maritima, in response to salt stress. *Biologia* **70**:467–477.

Soto G, Fox R, Ayub N, Alleva K, Guaimas F, Erijman EJ, Mazzella A, Amodeo G, Muschietti J. 2010. TIP5;1 is an aquaporin specifically targeted to pollen mitochondria and is probably involved in nitrogen remobilization in *Arabidopsis thaliana*. *The Plant Journal* **64**: 1038–1047.

Soto G, Stritzler M, Lisi C, Alleva K, Pagano ME, Ardila F, Mozzicafreddo M, Cuccioloni M, Angeletti M, Ayub ND. 2011. Acetoacetyl-CoA thiolase regulates the mevalonate pathway during abiotic stress adaptation. *Journal of Experimental Botany* **62**:5699–5711.

Steudle E. 2000. Water uptake by plant roots: an integration of views. *Plant and Soil* **226**:45–56.

Steudle E, Peterson CA. 1998. How does water get through roots? *Journal of Experimental Botany* **49**:775–788.

Suku S, Knipfer T, Fricke W. 2014. Do root hydraulic properties change during the early vegetative stage of plant development in barley (*Hordeum vulgare*)? *Annals of Botany* **113**:385–402.

Sutka M, Li G, Boudet J, Boursiac Y, Doumas P, Maurel C. 2011. Natural variation of root hydraulics in *Arabidopsis* grown in normal and salt-stressed conditions. *Plant Physiology* **155**:1264–1276.

Takase T, Ishikawa H, Murakami H, Kikuchi J, Sato-Nara K, Suzuki H. 2011. The circadian clock modulates water dynamics and aquaporin expression in *Arabidopsis* roots. *Plant and Cell Physiology* **52**:373–383.

Tardieu F, Davies WJ. 1993. Integration of hydraulic and chemical signaling in the control of stomatal conductance and water status of droughted plants. *Plant, Cell and Environment* **16**:341–349.

Tester M, Davenport R. 2003. Na^+ tolerance and Na^+ transport in higher plants. *Annals of Botany* **91**:503–527.

Turner NC. 1981. Techniques and experimental approaches for the measurement of plant water status. *Plant and Soil* **58**:339–366.

Tyerman SD, Niemietz CM, Bramley H. 2002. Plant aquaporins: multifunctional water and solute channels with expanding roles. *Plant, Cell and Environment* **25**:173–194.

Tyree MT. 2003. Hydraulic properties of roots. In: Visser EJW, De Kroon H, eds. *Root ecology*. Berlin: Springer, 125–150.

Vandeleur RK, Mayo G, Shelden MC, Gilliham M, Kaiser BN, Tyerman SD. 2009. The role of plasma membrane intrinsic protein aquaporins in water transport through roots: diurnal and drought stress responses reveal different strategies between isohydric and anisohydric cultivars of grapevine. *Plant Physiology* **149**: 445–460.

Vandeleur RK, Sullivan W, Athman A, Jordans C, Gilliham M, Kaiser BN, Tyerman SD. 2014. Rapid shoot-to-root signalling regulates root hydraulic conductance via aquaporins. *Plant, Cell and Environment* **37**:520–538.

Van den Honert TH. 1948. Water transport in plants as a catenary process. *Discussions of the Faraday Society* **3**:146–153.

Wan H, Zhao Z, Qian C, Sui Y, Malik AA, Chen J. 2010. Selection of appropriate reference genes for gene expression studies by quantitative real-time polymerase chain reaction in cucumber. *Analytical Biochemistry* **399**:257–261.

Wang W, Yang X, Zhang S, Sun Y. 2013. The root cortex cell hydraulic conductivity is enhanced with increasing chromosome ploidy in wheat. *Plant Physiology and Biochemistry* **68**:37–43.

Weatherley PE. 1950. Studies in the water relations of the cotton plant. I. The field measurement of water deficits in leaves. *New Phytologist* **49**:81–97.

12

Collaboration between grass seedlings and rhizobacteria to scavenge organic nitrogen in soils

James F. White*, Qiang Chen, Mónica S. Torres, Robert Mattera, Ivelisse Irizarry, Mariusz Tadych and Marshall Bergen

Department of Plant Biology and Pathology, Rutgers University, New Brunswick, NJ, USA

Associate Editor: James F. Cahill

Abstract. Plants require nitrogen (N) to make proteins, nucleic acids and other biological molecules. It is widely accepted that plants absorb inorganic forms of N to fill their needs. However, recently it has become clear that plants also have the capacity to absorb organic N from soils. In this paper we describe a new kind of symbiosis involving seed-vectored rhizobacteria and grasses that is targeted at enhancing acquisition of organic N from soils. Our proposal is based on results of experiments on seedlings of grass species *Festuca arundinacea* Schreb., *Lolium perenne* L. and *Poa annua* L. that suggest: (i) seed-vectored rhizobacteria colonize seedling roots and influence their development; (ii) reactive oxygen secretion by seedling roots plays a role in organic N procurement by denaturing microbial proteins in the vicinity of roots (daytime activity); and (iii) plant root and microbial proteases degrade denatured proteins prior to absorption by roots (night-time activity). This research involved the following types of studies: (i) seedling root development experiments with and without rhizobacteria on a variety of substrates in agarose media and (ii) isotopic N-tracking experiments to evaluate the absorption into seedlings of N obtained from degradation of proteins. We hypothesize that grasses, in particular, are adapted to scavenge organic N from soils through application of this 'oxidative nitrogen scavenging' symbiosis with rhizobacteria, and their soil-permeating root systems. This newly discovered symbiosis in grass species could lead to new ways to cultivate and manage grasses to enhance efficiency of N utilization and reduce applications of inorganic fertilizers.

Keywords: Grasses; microbiome; nitrogen use efficiency; oxidative nitrogen scavenging; plant growth-promoting rhizobacteria; symbiosis.

Introduction

All organisms require nitrogen (N) throughout development to make proteins, nucleic acids and other biological molecules. For plant growth, N is often the limiting nutrient. It has been believed that N available to plants was exclusively in the inorganic nitrate form. However, it has become clear that plants may obtain N from organic sources (Godlewski and Adamczyk 2007; Paungfoo-Lonhienne *et al.* 2008, 2012; Adamczyk *et al.* 2010). The need to push crop plants to higher yields has provoked heavy reliance on inorganic N fertilizers (Herridge *et al.* 2008). It has been estimated that only 30–50 % of applied inorganic N fertilizer is actually taken up by crop plants, with the rest lost in run-off, leaching or volatilization, increasing the probability for unintended environmental pollution (Mulvaney *et al.* 2009).

* Corresponding author's e-mail address: white@rci.rutgers.edu, jwhite3728@gmail.com

Agricultural scientists and plant breeders have targeted improvements in N use efficiency of plants as a way to reduce reliance on N applications (Garnett *et al.* 2009). Grasses associate with arbuscular mycorrhizae that function to absorb phosphorus and N from soils. Years of investigation on mycorrhizae have yielded volumes of information about the biology of mycorrhizal-plant associations (Gutjahr and Parniske 2013). However, comparatively little is known about other types of microbial endophytes or epiphytes that associate with healthy plants (Schulz and Boyle 2005). White *et al.* (2012) reported the occurrence of seed-vectored bacteria on many cool-season grasses. The bacteria, vectored on seed surfaces attached to adherent lemmas and paleas, rapidly colonize the roots of the seedlings after germination. The frequency of root colonizing bacteria on seeds of grasses suggested that bacteria could be involved in a nutritional symbiosis with plants; however, insufficient information was available to delineate clearly how the rhizobacteria benefited plants. In this paper we report studies conducted to evaluate the role of seed-vectored rhizobacteria in the nutrition of cool-season grass seedlings. In addition, we examine the role of grass root-secreted reactive oxygen in the process of protein degradation around roots. Some of the questions we pursued in this research include the following: Is there a survival/nutritional function for the seed-vectored bacteria on grass seeds? Do rhizobacteria facilitate organic N acquisition by grass seedlings? Does reactive oxygen secretion from grass roots relate to the function of seed-vectored bacteria?

Methods

Plant materials

Seeds used were selections of tall fescue (*Festuca arundinacea* Schreb.), perennial ryegrass (*Lolium perenne* L.) and annual bluegrass (*Poa annua* L.) obtained from the Rutgers University turf grass-breeding programme.

Bacteria isolation and identification

To isolate bacteria, 40 seeds of each grass species were washed for 5 min with constant agitation in sterile water, then plated onto yeast extract-sucrose agar (YES) and 10 % tryptic soy agar (TSA; Difco). YES agar contained 1 % yeast extract (Difco), 1 % sucrose (Sigma-Aldrich, St Louis, MO, USA) and 1.2 % agar (Difco). Twenty seeds were plated on each medium (five seeds per plate) and incubated 1 week at laboratory ambient temperature. The most common colonies appearing were selected for sequencing of the 16S rDNA region. Bacterial DNA was extracted using the UltraClean® Microbial DNA Isolation Kit (MoBio Laboratories, Carlsbad, CA, USA). The 16S rRNA gene was amplified using the universal primers 27F (5′-AGAGTTTGATCMTGGCTCAG) and 1492R (5′-CGGTTACC TTGTTACGACTT). Amplicons were sent to be sequenced by Genewiz (South Plainfield, NJ, USA). High-quality sequences were selected and bacteria were identified by searching for sequence similarity using BLAST (NCBI GenBank).

Removal of bacteria from seeds

To conduct seedling development studies, seeds free of bacteria were prepared by surface disinfection. This was accomplished by agitating seeds on a rotary shaker in a 3 % sodium hypochlorite solution for 35–40 min, after which seeds were rinsed twice for 5 min in sterile water. We optimized the surface disinfection conditions so that seed germination frequency was not reduced but rhizobacteria no longer could be seen on or isolated from resulting seedlings.

Preparation of media for seedling development experiments

Water agarose (0.7 %; Type 1 Low EEO, Sigma-Aldrich) was used as the base medium. This medium contained no added nutrient source and at 0.7 % agarose concentration was an ideal consistency for grass seedling root penetration. This base medium was amended using various vitamin or nutrient substrates (Tables 1–3). The following nutrient sources were employed in experiments: ammonium nitrate (Sigma-Aldrich), alanine (Sigma-Aldrich), biotin (Sigma-Aldrich), egg albumin (Sigma-Aldrich), catalase (Sigma-Aldrich), cellulase (Sigma-Aldrich), glycine (Sigma-Aldrich), lipase (Sigma-Aldrich), myo-inositol (Sigma-Aldrich), nicotinic acid (Sigma-Aldrich), pronase® (Behring Diagnostics), sodium ascorbate (Sigma-Aldrich) and yeast extract (Sigma-Aldrich). Nutrient sources or vitamins were added to media after autoclaving, except where the protein is identified as denatured (Tables 2 and 3). Media was poured into 6 cm plastic Petri plates.

Seedling development experiments

Seeds with and without native bacterial populations were placed onto plates. For each treatment 8 plates per treatment were used with 10–15 seeds on each plate. Plates were incubated for 10 days at laboratory ambient temperature in a 12-h alternating light/dark cycle. Seedlings were examined and assessed for the percentage of seedlings with roots growing downward into agarose and whether downward growing roots showed development of root hairs (Tables 1–3). Any inhibition of root growth was also noted.

Table 1. Annual bluegrass (*Poa annua*) seedling root hair growth with and without seed-vectored rhizobacteria.

Treatment	Bacteria present		Bacteria absent	
	% Roots vertical	Root hairs	% Roots vertical	Root hairs
Agarose	100.00	+	11.12	–
Yeast extract (0.01 %)	91.67	+	75.00	+
Glycine (0.01 %)	88.89	+	0.00	–
Biotin (0.01 %)	81.25	+	8.69	–
Nicotinic acid (0.01 %)	61.54	+	0.00	–
Myo-inositol (0.01 %)	83.33	+	10.00	–
Lipase (0.1 %)	N/A (root inhibition)		64.10	+
Albumin (0.1 %)	N/A (root inhibition)		52.94	+
Pronase (0.1 %)	N/A (root inhibition)		82.05	+
Catalase (0.01 %)	N/A (root inhibition)		77.08	+
Lipase (0.1 %) + 0.01 % ascorbate (pH 5.6)	N/A		N/A	–
Lipase (0.1 %) + 0.001 % ascorbate (pH 5.4)	N/A		N/A	+

Table 2. Tall fescue (*F. arundinacea*) seedling root growth with and without seed-vectored rhizobacteria.

Treatment	Bacteria present		Bacteria absent	
	% Roots vertical	Root hairs	% Roots vertical	Root hairs
Agarose	87.50	+	16.67	–
Yeast extract (0.01 %)	94.12	+	87.50	+
Alanine (0.01 %)	87.50	+	6.67	–
Glycine (0.01 %)	93.75	+	87.50	+
Denatured albumin (0.1 %)	65.20	+	93.75	+
Ammonium nitrate (0.01 %)	93.75	+	0.00	–

Table 3. Perennial ryegrass (*L. perenne*) seedling root growth with and without seed-vectored rhizobacteria.

Treatment	Bacteria present		Bacteria absent	
	% Roots vertical	Root hairs	% Roots vertical	Root hairs
Agarose	77.78	+	18.75	–
Yeast extract (0.01 %)	75.00	+	70.58	+
Alanine (0.01 %)	70.84	+	11.12	–
Glycine (0.01 %)	72.23	+	26.67	–
Denatured albumin (0.1 %)	Root suppression		81.82	+
Ammonium nitrate (0.01 %)	87.50	+	21.05	–

Microscopic examination of roots

For microscopic examinations (Figs 1–3), roots were stained for 8–10 h by flooding plates with 2.5 mM diaminobenzidine tetrachloride (DAB) and 5 purpurogallin units mL^{-1} of horseradish peroxidase (Type VI, Sigma-Aldrich) (White *et al.* 2012, 2014*b*). Stain was then poured off and plates rinsed several times with water. Seedling roots were examined microscopically using a compound

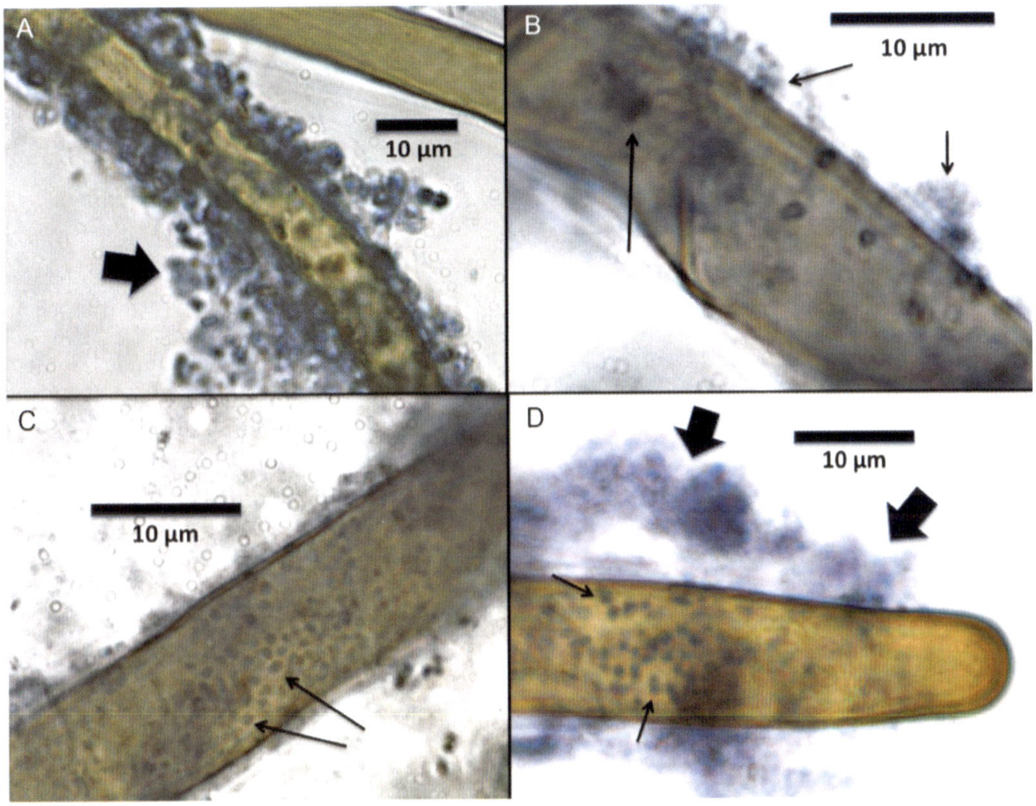

Figure 1. Bacteria on root hairs of cool-season grass seedlings; stained with DAB/peroxidase for reactive oxygen (brown) and counterstained with aniline blue/lacto-phenol for protein. (A) Bacteria (arrow) on surface of root hair of *L. perenne* seedling. (B) Bacteria and bacterial protein (arrows) on surface of root hair of *P. annua* seedling. (C) Bacteria (arrows) on surface of root hair of *P. annua* seedling. (D) Bacteria (small arrows) and denatured proteins (large arrows) on the surface of root hair of *P. annua* seedling.

light microscope through the reverse side of plates or by removing seedlings from agarose and placing in a drop of aniline blue/lacto-phenol stain on a slide (White *et al.* 2012, 2014*b*). For the evaluation of root effects on protein particles, examination was made only through the reverse of plates. DAB/horseradish peroxidase was used to stain H_2O_2 (red to brown colour); while aniline blue/lacto-phenol was used to stain proteins and bacterial cytoplasm (light blue to dense blue colour).

Production of ^{15}N-labelled protein mixture

To label bacterial proteins, *Bacillus amyloliquefaciens* (strain HF3) was inoculated into ^{15}N-labelling medium composed of 1 % sucrose, Murashige and Skoogs salt base without nitrogen (Sigma-Aldrich), 0.02 % yeast extract and 0.16 % ^{15}N-labelled glycine (Sigma-Aldrich) in 200 mL Erlenmeyer flasks. Flasks were then incubated at laboratory ambient temperature with constant agitation on a rotary shaker for 15 days. Cells were harvested from cultures by centrifugation and washed twice with phosphate-buffered saline solution. To extract proteins, cells were suspended in ice-cold acetone for 5 min, washed under nitrogen and then incubated in 1 % sodium dodecyl sulfate for 2 min. Extracted proteins were then freeze-dried to remove water; 0.16 g of labelled proteins was obtained. The labelled proteins were thoroughly mixed with 0.84 g of un-labelled egg albumin (Sigma-Aldrich) prior to use in media.

^{15}N-labelled protein absorption experiment

Plastic Petri plates (6 cm diameter) containing either 0.7 % agarose or 0.7 % agarose with 0.05 % of the labelled-protein mixture were prepared. The labelled-protein mixture was added to agarose after autoclaving with the agarose at ~70 °C and rapidly cooled to minimize additional protein denaturation. Twelve tall fescue seeds, either with seed-vectored rhizobacteria present or with rhizobacteria removed by surface disinfection as previously described, were placed onto plates of both media. The following treatments were prepared: (i) no protein in medium and seeds with rhizobacteria in 12-h alternating light/dark cycles; (ii) no protein in medium and no rhizobacteria on seeds in 12-h alternating light/dark cycles; (iii) protein in medium and rhizobacteria on seeds in 12-h alternating light/dark cycles; (iv) protein in medium but seeds without rhizobacteria on seeds in 12-h alternating light/dark cycles; (v) no protein in medium and no rhizobacteria on seeds maintained in total darkness; and

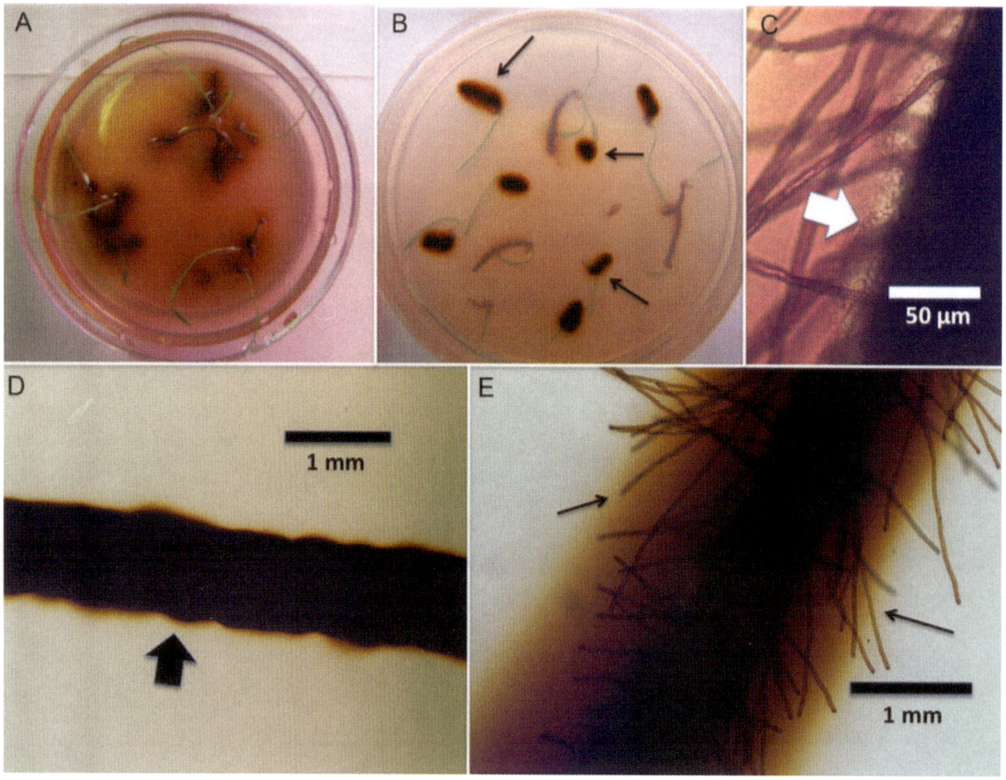

Figure 2. *Poa annua* seedlings showing reactive oxygen (H_2O_2) staining around roots. (A) Seedlings growing on 0.7 % agarose showing diffuse zones of reactive oxygen (brown) around roots. (B) Seedlings growing on 0.1 % albumin agarose showing dense zones of reactive oxygen (arrows) around roots. (C) Root surface showing root hairs and layer of bacteria (arrow). (D) Root without bacteria growing in 0.7 % water agarose medium, showing absence of root hairs. (E) Root with bacteria growing on 0.7 % water agarose medium, showing reactive oxygen zone and root hairs (arrows).

(vi) proteins in medium but seeds without rhizobacteria and maintained in total darkness. Three to six replicate plates were made for each treatment (Fig. 4). Plates in dark treatments were wrapped in aluminium foil to exclude light. All plates were placed in a clear plastic sealable canister and placed under fluorescent lights at laboratory ambient temperature for a 10-day period. After the incubation period, the shoots were excised from seedlings, with all shoots on a plate being combined and dried in an oven at 60 °C overnight. After drying, samples were further processed and analysed for $^{15}N/^{14}N$ ratio using mass spectroscopy at the Stable Isotope Laboratory at the Odum School of Ecology at the University of Georgia, Athens, GA, USA. Results were expressed as $\delta^{15}N$ vs. air measurements (Fig. 4). Statistical analysis of data involved application of the Duncan's multiple range test ($P < 0.05$).

Results

Identification of culturable bacteria

Culturable seed-vectored bacteria from tall fescue include *Pseudomonas* sp. (JX089400; 100 % identical) and *Pantoea agglomerans* (JX089401; 100 % identical); from perennial ryegrass *B. amyloliquefaciens* (KP053610; 99 % identical) and *Staphylococcus epidermidis* (KP053611; 99 % identical); and from annual bluegrass *Paenibacillus hordei* (KJ528493; 98 % identical) and *Pantoea* sp. (KJ528494; 99 % identical).

Seedling development experiments

Seedlings that possessed the rhizobacteria showed a majority of the seedling roots to be downward oriented with abundant root hair growth (Tables 1–3). Removal of rhizobacteria from surfaces of seed coats resulted in anomalous root development on agarose without nutrients. Without bacteria, the majority of the seedling roots grew on the surface of the agarose, and any roots that grew downward into the agarose lacked root hair development. Substances that were found to consistently restore normal root development in bacterial-free seedlings included yeast extract and various proteins (Tables 1–3). In tall fescue the amino acid glycine was seen to restore seedling development, but this was not seen in the other grasses (Tables 1–3). When rhizobacteria were present on seedlings, protein levels that stimulated proper seedling root development in seedlings without rhizobacteria were

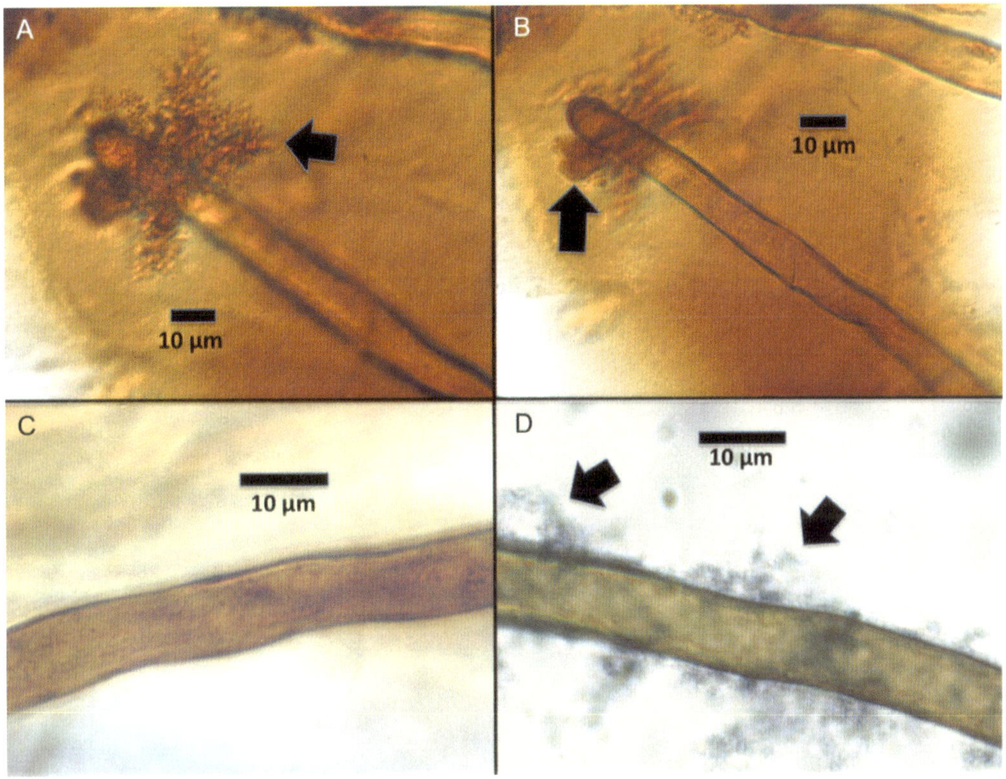

Figure 3. *Poa annua* root hairs stained for reactive oxygen. (A) Root hair showing denatured cellulase enzyme aggregation (arrow) on root hair with high reactive oxygen staining on protein as evidenced by darker brown colour. (B) Root hair showing the smoothing effect of the cellulase particle (arrow) in the area of close contact with the root hair. (C) Root hair without protein or bacteria. (D) Root hair showing adherent albumin protein (arrows).

instead inhibitory (Table 2). The addition of the reactive oxygen scavenger sodium ascorbate to agarose-containing proteins was seen to suppress the capacity of the proteins to trigger proper seedling root development in the absence of rhizobacteria (Table 1). The addition of ammonium nitrate to agarose was not sufficient to restore proper seedling root development in the absence of rhizobacteria (Tables 2 and 3).

Visualization of rhizobacteria and proteins on roots

Rhizobacteria could be visualized adhering closely to the surfaces of root hairs (Fig. 1A–D). Proteins could also be visualized in the vicinity of bacteria (Fig. 1B and D). DAB/horseradish peroxidase staining showed that H_2O_2 was abundantly secreted into agarose from roots (Fig. 2A). In media that contained protein H_2O_2 staining appeared to be more intense than in agarose without protein (Fig. 2B). H_2O_2 appeared to be secreted from the central axis of the seedling roots, rather than from root hairs (Fig. 2C–E). Visual evidence of root protease activity was seen in the smoothing of protein crystals that were in contact with root hairs (Fig. 3A and B). Seedlings grown on protein media frequently had protein accumulations on root hair surfaces (Fig. 3C and D).

^{15}N-labelled-protein absorption experiment

The protein mix itself had a $\delta^{15}N$ vs. air of 8869.57, indicating that it contained a substantial amount of ^{15}N-labelled proteins. The untreated tall fescue seed was 0.71 ± 0.04 $\delta^{15}N$ vs. air (all $\delta^{15}N$ vs. air values expressed as mean $\pm$ standard error of mean). In the experiment, the lowest absorption of ^{15}N from labelled protein (2488.49 ± 243.85 $\delta^{15}N$ vs. air) was seen in seedling shoots of tall fescue without rhizobacteria cultured in constant darkness in order to eliminate H_2O_2 secretion from roots (Fig. 4). The highest level of ^{15}N absorption from ^{15}N-labelled proteins (5415.46 ± 367.93 $\delta^{15}N$ vs. air) was seen in seedling shoots with rhizobacteria cultured under alternating light/dark conditions (Fig. 4). Seedling shoots without rhizobacteria grown in alternating light/dark conditions were intermediate (3817.49 ± 336.4 $\delta^{15}N$ vs. air) between the previous two treatments (Fig. 4). Minimal incorporation of ^{15}N into shoots was seen in the treatments without rhizobacteria grown on agarose without proteins (Fig. 4). Analysis of shoots of seedlings with rhizobacteria on water agarose free of any proteins showed several times more incorporation of the ^{15}N label (123.43 ± 14.4 $\delta^{15}N$ vs. air) compared with seedlings without

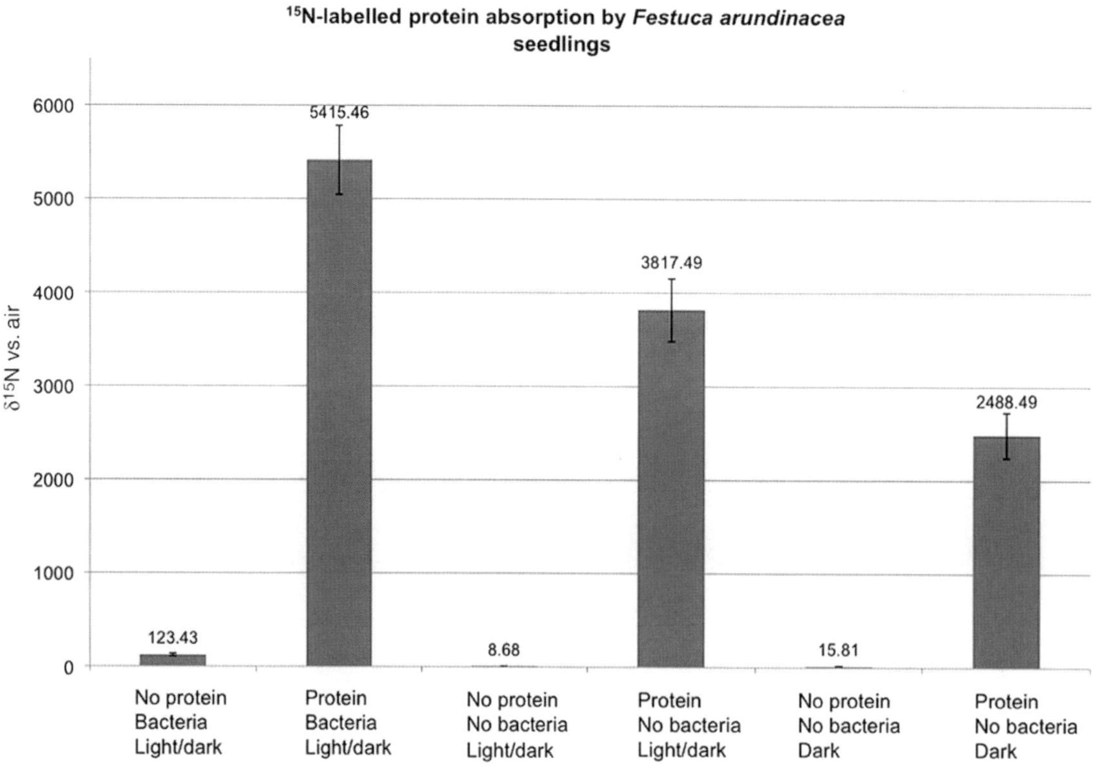

Figure 4. Results of ^{15}N-labelled-protein absorption experiment showing enhanced ^{15}N absorption due to rhizobacteria and reactive oxygen effects. All values expressed as mean ± standard error of mean; means with the same letter are not significantly different according to the Duncan's multiple range test ($P < 0.05$). The y-axis is the level of protein absorption (expressed as $\delta^{15}N$ vs. air) and the x-axis shows the six treatments. The numbers above bars are means of the $\delta^{15}N$ vs. air values and the bars show standard errors. The highest ^{15}N incorporation (5415.46) was seen in shoots of seedlings bearing bacteria grown on agarose containing ^{15}N-labelled protein in a 12-h alternating light/dark cycle (treatment = protein, bacteria, light/dark). Less ^{15}N incorporation (3817.49) was seen in shoots of seedlings grown under the same conditions but without bacteria (treatment = protein, no bacteria, light/dark). Slightly less ^{15}N incorporation (2488.49) was seen in shoots of seedlings grown on labelled proteins but without bacteria in total darkness to suppress reactive oxygen secretion (protein, no bacteria, dark). Minimal ^{15}N incorporation was seen into shoots of seedlings grown on agarose that did not contain ^{15}N-labelled protein (treatments = no protein, bacteria, light/dark; no protein, no bacteria, light/dark; no protein, no bacteria, dark).

bacteria under the same conditions (8.68 ± 1.31 $\delta^{15}N$ vs. air) (Fig. 4).

Discussion

Organic N absorption by roots

Previous research has shown that grasses, and plants in general, secrete proteases from their roots. Adamczyk *et al.* (2008, 2010) conducted studies on plant growth on proteins and presented evidence that plants actively degrade proteins without microbes. Further, Paungfoo-Lonhienne *et al.* (2008) showed that certain plants were capable of degrading proteins on the surfaces of root cells. Results in our seedling protein absorption experiment may reflect the activities of both plant and rhizobacterial proteases. The lowest ^{15}N absorption into seedling shoots on media with proteins was in the seedlings without rhizobacteria maintained in darkness (Fig. 4). In darkness, H_2O_2 secretion is minimized and absorption of organic N may reflect the activity of plant root proteases alone. A 53.4 % increase in incorporation of ^{15}N into shoots was observed in seedlings maintained under alternating light/dark conditions without microbes where H_2O_2 was secreted from roots in the light periods. The importance of reactive oxygen secretion from roots to the process of N scavenging may be seen in this 53.4 % increase in N absorption from protein over that seen in the seedlings without bacteria maintained in darkness where reactive oxygen was not secreted from roots. In our experiment, the real benefit of reactive oxygen may be lower than that seen in nature because our proteins were already partially denatured by the use of sodium dodecyl sulfate during the extraction process. The role of reactive oxygen in the process of protein degradation was further emphasized by results of the seedling development experiments (Table 1), where the addition of the antioxidant sodium ascorbate to agarose containing the enzyme lipase was seen to suppress development of root hairs. With reduced oxidative denaturing of lipase, the plant proteases alone were insufficient to degrade enough of the enzyme to trigger root hair development. Additional experiments will be needed to get a more accurate assessment of the real value of the pre-oxidation of proteins in the degradation process. The importance of the rhizobacteria to the process of N scavenging from proteins is evident in the 41.86 % increase in incorporation of ^{15}N into shoots of microbe bearing seedlings incubated in alternating light/dark conditions over that of seedlings without microbes under the same conditions. Secretion of H_2O_2 from roots and the added degradative capacity of the rhizobacteria increased organic N acquisition by seedlings ~119.74 % over what was seen in the seedlings without microbes or secreted H_2O_2. Evidence for the direct action of proteases produced by grass roots may be seen not only in accumulation of ^{15}N in shoots in the bacterial-free seedlings grown on the ^{15}N-labelled proteins, but also in the smoothing of protein crystals in direct contact with grass seedling root hairs (Fig. 3A and B). There is evidence that some plants (e.g. rice and wheat) under drought stress enhance their efforts to scavenge organic N from soils, where it has been shown that secreted proteases are up-regulated in plants under water stress (Kohli *et al.* 2012). The trigger for the shift to organic N use may be restriction of access to mineralized N due to reduced water flow through soils (Kohli *et al.* 2012).

Seedling development and rhizobacteria

The importance of the role of the rhizobacteria to nutrient acquisition by grass seedlings is emphasized in the results of seedling root development experiments with and without rhizobacteria on various substrates (Tables 1–3). Growth of seedlings with and without bacteria on nutrient-free agarose showed that bacteria stimulated roots to form root hairs and trigger the root gravitropic response where roots grew downward into the medium (Tables 1–3). In the absence of rhizobacteria, roots did not develop properly; roots tended to remain on the agarose surface and any roots that did penetrate into media did not produce root hairs. Seedling roots of all three species of grasses responded similarly when germinated on water agarose medium. It seems evident that grass seedlings interact with rhizobacteria in some way that triggers proper seedling root development. In seedlings without rhizobacteria, yeast extract and various proteins could consistently restore proper root development in triggering root hair formation and downward root growth. While we have no data that can clarify the molecular communication between rhizobacteria and grass seedling, the capacity of organic N to restore proper root development suggests that the function of rhizobacteria relates to organic N acquisition. This role is further emphasized because protein levels that were sufficient to stimulate root hair development without the presence of rhizobacteria were inhibitory to seedling growth when bacteria were present (Tables 1 and 3). Rhizobacteria appear to significantly enhance N scavenging capacity of grasses as is evident in results of the ^{15}N-labelled-protein absorption experiment (Fig. 4). Changes in plant development have also been shown for rice plants containing certain fungal endophytes (Rodriguez *et al.* 2009). It is entirely possible that improved growth in plants bearing other types of nonpathogenic microbes could also relate to increased supply of organic nutrients, with microbes functioning as collaborators of plants in scavenging organic N from soils or rhizosphere microbes.

Adaptations in grasses

The developmental dependence of grass seedling roots on the rhizobacteria suggests that grasses have evolved to rely on these seed-vectored rhizobacteria in order to maximize scavenging of organic N. Features of development of grasses seem consistent with this idea. Grasses possess lemmas and paleas that adhere tightly to the surface of the seed coat. We suggest that the adherent plant tissues are adaptations for vectoring the rhizobacteria on the seed surface. It is apparent that rhizobacteria are vectored on the seed surface because rigorous surface disinfection of seeds is sufficient to eliminate rhizobacteria from seedlings. Microbes vectored in and on seeds are likely to be essential elements of a plant's microbiome that must be transmitted reliably from parent to offspring. The bacteria that are vectored on seeds of the grasses that we examined are likely essential, since seedlings do not develop properly without them. It seems very likely that the rhizobacteria are important components of the 'functional microbiome' of grasses (Lakshmann *et al.* 2014; White *et al.* 2014*a*). Without competent rhizobacteria seedlings would likely be at a competitive disadvantage in terms of nitrogen supply.

Oxidative nitrogen scavenging

In a previous study we observed degradation of some of the bacteria on surfaces of grass seedling roots (White *et al.* 2012). We hypothesized that the direct oxidative degradation of bacteria and their proteins on roots could provide a source of N for plants, and we termed the process oxidative N scavenging (ONS). Our results here suggest that ONS involves the rhizobacteria integrally, and rather than being the exclusive target of ONS, they are crucial in its operation in providing additional proteases that efficiently degrade oxidatively denatured proteins. It seems likely that all microbial exoenzymes in the vicinity of roots would be vulnerable to degradation through ONS by grasses.

Diurnal model of ONS in grasses

We suggest a diurnal nature to ONS in grasses, with oxidative protein denaturation occurring only during the day and denatured protein degradation and absorption occurring at night (Fig. 5; White *et al.* 2012). This results in a temporal separation of oxidative and degradative processes. This temporal separation of oxidative and degradative processes is necessary because proteases of the plant and rhizobacteria are also vulnerable to oxidation by H_2O_2. In this cyclic process, grass seedlings collaborate with seed-vectored rhizobacteria to scavenge organic N in the vicinity of roots. Both grass and rhizobacteria have active roles in the decomposition of proteins around roots. Grass roots secrete reactive oxygen that denatures proteins, while rhizobacteria and grasses secrete proteases that further degrade proteins into smaller peptides or oligopeptides that may be absorbed by grass roots. Denaturation of proteins by reactive oxygen involves protein unfolding and increased exposure of peptide bonds to proteases (Fligiel *et al.* 1984; Davies 1987). Denatured proteins become highly susceptible to degradation by proteases (Davies 1987). Bongarzone *et al.* (1995) found enhanced susceptibility of peroxidized myelin proteins to degradation by *Bacillus* protease subtilisin. Several of the seed-vectored bacteria we isolated from grasses are known to produce secreted proteases, including *B. amyloliquefaciens*, *Paenibacillus hordei*, *Pantoea agglomerans* and *S. epidermidis* (Wells *et al.* 1983; Bott *et al.* 1988; Rodarte *et al.* 2011; Kim *et al.* 2013; Sugimoto *et al.* 2013). It is interesting to speculate that grasses may manage the rhizosphere to maximize protein production by rhizosphere microbes. The secretion of complex root exudates composed of carbohydrates and proteins into the rhizosphere could be targeted at encouraging secretion of microbial enzymes to degrade the exudate components. In essence, grasses may be cultivating microbes in the rhizosphere and harvesting their proteins to support growth (Walker *et al.* 2003). Validation of the diurnal oxidative nitrogen scavenging process could provide a key to understanding the mystery of why many plants grow faster during the night (Nusinow *et al.* 2012). The absorption of protein fragments during the evening hours could provide the rapidly acquired nitrogenous nutrients plants require to build enzymes needed for the growth spurt that occurs in the predawn hours.

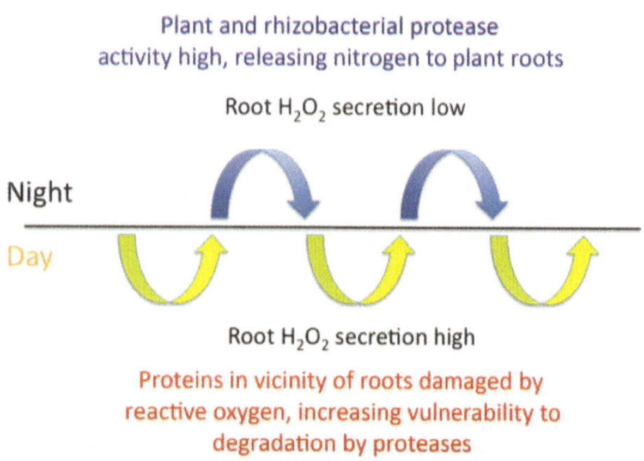

Figure 5. Proposed cyclic model of the oxidative nitrogen scavenging process in grasses. In daylight grasses secrete hydrogen peroxide from roots in order to denature proteins (microbial exoenzymes) around roots. At night plant and microbial proteases degrade oxidized proteins to form smaller peptides or oligopeptides that may be absorbed by roots.

Evidence for enhanced absorption of N from air

The results of the ^{15}N-labelled protein absorption experiment (Fig. 4) suggests that the seed-vectored rhizobacteria not only enhance absorption of organic N through increased degradation of proteins, but they may also increase N available to plants by scavenging it from air. In this experiment, Petri plates containing agarose media and seedlings were maintained together in a closed plastic canister. We hypothesize that the degradation of proteins by plants and microbes released ^{15}N-labelled ammonia or other compound into the air of the canister. Analysis of shoots of seedlings with rhizobacteria on water agarose free of any proteins showed a 14-fold increase in N absorption over that of seedlings grown under the same conditions but without rhizobacteria (Fig. 4). This difference can only be explained if rhizobacteria facilitated absorption of a ^{15}N-labelled volatile in the canister. It should be noted that seedlings without bacteria absorbed some ^{15}N as well. This is evident because seedlings without rhizobacteria or protein showed δ^{15}N vs. air values of 8.68 ± 1.31 (light) and 15.81 ± 2.23 (dark), while seed samples used in this experiment showed a 0.71 ± 0.04 δ^{15}N vs. air value and the agarose media showed no detectable N on analysis. It seems likely that the rhizobacteria are more efficient at capturing N than seedlings alone. Hurek *et al.* (1988) described a similar phenomenon in a study of bacteria associated with roots of kallar grass, where it was found that ammonia and nitrous oxide were being scavenged from the air by root-associated bacteria. Future experiments are needed to further evaluate the capacity of grass rhizobacteria to absorb and transfer to plants gaseous forms of N.

Conclusions

This paper presents observations and experiments that support occurrence of an organic N scavenging process that we hypothesize functions daily in grasses. It is necessary to rigorously evaluate whether ONS, as we have outlined it based on laboratory studies, is actually occurring in grasses in soils. This confirmation would likely involve larger scale N tracking experiments. It is necessary to elucidate the culturable and non-culturable rhizobacteria that assist plants to scavenge N. Each bacterium will need to be evaluated for its role in the process of N scavenging. It is possible that the bacteria that were most active in N scavenging were not cultured, or perhaps are not culturable. In most cases we can isolate multiple species of bacteria from grass seeds and seedling roots. It is unknown whether the rhizobacteria function in consortia, or if only a single species is functional. Finally, there are some important applications that can come from this research. If there are rhizobacteria that are more efficient at degrading soil proteins, it may be possible to use them to enhance the N scavenging efficiency of commercial grasses. This could have widespread positive effects, both economically and environmentally in reducing costs to apply fertilizers and reduce inorganic N applications that contaminate bodies of water. Further, Beltran *et al.* (2014) demonstrated that *Agave tequilana* plants are capable of absorption of N from ^{15}N-labelled bacteria, indicating that it is also capable of scavenging N from microbes. White *et al.* (2014*b*) demonstrated that reactive oxygen was associated with microbes in roots of seedlings of several species in other plant families, suggesting that the ONS process could be widespread in plants. A more complete understanding of which plants engage in N scavenging and exactly how it works in those plants could result in new approaches to cultivation and nourishment of crop plants.

Sources of Funding

This research was supported by funds from the John E. and Christina C. Craighead Foundation, the Rutgers Turfgrass Science Center, USDA-NIFA Multistate Project W3147 and the New Jersey Agricultural Experiment Station.

Contributions by the Authors

J.F.W. conceived the study; all authors contributed to experimental design; Q.C., I.I. and M.S.T. sequenced isolates, and M.T., R.M. and M.B. helped with experiments.

Acknowledgement

The authors are grateful to Drs Barbara Zilinskas and Chaim Frenkel for their ongoing encouragement and discussions regarding this research.

Literature Cited

Adamczyk B, Godlewski M, Zimny J, Zimny A. 2008. Wheat (*Triticum aestivum*) seedlings secrete proteases from roots and, after protein addition, grow well on medium without inorganic nitrogen. *Plant Biology* **10**:718–724.

Adamczyk B, Smolander A, Kitunen V, Godlewski M. 2010. Proteins as nitrogen source for plants, a short story about exudation of proteases by plant roots. *Plant Signaling & Behavior* **5**:817–819.

Beltran-Garcia MJ, White JF Jr, Prado FM, Prieto KR, Yamaguchi LF, Torres MS, Kato MJ, Medeiros MHG, Di Mascio P. 2014. Nitrogen acquisition in *Agave tequilana* from degradation of endophytic bacteria. *Scientific Reports* **4**:6938.

Bongarzone ER, Soto EF, Pasquini JM. 1995. Increased susceptibility to degradation by trypsin and subtilisin of in vitro peroxidized myelin proteins. *Neurochemical Research* **20**:421–426.

Bott R, Ultsch M, Kossiakoff A, Grayear T, Katz B, Power S. 1988. The three-dimensional structure of *Bacillus amyloliquefaciens* subtilisin at 1.8 Å and an analysis of the structural consequences of peroxide inactivation. *The Journal of Biological Chemistry* **263**: 7895–7906.

Davies KJA. 1987. Protein damage and degradation by oxygen radicals. *The Journal of Biological Chemistry* **262**:9895–9901.

Fligiel SEG, Lee EC, McCoy JP, Johnson KJ, Varani J. 1984. Protein degradation following treatment with hydrogen peroxide. *The American Journal of Pathology* **115**:418–425.

Garnett T, Conn V, Kaiser BN. 2009. Root based approaches to improving nitrogen use efficiency in plants. *Plant, Cell and Environment* **32**:1272–1283.

Godlewski M, Adamczyk B. 2007. The ability of plants to secrete proteases by roots. *Plant Physiology and Biochemistry* **45**:657–664.

Gutjahr C, Parniske M. 2013. Cell and developmental biology of arbuscular mycorrhiza symbiosis. *Annual Review of Cell and Developmental Biology* **29**:593–617.

Herridge DF, Peoples MB, Boddey RM. 2008. Global inputs of biological nitrogen fixation in agricultural systems. *Plant and Soil* **311**:1–18.

Hurek T, Reinhold B, Grimm B, Fendrik I, Niemann E-G. 1988. Occurrence of effective nitrogen-scavenging bacteria in the rhizosphere of kallar grass. *Plant and Soil* **110**:339–348.

Kim JM, Lee SH, Lee SH, Choi EJ, Jeon CO. 2013. *Paenibacillus hordei* sp. nov., isolated from naked barley in Korea. *Antonie van Leeuwenhoek* **103**:3–9.

Kohli A, Narciso JO, Miro B, Raorane M. 2012. Root proteases: reinforced links between nitrogen uptake and mobilization and drought tolerance. *Physiologia Plantarum* **145**:165–179.

Lakshmann V, Selvaraj G, Bais HP. 2014. Functional soil microbiome: belowground solutions to an aboveground problem. *Plant Physiology* **166**:689–700.

Mulvaney LR, Kahn SA, Elisworth TR. 2009. Synthetic nitrogen fertilizers deplete soil nitrogen: a global dilemma for sustainable cereal production. *Journal of Environmental Quality* **38**: 2295–2314.

Nusinow DA, Helfer A, Hamilton EA, King JJ, Imaizumi T, Schultz TF, Farré EM, Kay SA. 2012. The EF4-ELF3-LUX complex links the circadian clock to diurnal control of hypocotyl growth. *Nature* **475**: 398–402.

Paungfoo-Lonhienne C, Lonhienne TGA, Rentsch D, Robinson N, Christie M, Webb RI, Gamage HK, Carroll BJ, Schenk PM, Schmidt S. 2008. Plants can use protein as a nitrogen source without assistance of other organisms. *Proceedings of the National Academy of Sciences of the USA* **105**:4524–4529.

Paungfoo-Lonhienne C, Visser J, Lonhienne TGA, Schmidt S. 2012. Past, present and future of organic nutrients. *Plant and Soil* **359**:1–18.

Rodarte MP, Dias DR, Vilela DM, Schwan RF. 2011. Proteolytic activities of bacteria, yeasts and filamentous fungi isolated from coffee fruit (*Coffea arabica* L.). *Acta Scientiarum Agronomy* **33**:457–464.

Rodriguez RJ, Freeman DC, McArthur ED, Kim YO, Redman RS. 2009. Symbiotic regulation of plant growth, development and reproduction. *Communicative & Integrative Biology* **2**:141–143.

Schulz B, Boyle C. 2005. The endophytic continuum. *Mycological Research* **109**:661–686.

Sugimoto S, Iwamoto T, Takada K, Okuda K, Tajima A, Iwase T, Mizunoe Y. 2013. *Staphylococcus epidermidis* Esp. degrades specific proteins associated with *Staphylococcus aureus* biofilm formation and host-pathogen interaction. *Journal of Bacteriology* **195**:1645–1655.

Walker TS, Bais HP, Grotewold E, Vivanco JM. 2003. Root exudation and rhizosphere biology. *Plant Physiology* **132**:44–51.

Wells JA, Ferrari E, Henner DJ, Estell DA, Chen EY. 1983. Cloning, sequencing, and secretion of *Bacillus amyloliquefaciens* subtilisin in *Bacillus subtilis*. *Nucleic Acids Research* **11**: 7911–7925.

White JF Jr, Crawford H, Torres MS, Mattera R, Bergen M, Irisarry I. 2012. A proposed mechanism for nitrogen acquisition by grass seedlings through oxidation of symbiotic bacteria. *Symbiosis* **57**:161–171.

White JF Jr, Torres MS, Johnson H, Irizarry I, Tadych M. 2014*a*. A functional view of plant microbiomes: endosymbiotic systems that enhance plant growth and survival. In: Verma VC, Gange AC, eds. *Advances in endophytic research*. Bangalore, India: Springer, 425–439.

White JF Jr, Torres MS, Somu MP, Johnson H, Irizarry I, Chen Q, Zhang N, Walsh E, Tadych M, Bergen M. 2014*b*. Hydrogen peroxide staining to visualize bacterial infections of seedling root cells. *Microscopy Research and Technique* **77**:566–573.

13

Supplementary calcium ameliorates ammonium toxicity by improving water status in agriculturally important species

Elvia Hernández-Gómez[1], Luis A. Valdez-Aguilar[1*], Donita L. Cartmill[2], Andrew D. Cartmill[2] and Irán Alia-Tajacal[3]

[1] Programa de Agricultura Protegida, Universidad Autónoma Agraria Antonio Narro, Calzada Antonio Narro 1923, Buenavista, Saltillo, Coahuila 25315, México

[2] School of Agriculture, University of Wisconsin–Platteville, 1 University Plaza, Platteville, WI 53818, USA

[3] Facultad de Ciencias Agropecuarias, Universidad Autónoma del Estado de Morelos, Avenida Universidad 1001, Cuernavaca, Morelos 62209, México

Associate Editor: Wen-Hao Zhang

Abstract. Fertilization of agricultural plants with ammonium (NH_4^+) is often desirable because it is less susceptible to leaching than nitrate (NO_3^-), reducing environmental pollution, risk to human health and economic loss. However, a number of important agricultural species exhibit a reduction in growth when fertilized with NH_4^+, and increasing the tolerance to NH_4^+ may be of importance for the establishment of sustainable agricultural systems. The present study explored the feasibility of using calcium (Ca) to increase the tolerance of bell pepper (*Capsicum annuum*) to NH_4^+ fertilization. Although NH_4^+ at proportions $\geq$25 % of total nitrogen (N) decreased leaf dry mass (DM), supplementary Ca ameliorated this decrease. Increasing NH_4^+ resulted in decreased root hydraulic conductance (L_o) and root water content (RWC), suggesting that water uptake by roots was impaired. The NH_4^+-induced reductions in L_o and RWC were mitigated by supplementary Ca. Ammonium induced increased damage to the cell membranes through lipid peroxidation, causing increased electrolyte leakage; Ca did not reduce lipid peroxidation and resulted in increased electrolyte leakage, suggesting that the beneficial effects of Ca on the tolerance to NH_4^+ may be more of a reflection on its effect on the water status of the plant. Bell pepper plants that received NO_3^--N had a low concentration of NH_4^+ in the roots but a high concentration in the leaves, probably due to the high nitrate reductase activity observed. Ammonium nutrition depressed the uptake of potassium, Ca and magnesium, while increasing that of phosphorus. The results obtained in the present study indicate that NH_4^+ caused growth reduction, nutrient imbalance, membrane integrity impairment, increased activity of antioxidant enzymes and affected water relations. Supplementary Ca partially restored growth of leaves by improving root L_o and water relations, and our results suggest that it may be used as a tool to increase the tolerance to NH_4^+ fertilization.

Keywords: Antioxidant activity; electrolyte leakage; lipid peroxidation; nutrient imbalance; photosynthesis rate; root hydraulic conductance; stress tolerance.

* Corresponding author's e-mail address: luisalonso.valdez@uaaan.mx

Introduction

Nitrate (NO_3^-) and ammonium (NH_4^+) are the predominant forms in which nitrogen (N), a critical nutrient for the growth of agricultural species (Berger *et al.* 2013), is absorbed; nonetheless, the growth of cultivated plants may be affected by the predominant form of N present in the soil or nutrient solution. The use of NH_4^+ as the N source for crop production is particularly attractive when compared with NO_3^- as it is less susceptible to leaching from the soil profile (Fernández-Escobar *et al.* 2004), potentially increasing N use efficiency and reducing environmental pollution and the risk of human diseases, such as methemoglobinemia, cancers of the digestive tract, thyroid problems and diabetes (Knobeloch *et al.* 2000). Economic issues must also be considered as the production of NH_4^+-based fertilizers is less expensive than that of NO_3^--based fertilizers because a higher energy input is needed to oxidize NH_4^+ into NO_3^- (Havlin *et al.* 2013).

Plants may also benefit from NH_4^+ nutrition when provided at limited concentrations as its assimilation represents a lower energy cost due to its redox state, eliminating the requirement for reduction (Britto and Kronzucker 2002), which may result in increased plant growth and yield. However, high concentration of NH_4^+ may be toxic to some agricultural species (Borgognone *et al.* 2013), mainly due to a depletion of carbon supply caused by damage to the chloroplast ultrastructure (Takács and Técsi 1992), decreased photosynthesis rate (Setién *et al.* 2013) and an energy-demanding transmembrane NH_4^+ cycle in order to maintain low intracellular NH_4^+ concentrations (Britto *et al.* 2001). Other plant responses to high concentration of NH_4^+ include increased intracellular pH (Bittsánszky *et al.* 2015), damage to the cell membrane permeability (M'rah-Helali *et al.* 2010) and decrease in the uptake of essential cations such as potassium (K), calcium (Ca) and magnesium (Mg) (Hawkesford *et al.* 2012).

Ammonium at high concentrations induce a reduction in K uptake in several species; however, enhancing the external supply of K has been proposed as a tool for the alleviation of NH_4^+ toxicity (ten Hoopen *et al.* 2010). Potassium ameliorates NH_4^+ toxicity by reducing the free NH_4^+ concentration in the shoot and root tissues and by reducing the futile cycling of NH_4^+ at the plasma membrane (Balkos *et al.* 2010). Calcium concentration in plant tissues is reduced in cucumber (*Cucumis sativus*) (Roosta *et al.* 2009) and arabidopsis (*Arabidopsis thaliana*) plants (M'rah-Helali *et al.* 2010) grown at high concentration of NH_4^+, while tomato (*Solanum lycopersicum*) exhibits increased incidence of blossom end rot (a Ca-related disorder) (Borgognone *et al.* 2013). Improving plant performance through optimization of NH_4^+ nutrition and the use of supplementary levels of K, and other cations such as Ca, may be of substantial agronomic significance in global crop production (Balkos *et al.* 2010); however, little research has been conducted to elucidate the effect of supplementary concentration of Ca on the response of plants to NH_4^+ toxicity.

The objective of the present study was to determine the effect of NH_4^+ nutrition on growth, photosynthetic parameters, root hydraulic conductance (L_o), leaf water potential (ψ_w), antioxidant enzymes, nitrate reductase activity (NRA) and the nutrient status of *Capsicum annuum* cv. Darsena (bell pepper) plants in soilless culture, a species particularly sensitive to NH_4^+ and widely cultivated in protected agriculture and soilless systems in the southern portion of North America, and to explore the feasibility of using Ca to increase the tolerance to high NH_4^+ concentration.

Methods

Cultural conditions and plant material

The study was conducted under greenhouse conditions in Northeast México (25°27′LN, 101°02′LW, 1610 m above sea level); average minimum/maximum temperature and relative humidity for experiment duration were 28 °C/15 °C and 45 %/75 %, respectively, and average photosynthetically active radiation (PAR) measured at solar noon was 501 μmol m^{-2} s^{-1}.

Thirty-three-day-old bell pepper plants (15 cm height) were transplanted to a 39-L rigid plastic container with a drainage hole for retrieval of the nutrient solution. The container was filled with horticultural grade perlite [33 % (v/v) water holding capacity, 64 % air-filled pore space and 0.25 g cm^{-3} apparent density], with each container being considered as an experimental unit.

Nutrient solutions

Nutrient solutions with 14 meq L^{-1} total N were prepared with distilled water and varying proportions of NH_4^+: 0 % (control), 25% and 50 %; the remaining N was applied as NO_3^-. Potassium, phosphorus (P), Ca, Mg and sulfate were supplied at 6, 1, 10, 4 and 4 meq L^{-1}, respectively. Nutrient solution with 50 % of N as NH_4^+ was prepared with two concentrations of Ca: 9 and 14 meq L^{-1}. The pH of the nutrient solutions ranged from 5.8 to 6.0 and electrical conductivity (EC) from 2.0 to 2.4 dS m^{-1}. Fertigation was applied through a drip irrigation system (six emitters per experimental unit), designed to collect the leachate for reuse. Plants were irrigated for 10 min (~3.6 L) six times a day. Evapotranspirated water was replenished daily to each stock tank of nutrient solution with distilled water and the nutrient solutions were replaced every other day.

Gas exchange parameters and leaf water potential

Net photosynthetic rate, transpiration rate and stomatal conductance were measured (LI-COR Inc., LI-6200, Lincoln, NE, USA) at noon 40 days after planting on young leaves (the third fully developed leaf from top to bottom). Average PAR, CO_2 concentration and temperature were maintained at 400 μmol m^{-2} s^{-1}, 355 p.p.m. and 25.5 °C, respectively. Three measurements on each leaf from one plant per experimental unit were recorded. Water potential of two young leaves per plant was measured 30 days after planting (Scholander Pressure Chamber, Soil Moisture Equipment Corp., Santa Barbara, CA, USA).

Assessment of plant growth and nutrient status

Growth measurements were recorded at harvest (67 days after planting), including leaf area (LI-COR Inc., LI-3100) and dry mass (DM) of previously washed and dried stems, leaves and roots. Shoot and root water content (RWC) was calculated with the difference in fresh mass (FM) and DM {water content = [(shoot or root FM − shoot or root DM)/shoot or root DM] × 100}.

Concentration of N forms (NO_3^--N, NH_4^+-N and total N) was determined in roots and leaves of plants sampled at experiment termination. Samples were rinsed (2×) in distilled water, bagged and dried (75 °C for 72 h), and the dried material ground to pass a 40-mesh screen (Tekmar, model A-10, Cole-Parmer Instrument, Chicago, IL, USA). Nitrate and NH_4^+ concentration were determined following Cataldo's (Cataldo *et al.* 1975) and Nessler's (Krug *et al.* 1979) methods. Total N was determined with the Kjeldhal's procedure. Phosphorus, K, Ca and Mg were determined in digested leaf and root tissues (2 : 1 mixture of H_2SO_4 : $HClO_4$ and 2 mL of 30 % H_2O_2) with inductively coupled plasma emission spectrometer (ICP-AES, model Liberty, Varian, Santa Clara, CA, USA).

Lipid peroxidation, antioxidant enzymes and NRA

Lipid peroxidation was measured in leaves sampled 60 days after planting by quantifying the concentration of malondialdehyde using the thiobarbituric acid method (Kuk *et al.* 2003), whereas the superoxide dismutase (SOD) and ascorbate peroxidase (APX) activities were determined according to Bonnet *et al.* (2000) and Kuk *et al.* (2003), respectively.

Nitrate reductase activity was determined 60 days after planting using the procedure described by Ventura *et al.* (2010) with minor modifications. Briefly, fresh leaf samples (0.25 g) were collected at noon and vacuum infiltrated (pressure = 50 cmHg) for 1 min in a mixture containing 0.1 M KNO_3 and 0.1 % *n*-propanol in 50 mM KH_2PO_4 buffer (pH 7.0), in 1 : 20 (w : v) ratio. The reaction was allowed to proceed for 40 min at 30 °C in the dark. Samples were collected at the beginning and end of the assay, and the level of nitrite was determined with a spectrophotometer (Shimadzu UV—150-02) after the addition of 1 % sulfanilamide in 3 M HCl and 0.02 % *N*-1-naphthyl ethylenediamine mixture 1 : 1 (v : v) at 540 nm.

Electrolyte leakage and root hydraulic conductance

Electrolyte leakage was measured in 10 young leaf segments (1 cm^2) at 30 and 60 days after planting. The leaf segments were washed with distilled water to remove surface contamination and placed in stoppered vials containing 10 mL of distilled water. The vials were placed in a shaker at room temperature (27 °C) for 24 h at 100 r.p.m. Electrical conductivity (EC_1) of the bathing solution was read after the incubation period and then the samples were placed in an autoclave at 120 °C for 20 min; EC_2 was measured after cooling the bathing solution. Electrolyte leakage was calculated using the formula: 100 × (EC_1/EC_2).

Root L_o was measured 40 and 62 days after planting using a modified procedure as described by Joly (1989). Each plant was placed into a 100-mL beaker with distilled water. Shoots of plants were removed 1 cm above the root insertion and enclosed in a pressure chamber. A rubber tube was attached to the stump and protruded through a rubber gasket. The pressure in the chamber was increased gradually to 0.8 MPa and the sap flow captured in a glass vial for 30 min and weighed. The flow rate of sap exudate was determined as the average exudate weight per second (mg s^{-1}), and L_o was estimated as the exudate flow rate versus applied pressure per gram of root (mg g^{-1} s^{-1} MPa^{-1}).

Statistical design

Six replicates of each experimental unit (container with six plants) were distributed in a complete randomized block design. Significant effects of NH_4^+ and Ca were determined using ANOVA and Duncan's multiple mean comparison test ($P < 0.05$) with SAS (SAS v. 8.0, SAS Institute).

Results

Increasing NH_4^+ in the nutrient solution reduced leaf area and DM of stems and leaves, but increased DM of roots when NH_4^+ was at 50 % of total N (Table 1). Supplementary Ca at 14 meq L^{-1} partially reversed the toxicity in plants irrigated with 50 % of total N as NH_4^+ when compared with Ca at 9 meq L^{-1} as leaf DM was higher than that of plants with Ca at 9 meq L^{-1}.

Photosynthesis, transpiration rate and stomatal conductance were unaffected by the proportion of NH_4^+ in the nutrient solution (Table 2). Leaf ψ_w and RWC decreased as the proportion of NH_4^+ increased in the nutrient solution (Table 3); however, in plants treated with solutions containing 50 % of total N as NH_4^+, supplementary Ca

Table 1. Growth responses of bell pepper (*C. annuum* cv. Dársena) plants grown in perlite and irrigated with nutrient solutions of varying ammonium (NH_4^+) proportions and supplementary Ca. Total N concentration was constant at 14 meq L^{-1}. Degrees of freedom for numerator = 3; degrees of freedom for denominator = 15. Means within the same column followed by the same letter are not significantly different according to Duncan's multiple range test ($P < 0.05$).

NH_4^+ (%)	Ca (meq L^{-1})	Leaf area (cm^2 $plant^{-1}$)	Stem DM (g $plant^{-1}$)	Leaf DM (g $plant^{-1}$)	Shoot DM (g $plant^{-1}$)	Root DM (g $plant^{-1}$)
0	9	1056a	3.21a	3.29a	6.50a	3.07bc
25	9	503b	1.22c	1.61c	2.83c	2.75c
50	9	564b	1.87b	2.02c	3.90bc	3.87a
50	14	604b	1.63bc	2.56b	4.20b	3.52ab
ANOVA	*F* ratio	17.86	21.30	17.23	21.30	4.88
		$P < 0.001$	$P < 0.001$	$P < 0.001$	$P < 0.001$	$P < 0.015$

Table 2. Gas exchange parameters in bell pepper (*C. annuum* cv. Dársena) plants grown in perlite and irrigated with nutrient solutions of varying ammonium (NH_4^+) proportions and supplementary Ca. Total N concentration was constant at 14 meq L^{-1}. Degrees of freedom for numerator = 3; degrees of freedom for denominator = 15.

NH_4^+ (%)	Ca (meq L^{-1})	Net photosynthesis (μmol CO_2 m^{-2} s^{-1})	Transpiration rate (mmol m^{-2} s^{-1})	Stomatal conductance (mmol m^{-2} s^{-1})
0	9	9.03	7.67	0.64
25	9	8.98	8.59	0.75
50	9	9.72	7.41	0.63
50	14	9.43	7.38	0.56
ANOVA	*F* ratio	1.19	1.67	2.24
		$P < 0.347$	$P < 0.216$	$P < 0.126$

Table 3. Water relations in bell pepper (*C. annuum* cv. Dársena) plants grown in perlite and irrigated with nutrient solutions of varying ammonium (NH_4^+) proportions and supplementary Ca. Total N concentration was constant at 14 meq L^{-1}. dap, days after planting. Degrees of freedom for numerator = 3; degrees of freedom for denominator = 15. Means within the same column followed by the same letter are not significantly different according to Duncan's multiple range test ($P < 0.05$).

NH_4^+ (%)	Ca (meq L^{-1})	Shoot water content (%)	RWC (%)	Leaf ψ_w (MPa)	Root hydraulic conductance (mg g^{-1} s^{-1} MPa^{-1})	
					40 dap	60 dap
0	9	87.6	89.7a	−0.138a	12.09a	18.74a
25	9	87.9	84.8bc	−0.215ab	10.54ab	7.44c
50	9	87.3	84.9c	−0.428c	5.25c	4.56c
50	14	87.1	86.5b	−0.266b	8.86b	11.32b
ANOVA	*F* ratio	0.55	7.93	16.33	14.64	26.91
		$P = 0.656$	$P < 0.001$	$P < 0.001$	$P < 0.001$	$P < 0.001$

partially restored the water status of plants. Shoot water content was not affected by NH_4^+ or Ca concentration (Table 3). Root L_o was decreased by increasing the proportion of NH_4^+ in the nutrient solution, but in plants irrigated with solutions containing 50 % of total N as NH_4^+ (Table 3), supplementary Ca increased water conductance.

Electrolyte leakage was unaffected by the proportion of NH_4^+ at 30 days after planting; however, for the measurement conducted 60 days after planting, an increase in electrolyte loss was detected in plants with NH_4^+ at 50 % of total N (Table 4). Supplementary Ca was associated with a significant increase in electrolyte leakage

Table 4. Effect of varying ammonium (NH_4^+) proportions and supplementary Ca on lipid peroxidation, electrolyte leakage, SOD and APX activities in leaves of bell pepper (*C. annuum* cv. Dársena) plants grown in perlite. Total N concentration was constant at 14 meq L^{-1}. dap, days after planting. Means within the same column followed by the same letter are not significantly different according to Duncan's multiple range test ($P < 0.05$).

NH_4^+ (%)	Ca (meq L^{-1})	Lipid peroxidation ($\mu m\ g^{-1}$ FM)	Electrolyte leakage (%)		SOD activity ($U\ g^{-1}$)	APX activity ($U\ g^{-1}$)
			30 dap	60 dap		
0	9	9.23b	15.9b	13.9b	69.6b	1109b
25	9	14.61a	17.2b	14.8b	103.2a	1430ab
50	9	11.06ab	17.8b	15.4ab	86.4ab	1470a
50	14	11.66ab	27.2a	17.3a	72.2b	1480a
ANOVA	*F* ratio	3.16	9.82	4.44	7.04	3.41
	Degrees of freedom numerator, denominator	3, 18	3, 15	3, 15	3, 15	3, 18
		$P < 0.050$	$P < 0.001$	$P < 0.020$	$P < 0.004$	$P < 0.040$

Table 5. Total N, ammonium (NH_4^+) and nitrate (NH_4^+) concentration (mmol kg^{-1}) and NRA (nmol $NO_2\ g^{-1}$ fresh mass h^{-1}) in bell pepper (*C. annuum* cv. Dársena) plants grown in perlite and irrigated with nutrient solutions of varying NH_4^+ proportions and supplementary Ca. Total N concentration was constant at 14 meq L^{-1}. Means within the same column followed by the same letter are not significantly different according to Duncan's multiple range test ($P < 0.05$).

NH_4^+ (%)	Ca (meq L^{-1})	NRA	NH_4^+ concentration		NO_3^- concentration		Total N concentration	
			Leaf	Root	Leaf	Root	Leaf	Root
0	9	27 927a	233.0b	96.9c	108.70a	87.1ab	1284bc	1503b
25	9	25 707b	220.6b	207.6a	105.65a	102.3a	1187c	1579b
50	9	26 098b	211.1b	177.3b	39.91b	60.6bc	1388b	1492b
50	14	26 652ab	329.0a	221.8a	55.80b	40.5c	1907a	1856a
ANOVA	*F* ratio	3.10	14.63	53.14	20.69	4.97	52.98	8.77
	Degrees of freedom numerator, denominator	3, 21	3, 15	3, 15	3, 15	3, 15	3, 15	3, 15
		$P < 0.049$	$P < 0.001$	$P < 0.001$	$P < 0.001$	$P < 0.014$	$P < 0.001$	$P < 0.001$

in the 30- and 60-day measurements (Table 4). Increasing NH_4^+ caused an increase in lipid peroxidation and the activities of APX and SOD (Table 4), but reduced that of NRA (Table 5). However, when Ca was provided at higher concentration, the activity of the SOD was significantly decreased (Table 4).

Increasing proportions of NH_4^+ had no effect on NH_4^+ in the leaf tissue, while the root tissues exhibited increased NH_4^+ concentration (Table 5). Nitrate concentration was significantly decreased in both leaf and root tissues when 50 % of total N was provided as NH_4^+ (Table 5). Total leaf N concentration was increased as the proportion of NH_4^+ in the nutrient solution was increased, but root N concentration was unaffected (Table 5). Supplementary Ca was associated with a significant increase in total N and NH_4^+ concentrations in leaves and roots of plants irrigated with 50 % of the proportion of N in the nutrient solution provided as NH_4^+ (Table 5). Increasing NH_4^+ in the nutrient solution caused a decrease in leaf K, Ca and Mg concentrations. Phosphorus concentration in the leaf increased with increasing NH_4^+ in the nutrient solution (Table 6). Increasing NH_4^+ in the nutrient solution caused a decrease in root Ca and Mg concentration. Supplementary Ca partially counteracted the decreasing effect of NH_4^+ as leaf Mg and root Ca increased (Table 6).

Discussion

Bell pepper plants were sensitive to NH_4^+ proportions $\geq$25 %, being the shoot more affected than the roots as the stems and leaves exhibited up to a 62 and 51 % decrease in DM, respectively; in the roots, plants that received 25 % of N as NH_4^+ showed a 26 % decrease, whereas in plants with 50 % NH_4^+, root growth was

Table 6. Potassium, Ca, Mg and P concentrations in leaves and roots of bell pepper (*C. annuum* cv. Dársena) plants grown in perlite and irrigated with nutrient solutions of varying ammonium (NH_4^+) proportions and supplementary Ca. Total N concentration was constant at 14 meq L^{-1}. Degrees of freedom for numerator = 3; degrees of freedom for denominator = 15. Means within the same column followed by the same letter are not significantly different according to Duncan's multiple range test ($P < 0.05$).

NH_4^+ (%)	Ca (meq L^{-1})	K (mmol kg^{-1})		Ca (mmol kg^{-1})		Mg (mmol kg^{-1})		P (mmol kg^{-1})	
		Leaf	Root	Leaf	Root	Leaf	Root	Leaf	Root
0	9	1471a	943	170a	388a	232a	211a	34.8c	64.2
25	9	1273b	1007	110b	200b	139c	101b	57.9b	64.9
50	9	1316b	1143	114b	268b	190b	134b	64.5ab	54.5
50	14	1290b	1095	120b	361a	148c	121b	74.3a	77.4
ANOVA	*F* ratio	6.12	1.34	25.36	10.75	33.16	14.06	24.90	3.11
		$P < 0.006$	$P = 0.332$	$P < 0.001$	$P < 0.003$	$P < 0.001$	$P < 0.001$	$P < 0.001$	$P = 0.081$

enhanced by 26 %. The more pronounced effect on the shoots than on the roots has been reported in species sensitive to NH_4^+ (M'rah-Helali *et al.* 2010; Borgognone *et al.* 2013).

Calcium is reported to increase the tolerance of plants to abiotic stresses by regulating the response reactions and developmental processes (Steinhorst and Kudla 2013). Kaya *et al.* (2002) and Tuna *et al.* (2007) reported that increased Ca concentrations resulted in enhanced growth and yield of strawberry (*Fragaria* × *ananassa*) and tomato plants exposed to high salinity; however, to our knowledge, there is no information as to the potential role of Ca in enhancing the tolerance of agricultural species to high NH_4^+ stress. The present study demonstrates that supplementary Ca ameliorated to some extent the stress caused by high NH_4^+ proportions as suggested by the higher leaf DM than that of plants with typical Ca concentrations.

The observed reduction in leaf ψ_w, leaf expansion and shoot growth when NH_4^+ was >25 % may have been due to affected water transport to the shoot as demonstrated by the decrease in root L_o and RWC. Supplementary Ca improved to some extent root L_o, resulting in increased RWC and leaf ψ_w, and partially restored leaf DM accumulation. As L_o is largely determined by the activity of aquaporins (Li *et al.* 2015), probably the level of NH_4^+ used in the present study may have affected these water root channels, whereas Ca could have played a role in restoring it. Gao *et al.* (2010) reported that aquaporins activity in rice was higher when seedlings were supplied with N as NH_4^+ and indicated that the increase in L_o may be a consequence of the effect of NH_4^+ on root morphology (less development in the Casparian strip), growth, number of tips and surface area. However, our results showed that bell pepper plants responded differently and we have no evidence of modifications in root morphology (root DM of NO_3^--fed plants was comparable with that of plants irrigated with 50 % of total N as NH_4^+), suggesting that the increase in L_o may be the effect of the supplementary Ca.

Surprisingly, the restoring effect of Ca on L_o did not affect gas exchange parameters. However, our data showed that the decrease in transpiration rate and stomatal conductance was associated with an increase in root Ca and N concentration. Rothwell and Dodd (2014) provided evidence that shows that Ca added to the rhizosphere can decrease stomatal conductance and photosynthesis in beans (*Phaseoulus vulgaris*) and pea (*Pisum sativum*), resulting in increased xylem sap Ca concentration in beans. However, the higher Ca concentration in the roots observed in our study in plants with supplementary Ca suggests that Ca translocation rate was maintained as in control plants as leaf Ca concentration was not increased, which suggests that this nutrient was not affecting gas exchange parameters. Nonetheless, the decrease in stomatal conductance in plants supplemented with Ca was associated with an increase in total N in leaves, which in turn was associated with a significant increase in leaf NH_4^+ and NO_3^- concentrations. Similar effects of N nutrition of gas exchange parameters were reported by Hu *et al.* (2015) as stomatal conductance in Chinese cabbage (*Brassica pekinensis*) was decreased at proportions of 25 % of N in NH_4^+ form, compared with plants provided with lower NH_4^+, and by Du *et al.* (2015), that reported similar deleterious effects in NH_4^+-fed tea plants (*Camellia sinensis*) compared with NO_3^--fed plants. Therefore, the no relationship between stomatal conductance and the increased root hydraulic conductance in plants supplemented with increased Ca could have been related to the increase in leaf and root NH_4^+ and total N concentration.

Abiotic and biotic stresses are reported to primarily target the cell membranes, so that the maintenance of the integrity and stability of the plasmalemma are of foremost importance to stress tolerance (Bajji *et al.*

2002). In the present study, bell pepper plants exhibited increased damage in cell membranes as suggested by the higher lipid peroxidation observed when N was provided as NH_4^+, which in turn would explain the increase in electrolyte leakage as it has also been linked with impaired membrane stability (Bajji *et al.* 2002; Tuna *et al.* 2007). However, the supplementary concentration of Ca used in the present study did not reduce lipid peroxidation and resulted in increased electrolyte leakage, which may have been the result of NH_4^+ leakage, in addition to other ions such as K (Demidchik *et al.* 2014), as its concentration was markedly increased in leaves of plants with 50 % NH_4^+ at normal and supplementary Ca concentrations.

The potential damage of reactive oxygen species (ROS) may be counteracted by increasing the activity of a group of antioxidant compounds that protect the cell membranes (Domínguez-Valdivia *et al.* 2008). In the present study, the damage to cell membranes caused by increasing NH_4^+ resulted in increased SOD (+42 %) and APX (+32 %) activities, suggesting that bell pepper plants attempted to detoxify ROS. However, under high NH_4^+, supplementary Ca was associated with a reduction in SOD activity, suggesting that ROS were causing less damage. Calcium has been associated with increased tolerance to stress by alleviating the damage caused by ROS to cell membranes in tomato plants exposed to salinity (Tuna *et al.* 2007).

Bell pepper plants that received exclusively N in NO_3^- form exhibited a low concentration of NH_4^+ in the roots but a high concentration in the leaves, probably associated with the high reduction rate of NO_3^- as a result of the high NRA activity observed in the leaves. The decreased leaf and root NO_3^- concentration observed may have been caused by the lower availability of NO_3^- when external NH_4^+ was increased and due to its rapid reduction, as suggested by the limited effect observed on NRA activity under high NH_4^+.

Leaf and root NH_4^+ concentrations were not increased even when external NH_4^+ was at 50 %, suggesting that bell pepper plants may regulate internal NH_4^+ content, probably through efflux mechanisms across the plasma membrane (Britto *et al.* 2001). The higher leaf and root NH_4^+ concentrations when plants were exposed to high NH_4^+ and higher concentrations of Ca suggest that bell pepper was able to take up higher amounts of N in NH_4^+ form from the external solution and/or that the higher NRA activity rapidly reduced NO_3^- to NH_4^+.

Nitrogen nutrition with high levels of NH_4^+ is reported to cause nutrient imbalance as it depresses the uptake of K, Ca and Mg, while increases that of phosphate and sulfate (Roosta and Schjoerring 2007), which was confirmed in our study. The decrease in internal K concentration observed in our study was probably caused by competition with high levels of NH_4^+ (ten Hoopen *et al.* 2010), which in turn may be the cause of the decrease in leaf ψ_W, affecting leaf expansion and DM. Supplementary Ca in plants exposed to high NH_4^+ resulted in increased root Ca and decreased leaf Mg, probably due to the competition between Ca and Mg for uptake sites (Marschner 1995). The nutrient imbalance also explains the more reduced growth of plants with 25 % of N in NH_4^+ compared with plants with 50 % NH_4^+, as a higher root concentration of NH_4^+ directly affected leaf, stem and shoot growth, whereas a higher root NO_3^- affected root growth.

Conclusions

The results obtained in the present study indicate that NH_4^+ caused growth reduction, nutrient imbalance, membrane integrity impairment, increased activity of antioxidant enzymes and affected water relations in bell pepper plants; however, as supplementary Ca partially restored the growth of leaves by improving root L_o and water relations, it may be part of an integral management to enhance the tolerance of important agricultural species to excess NH_4^+.

Sources of Funding

Our work was funded by Universidad Autónoma Agraria Antonio Narro.

Contributions by the Authors

E.H.-G. and L.A.V.-A. designed and performed the study, D.L.C. and A.D.C. designed the experiment and wrote the document, and I.A.-T. performed biochemical and statistical analysis.

Literature Cited

Bajji M, Kinet J-M, Lutts S. 2002. The use of the electrolyte leakage method for assessing cell membrane stability as a water stress tolerance test in durum wheat. *Plant Growth Regulation* **36**: 61–70.

Balkos KD, Britto DT, Kronzucker HJ. 2010. Optimization of ammonium acquisition and metabolism by potassium in rice (*Oryza sativa* L. cv. IR-72). *Plant, Cell and Environment* **33**:23–34.

Berger B, Brock AK, Ruppel S. 2013. Nitrogen supply influences plant growth and transcriptional responses induced by *Enterobacter radicincitans* in *Solanum lycopersicum*. *Plant and Soil* **370**: 641–652.

Bittsánszky A, Pilinszky K, Gyulai G, Komives T. 2015. Overcoming ammonium toxicity. *Plant Science* **231**:184–190.

Bonnet M, Camares O, Veisseire P. 2000. Effects of zinc and influence of *Acremonium lolii* on growth parameters, chlorophyll a fluorescence and antioxidant enzyme activities of ryegrass (*Lolium perenne* L. cv Apollo). *Journal of Experimental Botany* **51**:945–953.

Borgognone D, Colla G, Rouphael Y, Cardarelli M, Rea E, Schwarz D. 2013. Effect of nitrogen form and nutrient solution pH on growth and mineral composition of self-grafted and grafted tomatoes. *Scientia Horticulturae* **149**:61–69.

Britto DT, Kronzucker HJ. 2002. NH_4^+ toxicity in higher plants: a critical review. *Journal of Plant Physiology* **159**:567–584.

Britto DT, Siddiqi MY, Glass ADM, Kronzucker HJ. 2001. Futile transmembrane NH_4^+ cycling: a cellular hypothesis to explain ammonium toxicity in plants. *Proceedings of the National Academy of Sciences of the USA* **98**:4255–4258.

Cataldo DA, Maroon M, Schrader LE, Youngs VL. 1975. Rapid colorimetric determination of nitrate in plant tissue by nitration of salicylic acid. *Communications in Soil Science and Plant Analysis* **6**:71–80.

Demidchik V, Straltsova D, Medvedev SS, Pozhvanov GA, Sokolik A, Yurin V. 2014. Stress-induced electrolyte leakage: the role of K^+-permeable channels and involvement in programmed cell death and metabolic adjustment. *Journal of Experimental Botany* **65**: 1259–1270.

Domínguez-Valdivia MD, Aparicio-Tejo PM, Lamsfus C, Cruz C, Martins-Loução MA, Moran JF. 2008. Nitrogen nutrition and antioxidant metabolism in ammonium-tolerant and -sensitive plants. *Physiologia Plantarum* **132**:359–369.

Du XH, Peng FR, Jiang J, Tan PP, Wu ZZ, Liang YW, Zhong ZK. 2015. Inorganic nitrogen fertilizers induce changes in ammonium assimilation and gas exchange in *Camellia sinensis* L. *Turkish Journal of Agriculture and Forestry* **39**:28–38.

Fernández-Escobar R, Benlloch M, Herrera E, García-Novelo JM. 2004. Effect of traditional and slow-release N fertilizers on growth of olive nursery plants and N losses by leaching. *Scientia Horticulturae* **101**:39–49.

Gao Y, Li Y, Yang X, Li H, Shen Q, Guo S. 2010. Ammonium nutrition increases water absorption in rice seedlings (*Oryza sativa* L.) under water stress. *Plant and Soil* **331**:193–201.

Havlin JL, Tisdale SL, Nelson WL, Beaton JD. 2013. *Soil fertility and fertilizers*, 8th edn. Upper Saddle River, NJ: Pearson.

Hawkesford M, Horst W, Kichey T, Lambers H, Schjoerring J, Møller IS, White P. 2012. Functions of macronutrients. In: Marschner P, ed. *Marschner's mineral nutrition of higher plants*, 3rd edn. Waltham: Academic Press, 135–189.

Hu L, Yu J, Liao W, Zhang G, Xie J, Lv J, Xiao X, Yang B, Zhou R, Bu R. 2015. Moderate ammonium: nitrate alleviates low light intensity stress in mini Chinese cabbage seedling by regulating root architecture and photosynthesis. *Scientia Horticulturae* **186**:143–153.

Joly RJ. 1989. Effects of sodium chloride on the hydraulic conductivity of soybean root systems. *Plant Physiology* **91**:1262–1265.

Kaya C, Kirnak H, Higgs D, Saltali K. 2002. Supplementary calcium enhances plant growth and fruit yield in strawberry cultivars grown at high (NaCl) salinity. *Scientia Horticulturae* **93**:65–74.

Knobeloch L, Salna B, Hogan A, Postle J, Anderson H. 2000. Blue babies and nitrate-contaminated well water. *Environmental Health Perspectives* **108**:675–678.

Krug FJ, Růžička J, Hansen EH. 1979. Determination of ammonia in low concentrations with Nessler's reagent by flow injection analysis. *The Analyst* **104**:47–54.

Kuk YI, Shin JS, Burgos NR, Hwang TE, Han O, Cho BH, Jung S, Guh JO. 2003. Antioxidative enzymes offer protection from chilling damage in rice plants. *Crop Science* **43**:2109–2117.

Li G, Boudsocq M, Hem S, Vialaret J, Rossignol M, Maurel C, Santoni V. 2015. The calcium-dependent protein kinase CPK7 acts on root hydraulic conductivity. *Plant, Cell and Environment* **38**: 1312–1320.

Marschner H. 1995. *Mineral nutrition of higher plants*, 2nd edn. London: Academic Press.

M'rah-Helali S, Nebli H, Kaddour R, Mahmoudi H, Lachaâl M, Ouerghi Z. 2010. Influence of nitrate–ammonium ratio on growth and nutrition of *Arabidopsis thaliana*. *Plant and Soil* **336**:65–74.

Roosta HR, Schjoerring JK. 2007. Effects of ammonium toxicity on nitrogen metabolism and elemental profile of cucumber plants. *Journal of Plant Nutrition* **30**:1933–1951.

Roosta HR, Sajjadinia A, Rahimi A, Schjoerring JK. 2009. Responses of cucumber plant to NH_4^+ and NO_3^- nutrition: the relative addition rate technique vs. cultivation at constant nitrogen concentration. *Scientia Horticulturae* **121**:397–403.

Rothwell SA, Dodd IC. 2014. Xylem sap calcium concentrations do not explain liming-induced inhibition of legume gas exchange. *Plant and Soil* **382**:17–30.

Setién I, Fuertes-Mendizabal T, González A, Aparicio-Tejo PMª, González-Murua C, González-Moro MB, Estavillo JM. 2013. High irradiance improves ammonium tolerance in wheat plants by increasing N assimilation. *Journal of Plant Physiology* **170**: 758–771.

Steinhorst L, Kudla J. 2013. Calcium and reactive oxygen species rule the waves of signaling. *Plant Physiology* **163**:471–485.

Takács E, Técsi L. 1992. Effects of NO_3^-/NH_4^+ ratio on photosynthetic rate, nitrate reductase activity and chloroplast ultrastructure in three cultivars of red pepper (*Capsicum annuum* L.). *Journal of Plant Physiology* **140**:298–305.

ten Hoopen F, Cuin TA, Pedas P, Hegelund JN, Shabala S, Schjoerring JK, Jahn TP. 2010. Competition between uptake of ammonium and potassium in barley and *Arabidopsis* roots: molecular mechanisms and physiological consequences. *Journal of Experimental Botany* **61**:2303–2315.

Tuna AL, Kaya C, Ashraf M, Altunlu H, Yokas I, Yagmur B. 2007. The effects of calcium sulphate on growth, membrane stability and nutrient uptake of tomato plants grown under salt stress. *Environmental and Experimental Botany* **59**:173–178.

Ventura Y, Wuddineh WA, Ephrath Y, Shpigel M, Sagi M. 2010. Molybdenum as an essential element for improving total yield in seawater-grown *Salicornia europaea* L. *Scientia Horticulturae* **126**:395–401.

14

Climate change reduces nectar secretion in two common Mediterranean plants

Krista Takkis*, Thomas Tscheulin, Panagiotis Tsalkatis and Theodora Petanidou
Laboratory of Biogeography and Ecology, Department of Geography, University of the Aegean, University Hill, GR-81100 Mytilene, Greece

Associate Editor: Colin M. Orians

Abstract. Global warming can lead to considerable impacts on natural plant communities, potentially inducing changes in plant physiology and the quantity and quality of floral rewards, especially nectar. Changes in nectar production can in turn strongly affect plant–pollinator interaction networks—pollinators may potentially benefit under moderate warming conditions, but suffer as resources reduce in availability as elevated temperatures become more extreme. Here, we studied the effect of elevated temperatures on nectar secretion of two Mediterranean Lamiaceae species—*Ballota acetabulosa* and *Teucrium divaricatum*. We measured nectar production (*viz.* volume per flower, sugar concentration per flower and sugar content per flower and per plant), number of open and empty flowers per plant, as well as biomass per flower under a range of temperatures selected *ad hoc* in a fully controlled climate chamber and under natural conditions outdoors. The average temperature in the climate chamber was increased every 3 days in 3 °C increments from 17.5 to 38.5 °C. Both study species showed a unimodal response of nectar production (volume per flower, sugar content per flower and per plant) to temperature. Optimal temperature for sugar content per flower was 25–26 °C for *B. acetabulosa* and 29–33 °C for *T. divaricatum*. According to our results, moderate climate warming predicted for the next few decades could benefit nectar secretion in *T. divaricatum* as long as the plants are not water stressed, but have a moderate negative effect on *B. acetabulosa*. Nevertheless, strong warming as predicted by climate change models for the end of the 21st century is expected to reduce nectar secretion in both species and can thus significantly reduce available resources for both wild bees and honeybees in Mediterranean systems.

Keywords: *Ballota acetabulosa*; elevated temperatures; global change; nectar production; nectar sugar content; phrygana; plant–pollinator interactions; *Teucrium divaricatum*.

Introduction

Global warming can have strong effects on plant species, their interactions and whole ecosystems (Tylianakis *et al.* 2008; Hegland *et al.* 2009; Traill *et al.* 2010). Elevated temperatures can, for instance, induce shifts in plant phenology across communities and change the interaction networks between plants and pollinators (Memmott *et al.* 2007; Hegland *et al.* 2009; Schweiger *et al.* 2010; Petanidou *et al.* 2014). One aspect of plant–pollinator interactions, which is particularly sensitive to climate change, is nectar production by plants (Scaven and Rafferty 2013). Plants have an optimum range of temperatures for nectar production, determined by their habitat and species characteristics (Jakobsen and Kristjánsson 1994). Often a moderate increase in average temperature can have a positive effect

* Corresponding author's e-mail address: k.takkis@geo.aegean.gr

on plant nectar production (Pacini and Nepi 2007; Nocentini *et al.* 2013) but beyond that plants experience temperature stress, which can induce changes in nectar production (Petanidou and Smets 1996; Pacini *et al.* 2003; Scaven and Rafferty 2013). A number of studies from different systems have found decreased nectar volumes and nectar production rates at higher temperatures under both experimental and natural conditions (Jakobsen and Kristjánsson 1994; Petanidou and Smets 1996; Keasar *et al.* 2008). At the same time, nectar sugar concentration is usually less dependent on external factors and more constant throughout the day and the flowering season (Southwick 1983; Villarreal and Freeman 1990; Nocentini *et al.* 2013). The amount of sugar produced per flower has nevertheless been shown to depend on nectar volume rather than concentration (Torres and Galetto 1998; Hoover *et al.* 2012) and thus can be strongly affected by changes in ambient temperatures.

A decrease in nectar quantity and sugar content in response to global warming reduces the amount of resources available for pollinators and thus can have negative effects on plant–pollinator interactions (Hegland *et al.* 2009; Hoover *et al.* 2012; Scaven and Rafferty 2013). Under temperature stress plants can even begin to produce flowers without any nectar (Petanidou and Smets 1996), which further reduces the amount of resources. A decrease in floral rewards, but also in the variation of nectar production patterns within a plant, can change pollinator behaviour patterns (Real and Rathcke 1988; Zimmerman 1988). Most often, pollinators are found to be risk-sensitive to greater variability of nectar volume (Shafir 2000; Keasar *et al.* 2008) and high variation can therefore decrease plant attractiveness to pollinators (Zimmerman 1988). Consequently, climate change can reduce plant reproductive success due to changed interaction patterns with pollinators and possibly threaten the persistence of both plant and pollinator populations (Scaven and Rafferty 2013).

The Mediterranean Basin is considered to be one of Europe's regions most threatened by climate change, due to the elevated drought risk during summer (Giorgi 2006; Hickler *et al.* 2012), caused by elevated temperatures in combination with decreased precipitation during summer (Giorgi and Lionello 2008; IPCC 2013). Different climate scenarios predict 4–5 °C warming for summer months by the end of the century, relative to the end of the 20th century (Giorgi and Lionello 2008; IPCC 2013). These changes in temperature and precipitation regime can generate increasingly severe conditions for plant communities and consequently increase the vulnerability of mutualistic interaction networks in the Mediterranean region. In the water-limited phryganic communities (East-Mediterranean low scrub), the nectar production is generally low, with only a few species producing large quantities (i.e. >0.5 μL) of nectar per flower (Petanidou and Smets 1995). In these systems, Lamiaceae are a dominant plant group constituting the main source of water and nutrients for pollinators (Herrera 1985; Petanidou and Vokou 1993; Petanidou and Smets 1995; Petanidou 2007). Their role is particularly vital during summer, when nectar production in this system is challenged by high ambient temperatures, whereas nectar-feeding insect diversity is high (Petanidou and Vokou 1993; Petanidou *et al.* 1995).

In this study, we explore the effect of temperature on the nectar production of two common phryganic species of Lamiaceae visited predominantly by bees, both honeybees and wild bees. In both a climate-controlled and natural setting, we study the patterns of flower and nectar production and their variation in response to a wide range of temperatures, including temperatures higher than current climatic means, thus allowing us to investigate the potential effects of climate change on nectar rewards available to pollinators in the future. Our hypothesis is that strongly elevated temperatures reduce flower and nectar production and affect the variation in nectar production patterns. Such results could indicate a possible threat to pollinators and to the persistence of both plant and pollinator populations in the Mediterranean systems. It may also affect negatively bee-keeping and honey production, which play an important role in the economy of this area.

Methods

Study species

We tested the effect of temperature on the nectar production of two perennial Mediterranean species of the Lamiaceae family, *Ballota acetabulosa* (L.) Benth. and *Teucrium divaricatum* Sieber ex Heldr. Both species inhabit the same phryganic communities, but prefer to grow in different microhabitats. While *B. acetabulosa* often grows near stony structures or walls (e.g. abandoned agricultural terraces) or in places partially shaded by taller vegetation (e.g. in dehesa type scrub) (Petanidou *et al.* 2000; T. Petanidou, pers. obs.), *T. divaricatum* clearly prefers well-sunlit open habitats. Both species flower at the same time of the year, from May to July (Davis 1982; Petanidou 1991). They are visited by a wide range of medium- to large-sized wild bees of the families Megachilidae and Apidae and are also important food plants for bee-keeping (Petanidou 1991; Dauber *et al.* 2010; T. Petanidou *et al.* unpubl. data).

Experiment design

Plants of *B. acetabulosa* used in the experiment were grown from seeds collected from the I. & A. Diomedes Botanical Garden of Athens University in 2005. Seeds were sown in October 2013, potted as seedlings, and grown outdoors until flowering. As for *T. divaricatum*, entire plants were collected from a natural population on Lesvos Island in October 2013, potted and grown outdoors until they were in bloom, ready to be used in the experiment.

The experiment was conducted at the University of the Aegean in Mytilene on Lesvos Island. Both study species were subjected to the same treatments simultaneously. For each species, two groups of plants were considered. One treatment group (15 plants per species) was tested under different temperature regimes in a climate chamber (Walk-in GRW-20 CMP 3/TBLIN, CDR Chryssagis™) and the second treatment group (six plants per species) was grown outdoors under naturally varying temperature and air humidity conditions. Comparing the two treatments allows us to estimate the effect of temperature separately from the effect of time and thus see, whether the trends are driven solely by temperature or whether they may be partially caused by the plants' intrinsic limitations (e.g. exhausted nutrient reserves), which plants naturally incur during their flowering period. Some of the initial plants (one for *B. acetabulosa* in the climate chamber, and 16 for *T. divaricatum*, of which 11 in the climate chamber and 5 outdoors) were replaced with new ones during the experiment when they got close to the end of their flowering period, in order to have the same number of flowering plants at all tested temperatures. Altogether, including the replacement plants, 22 *B. acetabulosa* and 38 *T. divaricatum* plants were employed. Each plant was labelled with a unique number and considered as a separate entity in the subsequent analyses. The experiment for both species lasted from 24 May to 17 June 2014 in the climate chamber and in the outdoors from 24 May to 8 July 2014, following the plants through the flowering period both in the climate chamber and outdoors.

In the climate chamber, a 14-h photoperiod was used for both species, corresponding to the natural day length at the time of the year. A mixture of plant growth fluorescent lamps (Gro-lux) and low-pressure sodium lamps were used, with a total light intensity of ~800 μmol m^{-2} s^{-1} (~43 000 lx) over the waveband 400–700 nm in the chamber. Light intensity in the chamber was somewhat lower than what the plants would experience in nature; however, the intensity was constant throughout the days, compensating for the lower intensity.

The climate chamber experiment for both species was started at the lowest temperatures, and the temperature was increased every 3 days by 3 °C increments. Tested 24 h average temperatures ranged from 17.5 to 38.5 °C (day temperatures 20–41 °C, night temperatures always 6 °C lower, corresponding to the approximate natural difference between day and night temperatures at the time of the year). Tested temperatures were chosen to cover a range around the long-term (1958–2001) June average temperature of ~26 °C according to Elefsis weather station near Athens, Greece, but including also elevated temperatures at least up to the temperatures predicted by climate change scenarios (IPCC 2013) to test for the effect of future climate warming. Relative air humidity was kept constant at 60 ± 5% during the day and 80 ± 5% during the night, corresponding approximately to current natural conditions. Plants were watered on the first day of every temperature step to avoid water stress.

Plants in the outdoor group were placed in an open area in full sunlight and were covered with large and airy tulle cages to prevent pollinator visits. The light intensity inside the covered cages was higher than in the climate chamber (~1400 μmol m^{-2} s^{-1} or ~75 000 lx). Data on ambient temperatures outdoors were obtained from the nearby (<300 m) climate station at the University of the Aegean in Mytilene (http://catastrophes.geo.aegean.gr/). The average temperature of the 3 days corresponding to each temperature step in the climate chamber was used in the analyses. The mean outdoor temperature during the test period ranged from 16.7 to 26.1 °C.

Nectar measurements both in the climate chamber and in the outdoor group were taken between 1230 and 1700 h on Day 3 of each temperature step, to give the plants in the climate chamber time to adjust to the changed conditions before sampling. Flowers were sampled on the first day of anthesis following Petanidou and Smets (1996). To ensure that only first day flowers were sampled, all open flowers were removed on Day 2, 24 h before nectar collection. Nectar was extracted from flowers with Drummond microcaps® (2 and 1 μL for *B. acetabulosa*; 0.5 μL for *T. divaricatum*), using a single flower per sample. Three flowers per plant, selected randomly from among the flowers directly exposed to light, were sampled for nectar volume and sugar concentration per flower. Nectar sugar concentration of each of the sampled flowers was measured by using special hand refractometers for small nectar volumes (Bellingham & Stanley LTD, Tunbridge Wells). After sampling, the same three flowers per plant that were sampled for nectar were dried at 50 °C for 12 h and their biomass per flower was weighed (calyx and corolla together). All flowers, which had opened during the previous 24 h were counted and removed after nectar sampling on Day 3 before proceeding to the next temperature set.

Data analysis

The sugar content per flower (micrograms) was calculated (volume per flower × concentration × density) with the density values taken from available sugar solution density tables (Dafni *et al.* 2005). In the very few cases (18/840 flowers of *B. acetabulosa* and 7/845 of

T. divaricatum) when the collected nectar volume was too small for the refractometer to measure sugar concentration, the average value of sugar concentration of the other two flowers from the same plant was used, in order to be able to calculate sugar content per flower. For data analysis, we used the average values of the three sampled flowers per plant. Sugar content per plant was calculated by multiplying the average sugar content per flower with the number of flowers per plant produced on Day 3. The proportion of empty flowers per plant (i.e. producing no nectar) was also calculated based on the three flowers sampled for nectar. Proportion of empty flowers per plant was used in models only in the case of *B. acetabulosa*, because *T. divaricatum* produced only few nectarless flowers. We calculated the coefficient of variation (CV) for each of the measured flower traits (nectar volume, concentration, sugar content and biomass per flower) to characterize trait variation within a plant. In the case of sugar concentration per flower, the CV was determined based on the original data, prior to calculating the sugar concentration for the measurements with too small nectar volumes for direct sugar concentration detection. If the mean trait value of a plant was zero, the CV of the corresponding trait was also defined to be zero for that plant. Nectar volume per flower, sugar content per flower, sugar content per plant, number of flowers per plant and biomass per flower data were log-transformed and nectar sugar concentration per flower was logit-transformed to deal with the constraints of percentage data.

We tested for the possible differences between the original and replacement plants of *T. divaricatum* in their response to time, using linear mixed models (LMM) with interaction terms ('time × replacement group'). There were no significant differences in the studied traits, except in sugar concentration per flower in the outdoor group ($time^2 \times$ replacement group interaction, $t = -2.013$, $P = 0.049$). Similarity of response indicates that the replacement plants had similar response patterns to the original plants and that the two plant groups can be analysed together in order to identify the general response patterns.

Data analysis was carried out in three consecutive steps. First, we used LMM models to identify the effect of temperature on plant traits in the climate chamber, testing both linear and quadratic effect of temperature on the traits and using plant number (plant ID) as a random factor. Analysis of variance (ANOVA) was used to compare the fit of the linear and quadratic models, based on the AIC values. However, it is essential to note that in the climate chamber we could not truly separate the effect of temperature and time on plants, since the temperature was increased linearly with time.

Secondly, we used the outdoor data to discern the separate effect of temperature on plant traits independent of time, fitting LMM models with two crossed random factors (plant ID and time from the beginning of the experiment) on the outdoor data. Naturally, temperature outdoors increased with time throughout the study period from May to July ($r = 0.65$, $P = 0.009$). However, this modelling approach enables us to separate the real effect of temperature on plant traits by taking into account the effect of time as a random factor.

As a third step, we identified the possible differences in response to time between the plants manipulated in the climate chamber and the plants grown outdoors. In these models we defined treatment group (climate chamber or outdoor) as an explanatory variable and used 'trait × group' interaction for identifying differences between the groups. No difference in trait response between the groups would therefore indicate that the trends in the climate chamber could be explained by the effect of time and that the effect of temperature was negligible or at least not significant. If the interaction terms in the models are significant, then the difference between the climate chamber and outdoor group can be attributed to the effect of elevated temperatures in the climate chamber. We did not perform the third set of models for the trait variation (CV) data because the effect of temperature on these traits was negligible in the first two sets of models.

In order to compare the response to time between the climate chamber and outdoor treatment groups, which had a different experiment duration, we used standardized time (mean = 0, SD = 1) in the models. Although having a different duration, the time span corresponds to the same ecological period for plants in both test groups, since the plants were examined from the beginning of full bloom until the end of the flowering period both in the climate chamber and outdoors.

All studied traits except the proportion of empty flowers per plant were modelled by following the description above. For modelling the proportion of empty flowers of *B. acetabulosa*, we followed a similar protocol, but used generalized linear mixed models (GLMM) with a negative binomial error distribution to account for the different distribution in the data. We also tested for the need to use zero-inflated models, but models with no zero-inflation parameter had a better fit (lower AIC values).

The analyses were conducted in R 3.1.1 (R Core Team 2014). Linear mixed models were fitted using the function *lmer* in the *lme4* package (Bates *et al.* 2014), the GLMM models were built using the function *glmmadmb* in the package *glmmADMB* (Fournier *et al.* 2012; Skaug *et al.* 2015). Additional *P*-values for the *t*-values in LMM models were calculated using the package *lmerTest* (Kuznetsova et al. 2014) and the conditional and marginal coefficients of determination (R^2c and R^2m) for the LMM models were calculated with the function *r.squaredGLMM* in the package *MuMIn* (Barton 2015). R^2c shows the model variance explained by both fixed and random factors, while R^2m represents the variance explained by fixed factors alone. Graphs were compiled using the function *ggplot* in the *ggplot2* package (Wickham 2009) using a smoothing function to plot the relationships.

Results

Ballota acetabulosa

In the climate chamber, the traits of *B. acetabulosa* showed a significant dependence on temperature through time (henceforth simply 'temperature'; Table 1 and **[Supporting Information—Table S1]**). All tested traits had a positive unimodal relationship to temperature, except sugar concentration per flower, which showed a negative unimodal response (Table 1, Fig. 1). Optimum temperature for nectar volume per flower and sugar content per flower and per plant was at ~25–26 °C (Fig. 1A, C and D), even though nectar sugar concentration per flower was the lowest at these temperatures (Fig. 1B). Number of flowers per plant per day was also highest at the same favourable temperatures (Fig. 1E), but biomass per flower declined unimodally through the test period (Fig. 1F). Contrarily, trait variation (CV) within a plant did not show significant changes in response to elevated temperatures in most of the tested traits, except CV of sugar concentration **[see Supporting Information—Table S2]**. Elevated temperatures in the climate chamber also accelerated flower production and thus resulted in shortening the flowering period nearly 2-fold compared with the outdoor group (duration of the experiment was 24 days in the climate chamber vs 45 days outdoors).

Under this experimental design we were, however, unable to distinguish between the effect of temperature and time in the climate chamber as the relationships in this group could be due to both temperature and time. Under naturally varying conditions in the outdoor group we were able to model the effect of temperature separately, although the observed temperature range was narrower than the one tested in the climate chamber. For the outdoor group of *B. acetabulosa*, the linear models were favoured over the quadratic ones in all traits. Nevertheless, only sugar content per plant and biomass per flower had a significant relationship with temperature, showing a negative response to higher average temperatures (Table 2). Trait variation (CV) did not show a response to temperature, similarly to the climate chamber data, except in the case of CV of sugar concentration per flower, which was positively unimodally related to temperature (R^2m = 0.15, R^2c = 0.29) **[see Supporting Information—Table S3]**.

Comparison between the climate chamber and the outdoor group revealed a difference in the response to time in all *B. acetabulosa* traits, except sugar concentration per flower (Table 3 and **[Supporting Information—Fig. S1]**). The difference between the responses indicates an additional effect of temperature on plant traits in the climate chamber. Since the outdoor models indicated that temperature had no separate significant effect on nectar volume per flower, sugar content per flower and number of flowers per plant (Table 2), we can conclude that the difference observed between these trait responses is caused by the effect of elevated temperatures **[see Supporting Information—Fig. S1A, C and E]**.

At the moderately warm temperatures in the climate chamber (~26.5–29.5 °C) the plants had no empty flowers (producing no nectar) per plant, while at lower and higher temperatures the proportion of empty flowers increased. Nevertheless, modelling showed that the proportion of empty flowers per plant was not significantly related to temperature and the comparison between climate chamber and outdoor data was also not significant **[see Supporting Information—Table S4]**.

Teucrium divaricatum

All tested traits of *T. divaricatum*, except sugar concentration per flower, were significantly positively unimodally related to temperature in combination with time in the climate chamber (Table 1, Fig. 2). Nectar volume and flower sugar content per flower of *T. divaricatum* peaked at higher average temperatures (~29–33 °C; Fig. 2A and C) than the number of flowers per plant produced per day (~24–25 °C; Fig. 2E). Sugar content per plant was therefore greatest at ~26–27 °C (Fig. 2D). Trait variation (CV) of flower traits was not related to elevated temperatures in most of the studied traits, except CV of sugar concentration [**see Supporting Information—Table S2**]. In the outdoor group, only two traits showed a significant relationship to average temperature; the number of flowers per plant decreased with increasing temperatures and biomass per flower was positively unimodally related to temperature (Table 2). Most of the outdoor models also had very low marginal coefficients of determination (R^2m), showing that much of the variation was explained by differences between individual plants and not by temperature (Table 2).

Comparing the response between the climate chamber and the outdoor group in response to time showed a difference in nectar volume per flower, number of flowers per plant and biomass per flower (Table 3 and **[Supporting Information—Fig. S2]**). Number of flowers per plant and biomass per flower also showed a separate response

Table 1. Effect of temperature (simple and quadratic effect of temperature, 'T' and 'T^2', respectively) on nectar and flower traits in the climate chamber. 'I' represents model intercept, 'R^2m' and 'R^2c' denote marginal and conditional coefficients of determination, indicating the variation explained by fixed factors (R^2m) and the whole model (R^2c; Barton 2015). Statistically significant ($P < 0.05$) results are presented in bold.

Species	Modelled trait		Estimate	SE	t	P	R^2m	R^2c
B. acetabulosa	Nectar volume per flower	I	−0.748	0.162	−4.614	**<0.001**	0.31	0.47
		T	0.079	0.012	6.441	**<0.001**		
		T^2	−0.002	0.0002	−6.942	**<0.001**		
	Sugar concentration per flower	I	0.945	0.608	1.554	0.124	0.03	0.53
		T	−0.118	0.046	−2.586	**0.011**		
		T^2	0.002	0.001	2.656	**0.009**		
	Sugar content per flower	I	−2.578	0.624	−4.134	**<0.001**	0.37	0.50
		T	0.320	0.047	6.863	**<0.001**		
		T^2	−0.006	0.001	−7.477	**<0.001**		
	Sugar content per plant	I	−6.153	1.181	−5.208	**<0.001**	0.51	0.58
		T	0.740	0.088	8.389	**<0.001**		
		T^2	−0.015	0.002	−9.239	**<0.001**		
	Number of flowers per plant	I	−1.436	0.321	−7.598	**<0.001**	0.82	0.84
		T	0.346	0.024	14.500	**<0.001**		
		T^2	−0.007	0.0004	−16.561	**<0.001**		
	Biomass per flower	I	−1.815	0.046	−39.637	**<0.001**	0.75	0.92
		T	0.005	0.003	1.622	0.108		
		T^2	−0.0003	0.0001	−5.194	**<0.001**		
T. divaricatum	Nectar volume per flower	I	−0.074	0.053	−1.383	0.170	0.08	0.41
		T	0.011	0.004	2.748	**0.007**		
		T^2	−0.0002	0.0001	−2.428	**0.017**		
	Sugar concentration per flower	I	0.676	0.104	6.510	**<0.001**	0.02	0.53
		T	−0.005	0.004	−1.392	0.167		
	Sugar content per flower	I	0.088	0.257	0.341	0.734	0.08	0.60
		T	0.085	0.020	4.313	**<0.001**		
		T^2	−0.002	0.0004	−4.055	**<0.001**		
	Sugar content per plant	I	−0.217	0.678	−0.320	0.749	0.42	0.82
		T	0.327	0.052	6.317	**<0.001**		
		T^2	−0.007	0.001	−7.524	**<0.001**		
	Number of flowers per plant	I	−0.202	0.611	−0.330	0.742	0.54	0.87
		T	0.254	0.045	5.680	**<0.001**		
		T^2	−0.006	0.001	−7.851	**<0.001**		
	Biomass per flower	I	−1.965	0.060	−32.871	**<0.001**	0.45	0.93
		T	0.004	0.005	0.934	0.353		
		T^2	−0.0003	0.0001	−3.447	**<0.001**		

to temperature in the outdoor models (Table 2), thus it is more difficult to say how much of the difference could be caused by the elevated temperatures in the climate chamber **[see Supporting Information—Fig. S2E and F]**. However, the difference in the case of nectar volume per flower is likely caused by the effect of elevated temperatures in the climate chamber **[see Supporting Information—Fig. S2A]**.

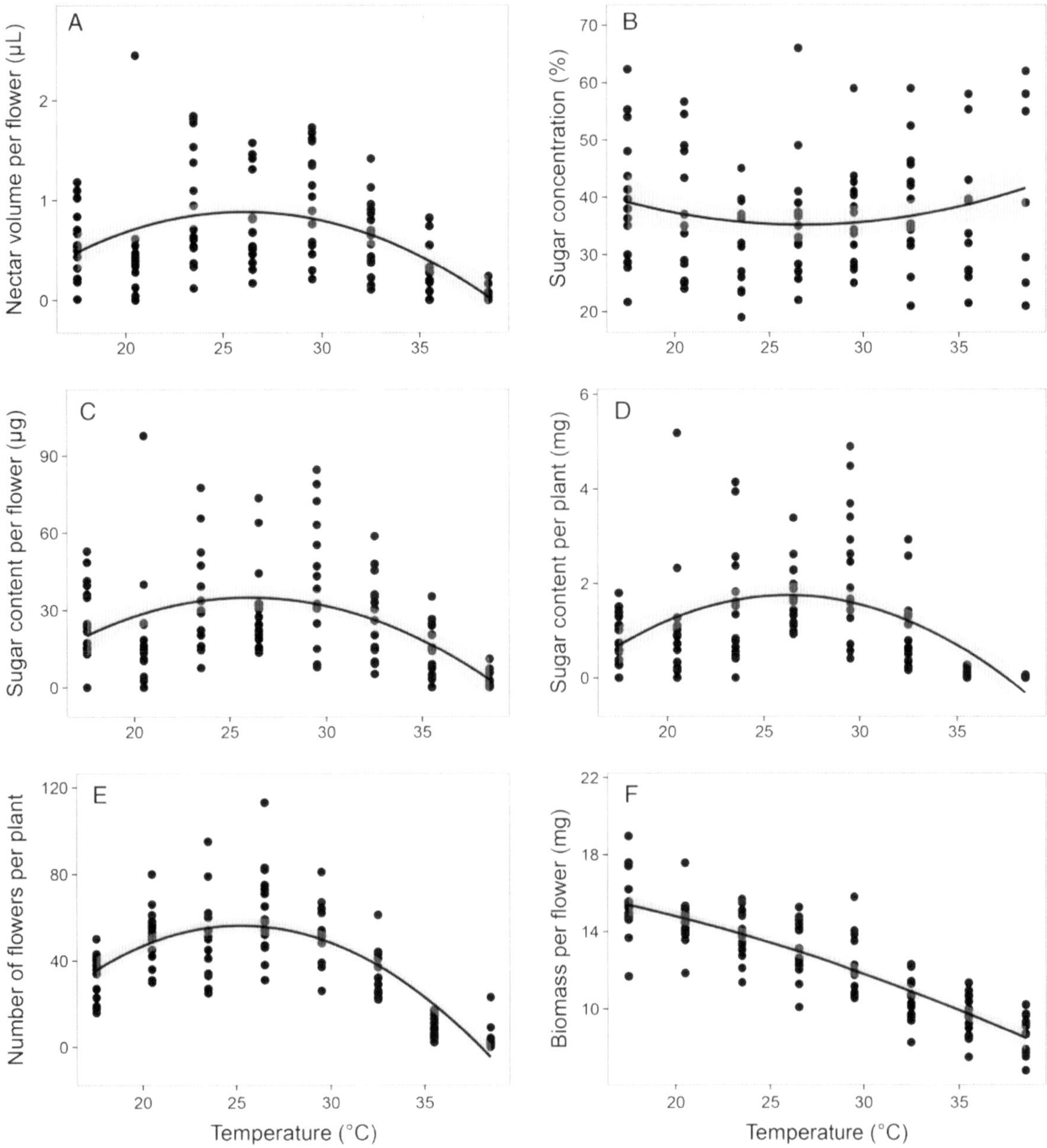

Figure 1. *Ballota acetabulosa* trait response to temperature in the climate chamber. Grey areas represent 95% confidence intervals.

Discussion

Nectar secretion in the Mediterranean under global warming

We studied the effect of a range of temperatures on two phryganic Lamiaceae species in Greece to evaluate the potential threat of climate change to floral nectar production and plant–pollinator interactions in the Mediterranean. The Mediterranean region is considered to be particularly threatened by global warming in the future due to strongly elevated temperatures and a heightened probability of drought during the summer (Giorgi 2006; Hickler *et al.* 2012). The average summer temperatures across the Mediterranean between 1961 and 1990 were between ~22.5 and 30 °C (New *et al.* 1999) but the climate warming scenarios now predict a 0.9–1.2 °C of warming for the summer months over the next two decades and up to 4–5 °C temperature increase before the end of the century (Giorgi and Lionello 2008; Giannakopoulos *et al.* 2009; IPCC 2013).

Strongly elevated temperatures, predicted to prevail in the Mediterranean region in the end of the century, would likely decrease nectar secretion and sugar production in both *B. acetabulosa* and *T. divaricatum*, which based on

Table 2. Effect of temperature (simple and quadratic effect of temperature, '*T*' and 'T^2', respectively) on nectar and flower traits in the outdoor group. '*I*' represents model intercept, 'R^2m' and 'R^2c' denote marginal and conditional coefficients of determination, indicating the variation explained by fixed factors (R^2m) and the whole model (R^2c; Barton 2015). Statistically significant ($P < 0.05$) results are presented in bold.

Species	Modelled trait		Estimate	SE	t	P	R^2m	R^2c
B. acetabulosa	Nectar volume per flower	*I*	0.819	0.317	2.583	**0.023**	0.14	0.63
		T	−0.027	0.014	−2.025	0.064		
	Sugar concentration per flower	*I*	−1.568	0.960	−1.634	0.126	0.03	0.27
		T	0.042	0.041	1.025	0.324		
	Sugar content per flower	*I*	3.416	1.076	3.175	**0.007**	0.13	0.53
		T	−0.093	0.046	−2.022	0.065		
	Sugar content per plant	*I*	8.532	2.690	3.172	**0.008**	0.23	0.75
		T	−0.262	0.115	−2.289	**0.041**		
	Number of flowers per plant	*I*	3.113	1.222	2.547	**0.022**	0.11	0.85
		T	−0.085	0.054	−1.587	0.134		
	Biomass per flower	*I*	−1.411	0.180	−7.854	**<0.001**	0.35	0.95
		T	−0.024	0.008	−3.058	**0.008**		
T. divaricatum	Nectar volume per flower	*I*	−0.073	0.197	−0.369	0.718	0.03	0.51
		T	0.009	0.008	1.006	0.333		
	Sugar concentration per flower	*I*	−0.231	0.788	−0.293	0.774	0.01	0.55
		T	0.015	0.034	0.450	0.660		
	Sugar content per flower	*I*	0.025	0.717	0.035	0.973	0.07	0.50
		T	0.053	0.031	1.743	0.113		
	Sugar content per plant	*I*	5.690	2.296	2.478	**0.028**	0.06	0.41
		T	−0.129	0.098	−1.313	0.213		
	Number of flowers per plant	*I*	3.803	1.124	3.383	**0.004**	0.15	0.57
		T	−0.107	0.049	−2.162	**0.048**		
	Biomass per flower	*I*	−3.375	0.441	−7.648	**<0.001**	0.09	0.75
		T	0.119	0.042	2.825	**0.021**		
		T^2	−0.003	0.001	−2.727	**0.025**		

our results, currently grow in the wild at or close to their optimum temperature range for nectar secretion. This effect can be particularly strong in regions of the Mediterranean where the current average temperatures are higher. Strongly elevated temperatures in the climate chamber decreased sugar content per plant of both *B. acetabulosa* and *T. divaricatum* (Figs 1D and 2D) by decreasing the volume of nectar produced per flower (Figs 1C and 2C) and the number of flowers per plant (Figs 1E and 2E). High temperatures induce physiological stress in plants, which decreases nectar production per flower (Petanidou and Smets 1996; Pacini and Nepi 2007; Scaven and Rafferty 2013) and can cause plants to produce fewer flowers (Saavedra *et al.* 2003). Extreme climate warming could consequently have a negative effect on nectar production and hence on the phryganic communities, where Lamiaceae are a dominant plant group (Petanidou and Vokou 1993).

Moderate warming predicted for the next two decades, however, should have only a moderate effect on *B. acetabulosa* nectar secretion, remaining close to this species' optimum range, and even promote nectar production of *T. divaricatum*, which demonstrated a relatively high optimal temperature for nectar volume per flower (30–35 °C; Fig. 2A) and sugar content per flower (~29 °C; Fig. 2C). Similarly high temperature optimum (32.5 °C during the day, analogous to the optimal daily temperatures of *T. divaricatum*) has also been found for the sugar content per flower of *Thymus capitatus*, another phryganic species (Petanidou and Smets 1996). In species with similarly high temperature optima, climate change can potentially increase nectar production, at least in

Table 3. Comparison models testing the difference of the effect of time (simple and quadratic effect, 'Time' and 'Time^2', respectively) on nectar and flower traits between the climate chamber and outdoor treatment ('group'). Only interaction terms are presented here from the model full results. Statistically significant ($P < 0.05$) results are presented in bold.

Species	Modelled trait		*t*	*P*
B. acetabulosa	Nectar volume per flower	Time × group	−0.757	0.450
		Time^2 × group	3.543	**<0.001**
	Sugar concentration per flower	Time × group	0.310	0.757
		Time^2 × group	−1.841	0.069
	Sugar content per flower	Time × group	0.554	0.581
		Time^2 × group	3.880	**0.002**
	Sugar content per plant	Time × group	−1.209	0.228
		Time^2 × group	3.007	**0.003**
	Number of flowers per plant	Time × group	−3.824	**<0.001**
		Time^2 × group	1.027	0.306
	Biomass per flower	Time × group	−4.379	**<0.001**
		Time^2 × group	0.415	0.679
T. divaricatum	Nectar volume per flower	Time × group	0.481	0.632
		Time^2 × group	−2.540	**0.012**
	Sugar concentration per flower	Time × group	−0.606	0.546
		Time^2 × group	1.868	0.064
	Sugar content per flower	Time × group	1.058	0.292
		Time^2 × group	−0.997	0.320
	Sugar content per plant	Time × group	−1.041	0.300
		Time^2 × group	−0.066	0.948
	Number of flowers per plant	Time × group	−0.418	0.676
		Time^2 × group	2.679	**0.008**
	Biomass per flower	Time × group	4.510	**<0.001**
		Time^2 × group	3.622	**<0.001**

the case of moderate warming. However, sugar content per plant of *T. divaricatum* peaked at lower temperatures (optimum ~26–27 °C; Fig. 2D) than sugar content per flower, indicating that the nectar production per plant could still be negatively affected by even moderate warming, even if sugar content per flower can still increase under higher temperatures. The effect of moderate warming on the sugar content per plant of *T. divaricatum*, as well as *B. acetabulosa* sugar content per flower (Fig. 1C) and sugar content per plant (optimum ~25–27 °C; Fig. 1D), might in the near future be promoted in some regions of the Mediterranean with lower average temperatures, or be moderately decreased in regions with higher temperatures.

In contrast to nectar and sugar production, trait variation (CV) within a plant was not related to temperature in most of the tested traits. Only CV of sugar concentration was positively related to temperature indoors (in both species) and positively unimodally related to temperature outdoors (in *B. acetabulosa*). Changes in the variation of nectar production can decrease plant attractiveness to pollinators (Real and Rathcke 1988; Zimmerman 1988; Shafir 2000), thus having a negative impact on plant pollination and consequently on plant reproduction and population persistence (Scaven and Rafferty 2013). Nevertheless, the stability of trait variation patterns in our experiment signifies that in these two species this aspect of plant–pollinator interactions could be relatively unaffected by elevated temperatures and not cause additional alterations under climate warming.

Moderate warming can be beneficial for nectar production at least as long as the plants are not water stressed. The Mediterranean plants are indeed well adapted to cope with the hot and dry conditions during the flowering period (Petanidou 2007). However, different scenarios of

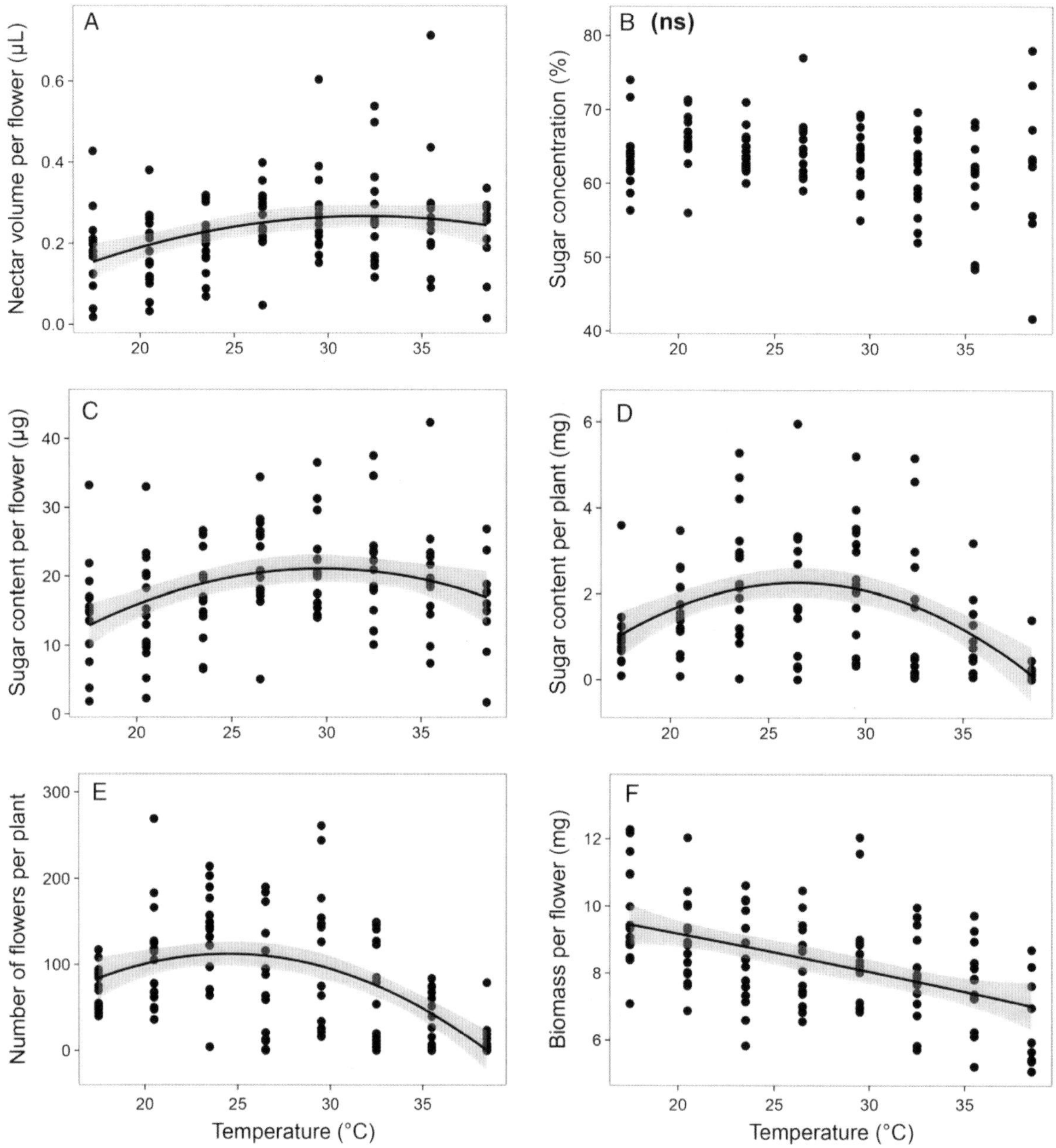

Figure 2. *Teucrium divaricatum* trait response to temperature in the climate chamber. Grey areas represent 95% confidence intervals. Non-significant ($P > 0.05$) relationships are marked with 'ns'.

climate change predict a substantial decrease in summer precipitation in the Mediterranean region in addition to increased temperatures (Giorgi and Lionello 2008; Giannakopoulos *et al.* 2009; IPCC 2013). Consequently, the actual effect of climate warming with additional water stress (Villarreal and Freeman 1990; Carroll *et al.* 2001) could further increase the negative effect of elevated temperatures on nectar production in the phryganic systems. The combined effect of temperature and drought on plants and their ability to adapt to future climate changes in this region still needs further study.

Differences in species' response to global warming

Our two study species responded to the elevated temperatures somewhat differently, which suggests possible disparate responses to climate change in phryganic species. For one, *B. acetabulosa* optimal temperatures for nectar and sugar production were in general lower, close to the current average temperatures, and the species therefore more sensitive to climate warming than *T. divaricatum*, which had higher optimal temperatures for nectar secretion (Figs 1 and 2). Additionally, several traits of *B. acetabulosa* exhibited dependence on elevated temperatures (nectar volume per flower, sugar content per flower and number of flowers per plant) in addition to the effect of time through the flowering period. On the other hand, in *T. divaricatum* it was only nectar volume per flower that clearly depended on elevated temperatures. Strong dependence on temperature implies that *B. acetabulosa* could be rather sensitive to temperature changes, whereas *T. divaricatum* might be more moderately affected by climate warming.

The differences in species' nectar production patterns could be caused by the plants' adaptations to the different phryganic microhabitats, which the two species inhabit—*B. acetabulosa* prefers modified or somewhat protected microhabitats (understorey, partially shaded, neighbouring to different structures), while *T. divaricatum* grows under full sun (Petanidou and Smets 1996; Petanidou *et al.* 2000). As a plant of open natural areas, *T. divaricatum* is adapted to particularly high temperatures and does not experience physiological stress at moderately higher than average temperatures, akin to other open phryganic species, such as *Thymus capitatus*, which has a similarly high optimal temperature range for nectar sugar production (Petanidou and Smets 1996). At the same time, *B. acetabulosa* as a plant of less exposed habitats could be less adapted to the heat and considerably more sensitive to temperature rise. This difference suggests a stronger effect of climate change on understorey plants in the phryganic systems, which are less adapted to high temperatures and the effect of drought. At the same time, species that are adapted to the harsh conditions of the open phrygana might be able to cope better with the forthcoming changes.

Effect of global warming on Mediterranean pollinators

Our results demonstrate that in case of moderate warming, as predicted in the Mediterranean region for the next two decades, nectar sugar production and consequently the amount of resources available for pollinators could be moderately adversely affected in the case of some Mediterranean species (e.g. *B. acetabulosa*) and might even be benefitted in others (e.g. *T. divaricatum*). This difference in species responses suggests that at the community level species nectar production might be able to balance out, at least for the more generalized pollinators (Scaven and Rafferty 2013). Nevertheless, in the case of more extreme warming, as predicted for the end of the century, nectar production of both species is expected to decrease.

Both our study species are among the highest nectar producers in phrygana (Petanidou and Smets 1995) and are therefore an essential resource for a number of pollinator species (Petanidou and Vokou 1993; T. Petanidou *et al.*, unpubl.). Due to the greater sensitivity of *B. acetabulosa* to elevated temperatures, the effect of climate warming could be more pronounced on the pollinators of this species already during the next few decades under the effect of moderate warming. Although plant–pollinator networks, especially generalist interactions, are expected to be rather robust to the effect of climate change (Devoto *et al.* 2007; Schweiger *et al.* 2010; DeLucia *et al.* 2012; Burkle *et al.* 2013), the loss of a generalist plant species can still be a considerable risk for the population persistence of pollinators (Memmott *et al.* 2004). Even if the plant and pollinator populations are able to persist for some time, the strength of interactions can be changed due to alterations in plant resources and pollinator behaviour and thus still significantly affect the mutualistic interaction networks due to climate warming (Memmott *et al.* 2007; Scaven and Rafferty 2013).

Moreover, experimental warming in the climate chamber accelerated plant flowering and thus reduced the length of flowering period under elevated temperatures. Under natural conditions, shorter flowering time in consequence of global warming could increase the probability of creating temporal mismatches with pollinators (Memmott *et al.* 2007; Hegland *et al.* 2009; DeLucia *et al.* 2012). Since Lamiaceae are the most essential group for the pollinators in the phryganic systems in summer (Herrera 1985; Petanidou and Smets 1995; Petanidou 2007), the reduced nectar secretion and temporal mismatches with these species at strongly elevated temperatures could have a considerable impact on the phryganic pollinator fauna. In addition, it could also negatively affect apiculture in the Mediterranean region, which in the phryganic systems is strongly dependent on the abundance and nectar production of different Lamiaceae species (Petanidou and Smets 1995).

Conclusions

Mediterranean ecosystems may be able to endure moderate climate warming without major changes in plant–pollinator interactions, at least as long as the plant communities are not overly water stressed. Additional water stress due to decreased rainfall predicted by climate change scenarios could, however, induce stronger and more rapid changes in nectar production and plant–pollinator interactions. More extensive changes in Mediterranean communities can be expected towards the end of the century due to more extreme warming, when even species adapted to the severe conditions of open phrygana, such as *T. divaricatum*, might be excessively stressed by the elevated temperatures. Consequent changes in plant–pollinator interactions can include weakening or disruption of interaction networks and can eventually lead to a possible loss of pollinator species in the Mediterranean systems (Hegland *et al.* 2009; Scaven and Rafferty 2013; Petanidou *et al.* 2014).

Sources of Funding

The research has been co-financed by the European Union (European Social Fund—ESF) and Greek national funds through the Operational Program 'Education and Lifelong Learning' of the National Strategic Reference Framework (NSRF)—Research Funding Program: THALES. Investing in knowledge society through the European Social Fund.

Contributions by the Authors

T.P. conceived the idea, P.T. and K.T. conducted the experiment, T.P. and T.T. supervised and supported the experimental work. K.T. analysed the data and led the writing with the assistance of T.P. and T.T.

Acknowledgements

We thank Eirini Vallianatou for providing seeds of *Ballota acetabulosa* from the I. & A. Diomedes Botanical Garden seed repository, Lazaros Neokosmidis for helping with *Teucrium divaricatum* collection in the field and Prof. Konstantinos Kalabokidis and Palaiologos Palaiologou for providing the Mytilene climate station data. We are grateful to the Associate Editor and the anonymous reviewer for their very helpful comments on the manuscript.

Supporting Information

The following additional information is available in the online version of this article -

Table S1. Nectar secretion values of the study plants under different temperature regimes in the climate chamber.

Table S2. Effect of temperature on the CV of flower traits in the climate chamber.

Table S3. Effect of temperature on the CV of flower traits in the outdoor group.

Table S4. Proportion of empty flowers of *Ballota acetabulosa* in relation to temperature and comparison models testing the difference of the effect of time between the climate chamber and the outdoor treatment.

Figure S1. Comparison of *Ballota acetabulosa* trait response to time between the climate chamber and outdoor group.

Figure S2. Comparison of *Teucrium divaricatum* trait response to time between the climate chamber and outdoor group.

Literature Cited

Barton K. 2015. *MuMIn: multi-model inference*. R package version 1.13.4. http://CRAN.R-project.org/package=MuMIn (20 March 2015).

Bates D, Maechler M, Bolker B, Walker S. 2014. *lme4: linear mixed-effects models using Eigen and S4*. R package version 1.1-7. http://CRAN.R-project.org/package=lme4 (20 March 2015).

Burkle LA, Marlin JC, Knight TM. 2013. Plant-pollinator interactions over 120 years: loss of species, co-occurrence, and function. *Science* **339**:1611-1615.

Carroll AB, Pallardy SG, Galen C. 2001. Drought stress, plant water status, and floral trait expression in fireweed, *Epilobium angustifolium* (Onagraceae). *American Journal of Botany* **88**:438-446.

Dafni A, Kevan PG, Husband BC. 2005. *Practical pollination biology*. Cambridge, ON: Enviroquest, Ltd.

Dauber J, Biesmeijer JC, Gabriel D, Kunin WE, Lamborn E, Meyer B, Nielsen A, Potts SG, Roberts SPM, Sõber V, Settele J, Steffan-Dewenter I, Stout JC, Teder T, Tscheulin T, Vivarelli D, Petanidou T. 2010. Effects of patch size and density on flower visitation and seed set of wild plants: a pan-European approach. *Journal of Ecology* **98**:188-196.

Davis PH, ed. 1982. *Flora of Turkey and the East Aegean Islands*, Vol. 7. Edinburgh: University Press.

DeLucia EH, Nabity PD, Zavala JA, Berenbaum MR. 2012. Climate change: resetting plant-insect interactions. *Plant Physiology* **160**: 1677-1685.

Devoto M, Zimmermann M, Medan D. 2007. Robustness of plant-flower visitor webs to simulated climate change. *Ecología Austral* **17**:37-50.

Fournier DA, Skaug HJ, Ancheta J, Ianelli J, Magnusson A, Maunder MN, Nielsen A, Sibert J. 2012. AD model builder: using automatic differentiation for statistical inference of highly parameterized complex nonlinear models. *Optimization Methods and Software* **27**:233-249.

Giannakopoulos C, Le Sager P, Bindi M, Moriondo M, Kostopoulou E, Goodess CM. 2009. Climatic changes and associated impacts in the Mediterranean resulting from a 2°C global warming. *Global and Planetary Change* **68**:209-224.

Giorgi F. 2006. Climate change hot-spots. *Geophysical Research Letters* **33**:L08707.

Giorgi F, Lionello P. 2008. Climate change projections for the Mediterranean region. *Global and Planetary Change* **63**:90-104.

Hegland SJ, Nielsen A, Lázaro A, Bjerknes A-L, Totland Ø. 2009. How does climate warming affect plant-pollinator interactions? *Ecology Letters* **12**:184-195.

Herrera J. 1985. Nectar secretion patterns in Southern Spanish Mediterranean scrublands. *Israel Journal of Botany* **34**:47-58.

Hickler T, Vohland K, Feehan J, Miller PA, Smith B, Costa L, Giesecke T, Fronzek S, Carter TR, Cramer W, Kühn I, Sykes MT. 2012. Projecting the future distribution of European potential natural vegetation zones with a generalized, tree species-based dynamic vegetation model. *Global Ecology and Biogeography* **21**:50-63.

Hoover SER, Ladley JJ, Shchepetkina AA, Tisch M, Gieseg SP, Tylianakis JM. 2012. Warming, CO_2, and nitrogen deposition interactively affect a plant-pollinator mutualism. *Ecology Letters* **15**:227-234.

IPCC. 2013. *Climate change 2013: the physical science basis. Contribution of Working group I to the Fifth assessment report of the Intergovernmental panel on climate change* (Stocker TF, Qin D, Plattner G-K, Tignor M, Allen SK, Boschung J, Nauels A, Xia Y, Bex V, Midgley PM, eds.). Cambridge, UK: Cambridge University Press.

Jakobsen HB, Kristjánsson K. 1994. Influence of temperature and floret age on nectar secretion in *Trifolium repens* L. *Annals of Botany* **74**:327–334.

Keasar T, Sadeh A, Shmida A. 2008. Variability in nectar production and standing crop, and their relation to pollinator visits in a Mediterranean shrub. *Arthropod-Plant Interactions* **2**: 117–123.

Kuznetsova A, Brockhoff PB, Christensen RHB. 2014. *lmerTest: tests in linear mixed effects models*. R package version 2.0-20. http://CRAN.R-project.org/package=lmerTest (20 March 2015).

Memmott J, Waser NM, Price MV. 2004. Tolerance of pollination networks to species extinctions. *Proceedings of the Royal Society B: Biological Sciences* **271**:2605–2611.

Memmott J, Craze PG, Waser NM, Price MV. 2007. Global warming and the disruption of plant–pollinator interactions. *Ecology Letters* **10**:710–717.

New M, Hulme M, Jones P. 1999. Representing twentieth-century space–time climate variability. Part I: development of a 1961–90 mean monthly terrestrial climatology. *Journal of Climate* **12**: 829–856.

Nocentini D, Pacini E, Guarnieri M, Martelli D, Nepi M. 2013. Intrapopulation heterogeneity in floral nectar attributes and foraging insects of an ecotonal Mediterranean species. *Plant Ecology* **214**:799–809.

Pacini E, Nepi M. 2007. Nectar production and presentation. In: Nicolson SW, Nepi M, Pacini E, eds. *Nectaries and nectar*. Dordrecht, The Netherlands: Springer, 167–214.

Pacini E, Nepi M, Vesprini JL. 2003. Nectar biodiversity: a short review. *Plant Systematics and Evolution* **238**:7–21.

Petanidou T. 1991. Pollination ecology in a phryganic ecosystem. PhD Thesis (in Greek, with English summary). Aristotle University of Thessaloniki, Thessaloniki, Greece.

Petanidou T. 2007. Ecological and evolutionary aspects of floral nectars in Mediterranean habitats. In: Nicolson SW, Nepi M, Pacini E, eds. *Nectaries and nectar*. Dordrecht, The Netherlands: Springer, 343–375.

Petanidou T, Smets E. 1995. The potential of marginal lands for bees and apiculture: nectar secretion in Mediterranean shrublands. *Apidologie* **26**:39–52.

Petanidou T, Smets E. 1996. Does temperature stress induce nectar secretion in Mediterranean plants? *New Phytologist* **133**: 513–518.

Petanidou T, Vokou D. 1993. Pollination ecology of Labiatae in a phryganic (East Mediterranean) ecosystem. *American Journal of Botany* **80**:892–899.

Petanidou T, Ellis WN, Margaris NS, Vokou D. 1995. Constraints on flowering phenology in a phryganic (East Mediterranean shrub) community. *American Journal of Botany* **82**:607–620.

Petanidou T, Goethals V, Smets E. 2000. Nectary structure of Labiatae in relation to their nectar secretion and characteristics in a Mediterranean shrub community—does flowering time matter? *Plant Systematics and Evolution* **225**:103–118.

Petanidou T, Kallimanis AS, Sgardelis SP, Mazaris AD, Pantis JD, Waser NM. 2014. Variable flowering phenology and pollinator use in a community suggest future phenological mismatch. *Acta Oecologica* **59**:104–111.

R Core Team. 2014. *R: a language and environment for statistical computing*. Vienna, Austria: R foundation for statistical computing. http://www.R-project.org/ (20 March 2015).

Real L, Rathcke BJ. 1988. Patterns of individual variability in floral resources. *Ecology* **69**:728–735.

Saavedra F, Inouye DW, Price MV, Harte J. 2003. Changes in flowering and abundance of *Delphinium nuttallianum* (Ranunculaceae) in response to a subalpine climate warming experiment. *Global Change Biology* **9**:885–894.

Scaven VL, Rafferty NE. 2013. Physiological effects of climate warming on flowering plants and insect pollinators and potential consequences for their interactions. *Current Zoology* **59**:418–426.

Schweiger O, Biesmeijer JC, Bommarco R, Hickler T, Hulme PE, Klotz S, Kühn I, Moora M, Nielsen A, Ohlemüller R, Petanidou T, Potts SG, Pyšek P, Stout JC, Sykes MT, Tscheulin T, Vilà M, Walther G-R, Westphal C, Winter M, Zobel M, Settele J. 2010. Multiple stressors on biotic interactions: how climate change and alien species interact to affect pollination. *Biological Reviews* **85**:777–795.

Shafir S. 2000. Risk-sensitive foraging: the effect of relative variability. *Oikos* **88**:663–669.

Skaug H, Fournier D, Bolker B, Magnusson A, Nielsen A. 2015. *Generalized linear mixed models using AD model builder*. R package version 0.8.1.

Southwick EE. 1983. Nectar biology and nectar feeders of common milkweed, *Asclepias syriaca* L. *Bulletin of the Torrey Botanical Club* **110**:324–334.

Torres C, Galetto L. 1998. Patterns and implications of floral nectar secretion, chemical composition, removal effects and standing crop in *Mandevilla pentlandiana* (Apocynaceae). *Botanical Journal of the Linnean Society* **127**:207–223.

Traill LW, Lim MLM, Sodhi NS, Bradshaw CJA. 2010. Mechanisms driving change: altered species interactions and ecosystem function through global warming. *Journal of Animal Ecology* **79**:937–947.

Tylianakis JM, Didham RK, Bascompte J, Wardle DA. 2008. Global change and species interactions in terrestrial ecosystems. *Ecology Letters* **11**:1351–1363.

Villarreal AG, Freeman CE. 1990. Effects of temperature and water stress on some floral nectar characteristics in *Ipomopsis longiflora* (Polemoniaceae) under controlled conditions. *Botanical Gazette* **151**:5–9.

Wickham H. 2009. *ggplot2: elegant graphics for data analysis*. New York: Springer.

Zimmerman M. 1988. Nectar production, flowering phenology, and strategies for pollination. In: Lovett-Doust J, Lovett-Doust L, eds. *Plant reproductive ecology: patterns and strategies*. New York: Oxford University Press, 157–178.

15

Natural populations of *Arabidopsis thaliana* differ in seedling responses to high-temperature stress

Nana Zhang[1], Brian Belsterling[1], Jesse Raszewski[1] and Stephen J. Tonsor[1,2*]

[1] Department of Biological Sciences, University of Pittsburgh, 4249 Fifth Ave., Pittsburgh, PA 15260, USA

[2] Carnegie Museum of Natural History, 4400 Forbes Ave., Pittsburgh, PA 15213, USA

Associate Editor: F. Xavier Picó

Abstract. Little is known about adaptive within-species variation in thermotolerance in wild plants despite its likely role in both functional adaptation at range limits and in predicting response to climate change. Heat shock protein Hsp101, rapidly heat induced in *Arabidopsis thaliana*, plays a central role in thermotolerance in laboratory studies, yet little is known about variation in its expression in natural populations. We explored variation in thermotolerance and Hsp101 expression in seedlings from 16 natural populations of *A. thaliana* sampled along an elevation and climate gradient. We tested both naive controls (maintained at 22 °C until heat stress) and thermally pre-acclimated plants (exposed to a 38 °C 3-h acclimation treatment). After acclimation, seedlings were exposed to one of two heat stresses: 42 or 45 °C. Thermotolerance was measured as post-stress seedling survival and root growth. When stressed at 45 °C, both thermotolerance and Hsp101 expression were significantly increased by pre-acclimation. However, thermotolerance did not differ between pre-acclimation and control when followed by a 42 °C stress. Immediately after heat stress, pre-acclimated seedlings contained significantly more Hsp101 than control seedlings. At 45 °C, Hsp101 expression was positively associated with survival ($r^2 = 0.37$) and post-stress root growth ($r^2 = 0.15$). Importantly, seedling survival, post-stress root growth at 45 °C and Hsp101 expression at 42 °C were significantly correlated with the home sites' first principal component of climate variation. This climate gradient mainly reflects a temperature and precipitation gradient. Thus, the extent of Hsp101 expression modulation and thermotolerance appear to be interrelated and to evolve adaptively in natural populations of *A. thaliana*.

Keywords: Adaptation; *Arabidopsis*; climates; Hsp101; natural variation; thermotolerance.

Introduction

Increasing temperatures in many regions of the world may be a defining environmental change in the 21st century. Increasing heat stress is highly likely to lead to shifts in geographic distributions or even to complete extinction of many species (Field *et al.* 2014). Even without current human-driven rapid changes in thermal environments, populations at the warmer edge of their species' geographic range often face strong stresses from high-temperature events (Gaston 2009). Variation in thermal environment among sites within species' ranges may result in adaptive variation in thermotolerance. However, little is known about the within-species ability to evolve thermotolerance responses (Barua *et al.* 2008; Tonsor *et al.* 2008; Paul *et al.* 2011). In this study, we examine variation among natural populations of *Arabidopsis thaliana* in

* Corresponding author's e-mail address: tonsors@carnegiemnh.org

their ability to survive and recover from heat stress. We also test the extent to which survival and recovery from heat stress is correlated with the expression of heat shock protein (Hsp).

Exposure to high temperature can cause severe, irreversible cellular damage and loss of cellular function (Shabala 2012). The threshold temperature at which irreversible damage begins can vary among populations from contrasting in climates (Barua *et al.* 2008). The threshold temperature for damage can also be plastic in response to conditions experienced prior to high-temperature exposure (Wehner *et al.* 1985). When temperatures increase gradually or when plants experience a prior exposure to moderately high temperatures, changes in gene expression ensue (Larkindale *et al.* 2005; Hannah *et al.* 2006; Larkindale and Vierling 2008), leading to greater thermotolerance. This reprogramming and subsequent increase in thermotolerance is termed acclimation or acquired thermotolerance.

Acquired thermotolerance has been shown to depend, in part, on the rapid expression of Hsps (Lindquist 1986; Wang *et al.* 2004). Heat causes protein denaturation, disrupting normal protein function. The Hsps act as molecular chaperons to prevent protein aggregation, repair protein damage and maintain cellular homeostasis. Hsps stabilize protein form and prevent aggregation. The Hsps together recover or protect normal cellular function provided that the heat stress is not too extreme (Hong and Vierling 2001). While some other Hsps have been shown to have important roles in the heat shock response and in thermotolerance, to date only Hsp101 has been shown to be essential for acquired thermotolerance in plants. Hsp101 has a specific role as a protein machine that disaggregates misfolded proteins (Doyle and Wickner 2009).

This has been demonstrated in both *A. thaliana* (Hong and Vierling 2001) and maize (Nieto-Sotelo *et al.* 2002). Hsp101, the cytosol-expressed homologue of the Hsp100/ClpB gene family in *A. thaliana*, re-solubilizes and refolds denatured protein. Hsp101 is present in a single copy, Athsp101, in the *A. thaliana* genome. To date, nearly all plant studies of Hsp100/ClpB gene products have focussed on understanding Hsp101 genetics and biochemistry using genetic derivatives of the most commonly studied genotype of *A. thaliana*, Columbia (Col-0), in laboratory settings (Larkindale and Knight 2002; Hong *et al.* 2003; Larkindale *et al.* 2005; Tonsor *et al.* 2008).

Only a little is known about the extent to which natural populations evolve different patterns of Hsp101 expression in response to the differing patterns of heat stress they have experienced in nature. Natural populations of wild species can be important tools for studying the evolution of thermotolerance because they can carry adaptive signatures of selection imposed by their climates of origin. We know of only four studies that directly investigate the relationship between Hsps and thermotolerance in natural plant populations. Barua *et al.* (2003) demonstrated variation among ecotypes of *Chenopodium album* in thermotolerance of photosynthetic electron transport. This thermotolerance variation was strongly associated with variation in chloroplast small Hsp expression and with the thermal environments from which the genotypes were collected (Barua *et al.* 2003). Barua *et al.* (2008) further characterized geographically based variation among genotypes of *C. album* in Hsp60, Hsp70 and small Hsp expression, a finding that expression was greatest in genotypes from habitats that were subject to the greatest fluctuations in temperature. Habitats with high mean and maximum temperatures did not show greater Hsp expression (Barua *et al.* 2008). Similarly, Tonsor *et al.* (2008) tested standard accessions of *A. thaliana* whose sites of origin came from a broad latitudinal range. These accessions showed co-variation between latitude of origin and leaf Hsp101 content, with genotypes from warmer latitudes exhibiting lower content (Tonsor *et al.* 2008). Amano *et al.* (2012) compared two species of *Potamogeton* (Potamogetonaceae); one species was heat tolerant, and the other was heat intolerant. *Potamogeton malainus* from shallow waters with high-temperature fluctuations showed higher thermotolerance, while *P. perfoliatus* from deeper waters with cooler, more constant temperatures had lower basal thermotolerance and could not develop acquired thermotolerance because of changes in some of the heat shock element binding sites for HSFA2 (Amano *et al.* 2012). Taken together, these studies suggest that environments that differ in temperature regimes may lead, in general, to evolved differences in heat shock responses. The studies further suggest that evolved responses are likely the result of complex interactions of plants with their local climates. Predictions as to the expected pattern of response are, therefore, not likely to be accurate. Instead, we chose to simply ask: what are the extant patterns of responses?

Population-level studies exploring the relationship among the climate of origin, differentiation in thermotolerance and the role of changes in gene expression in acquired thermotolerance are needed to improve understanding of optimal heat stress responses in plants. Studies at the population level can provide insights on how populations, from common genetic background, adapted in the past to the novel climates in which they found themselves. Such studies can also help us predict more finely on how plants will respond to future climate shifts. Evolution of Hsps and heat shock response at large geographic scale has been largely conducted in animal species (Feder 1999). In fruit flies, *Drosophila virilis* from low latitude showed higher thermotolerance and Hsp70 accumulation after 40–41 °C heat shock (Garbuz *et al.* 2003). In a marine system, differential response to heat stress in invasive vs. native blue mussels showed the importance of heat shock response to invasive success (Lockwood *et al.* 2010). Heat shock proteins were found to have different roles in heat stress response, but to have generally been up-regulated (Sørensen *et al.* 2001, 2005; but see Jensen *et al.* 2010). Here, we identify adaptively differing patterns of Hsp expression that correspond with adaptation to contrasting thermal climates in the genomically highly enabled study system, *A. thaliana*. These results provide fundamental insights needed for future studies of the genetics, biochemistry and evolutionary ecology of variation in Hsp expression and thermotolerance in natural populations.

Many climate variables are correlated with elevation (Körner 2007), and elevation gradients create climate gradients across short distances that substantially mimic the broader gradients incorporated in many species' full geographic ranges. Thus, studying plants along an elevation gradient can simplify the logistics of climate–plant adaptation studies (Montesinos-Navarro *et al.* 2011). With F. Xavier Picó (La Doñana Biological Station, Seville, Spain), we collected 16 natural populations of *A. thaliana* along a climate and elevation gradient in southern Europe extending from near sea level at the Mediterranean coast to near treeline (~2200 m above sea level (a.s.l.)) in the Pyrenee Mountains. Low-elevation sites are hotter and dryer overall, while high-elevation sites are cooler and wetter (Montesinos-Navarro *et al.* 2009, 2011, 2012; Wolfe and Tonsor 2014). Because the collection locations were precisely geo-referenced, we obtained accurate local climate data from geo-referenced climate databases. In addition, these populations appear to share common ancestry. Their co-ancestry results from dispersal across the landscape of northern Spain after emergence from a glacial refugium (Pico *et al.* 2008). These natural populations, therefore, allow the exploration of a series of important ecological and evolutionary questions regarding adaptive divergence from common ancestry. These local populations have evolved functional differences that appear to adapt them to local climates. Traits associated with life history, such as biomass allocation, fecundity, developmental time and demographic traits, such as seed dormancy and timing of seed germination, showed clinal variation along the elevation gradient (Montesinos-Navarro *et al.* 2009, 2011, 2012). Low-elevation populations evolved an early flowering strategy that adapts them to spring heat and drought (Wolfe and Tonsor 2014). These populations also showed variation in genetic diversity, with high-elevation populations being more genetically diverse than low-elevation populations (Gomaa *et al.* 2011). In this study, we address local population evolutionary adaptive divergence in heat shock response and Hsps across a climate gradient. Our study examines mechanisms that may underpin adaptive variation and contributes to growing knowledge on abiotic stress responses in natural plant populations.

In this study, we characterize genetically based differences among 16 wild-collected populations in their responses to heat challenges, focussing on evolutionary divergence in thermotolerance (survival and post-stress growth) and Hsp101 expression.

We addressed the following hypotheses:

- Hypothesis 1: Populations of *A. thaliana* vary in their responses to high temperature when tested in experimentally controlled heat treatments;
- Hypothesis 2: Hsp101 expression variation is positively related with observed variation in thermotolerance;
- Hypothesis 3: The observed variation in thermotolerance and Hsp101 expression is associated with the climate of population origin and is therefore adaptive.

Methods

Source populations

Seeds of *A. thaliana* were field-collected and subjected to at least two rounds of single seed descent in controlled environment chambers. Sixty-four genotypes, 4 from each of 16 population locations (Table 1) [also **see Supporting Information—Fig. S1**], were chosen for study. The original seed was collected in collaboration with F. Xavier Picó (Montesinos-Navarro *et al.* 2009, 2011). Dr Picó states that no specific permissions were necessary for any of the 16 collection sites (F. Xavier Picó, pers. comm. to S.J.T., 30 June 2014). The population locations are listed in Table 1. All genotypes have been donated to the *Arabidopsis* Biological Resource Center at Ohio State University (stock CS78884).

For all collections, we quantified Hsp101 expression directly after heat stress (detailed below). In contrast, we quantified thermotolerance 1 week after heat stress. To obtain these disparate data, we planted two full sets of seeds/populations for each of the four treatments, one destined for Hsp101 quantification immediately following treatment, and the other was grown for 1 week following heat treatment and assayed for seedling survival and root growth.

Planting designs

Our hypotheses are focussed mainly on understanding population-level differences. We maximized the accuracy

Table 1. Population locations and climate PCs. Condensed from Tables S1 and S3 in Wolfe and Tonsor (2014).

Population	Longitude	Latitude	Elevation (m a.s.l.)	ClimatePC1	ClimatePC2
PIN	2.6591	41.6592	109	−3.6458	−1.7597
RAB	3.0537	42.3781	110	−4.4803	0.8848
SPE	2.9162	41.9275	332	−2.9213	−1.6334
BAR	2.1278	41.4322	340	−4.58	−1.1973
HOR	2.6202	41.6645	351	−2.7083	−1.8691
ARB	2.4941	41.8143	440	−2.1884	−0.6771
COC	3.1653	42.3263	519	−3.2761	−0.087
POB	1.0235	41.3526	597	−2.0028	3.7342
BOS	0.6913	42.7836	719	2.6056	2.6257
MUR	2.0002	41.6757	836	−2.0033	0.2208
VDM	1.0249	42.0292	912	1.2933	2.6184
ALE	1.3187	42.4105	1163	2.3143	1.2497
BIS	0.5334	42.4878	1397	5.8196	−2.246
PAL	1.2926	42.4312	1491	4.2579	0.6184
VIE	0.7606	42.6256	1538	5.7257	−1.1941
PAN	−0.231	42.762	1664	5.7899	−1.2883

of our population mean estimates within the constraints of total size of the experiment. Estimates of individual line performance were less important, especially given the relatively low within-population genetic variance in this highly selfing species.

Thermotolerance assay planting design. Four complete replicate sets of agar plates were prepared. Each set included two seedlings of each of the 64 Spanish lines, 128 seedlings total. This gave us eight replicate measures of each population's characteristics. Six seeds were planted per plate; the 128 seeds of a set were distributed across 22 plates. Seeds were randomly assigned to plates, but no plate contained two seeds from any one genotype or population. Because the design does not completely fill $6 \times 22 = 132$ locations, we were able to also include seeds for a pilot study that are not included in this study in some randomly chosen locations. These pilot study seeds will not be mentioned further.

Hsp101 assay planting design. Population-level estimates of Hsp101 expression were based on a single randomly chosen genotype from each population. This was justified because *A. thaliana* is highly selfing (Abbott and Gomes 1989; Nordborg *et al.* 2002) and highly selfing populations tend towards low amounts of internal genetic variation (Loveless and Hamrick 1984; Hamrick and Godt 1996; Duminil *et al.* 2007 but see Platt *et al.* 2010). Each genotype was represented by two replicate plantings, each containing 15–20 seeds; four genotypes were randomly assigned to a plate. This number of seedlings assured sufficient tissue for Hsp101 extraction and quantification.

Heat stress treatments

The most informative heat stress temperature was unknown at the start of the experiment. Virtually all prior heat stress experiments with *A. thaliana* have been conducted on mid-northern European genetic lines. However, Tonsor *et al.* demonstrated a cline in Hsp101 expression with latitude, suggesting that populations will differ in the temperatures that represent a heat stress (Tonsor *et al.* 2008). Our study populations are from southern European latitudes (Montesinos-Navarro *et al.* 2009). Temperatures in their sites of origin range from cooler to warmer, unlike temperatures in sites of the standard laboratory lines. Given the unknown nature of temperature responses among our lines, we, therefore, imposed two heat challenges, 42 and 45 °C. Our approach was to heat stress 10-day-old seedlings grown on agar plates, and then measure survival, post-stress root growth and quantify Hsp101 expression.

Arabidopsis seedlings are known to acquire greater heat tolerance with acclimation. The standard acclimation treatment (AT) in prior published studies was 3 h at 38 °C (Hong *et al.* 2003). We, therefore, compared responses at both 42 and 45 °C for two sets of replicates: those with 38 °C (AT) and thermally naive control treatment (CT) seedlings maintained at 22 °C prior to heat stress. Thus, we employed four thermal treatments. Each of these seedlings were grown at 22 °C prior to and following any heat treatments. Our four thermal treatments were (i) CT42: 42 °C for 3 h; (ii) CT45: 45 °C for 3 h; (iii) AT42: 3 h at 38 °C, recovery at 22 °C for 3 h, then 3 h at 42 °C; (iv) AT45: 3 h at 38 °C, recovery at 22 °C for 3 h, then 3 h at 45 °C.

Growth protocols and heat stress assays

Seeds were surface-sterilized by exposure to chlorine gas for 3 h. Seeds were grown on gridded square agar plates containing Mirashige and Skoog nutrient solution. For the 'thermotolerance assay', seeds were placed on the top grid line and plates were vertically oriented to allow measurement of root growth. For the 'Hsp101 assay', the plates were divided into four quadrats and seeds of a randomly chosen genotype were placed in the middle of one of the four quadrats. Plates were oriented horizontally. For all plates in both planting designs, after 5 days of stratification at 4 °C, plates were placed under fluorescent lights and maintained at ~22 °C with 16 h light, 8 h dark, except during any of the experimental heat treatments. After seed placement, all plates were sealed for the duration of the experiment. Fifteen days after planting, 10 days after emerging from stratification, each replicate heat stress set was exposed to one of the four heat stress treatments as mentioned above. The plates were vertically positioned on racks in a large forced-air drying oven to guarantee rapid air movement and conductive heat transfer. Although we did not measure the temperature of individual plates, heat transfer in the oven is very fast and even. In combination with the low mass of the agar plates, we are certain that plate and agar temperature equilibrated with oven air temperature within a few minutes of placement in the oven. Those sets intended for the thermotolerance assay were returned to 22 °C with 16 h light, 8 h dark and grown for 1 week after heat stress. The sets of plates destined for Hsp101 quantification were harvested immediately after heat stress.

Thermotolerance measures

We measured seedling survival and post-stress root growth, two widely used measures of heat stress effects (Yeh *et al.* 2012). A seedling was counted as dead if all the leaves turned from green to white. Per cent seedling survival was calculated as the percentage of living seedlings 1-week post-stress for each population. Plates were photographed just prior to heat stress and 1-week post-treatment. Root length was traced using NeuronJ (a macro in NIH ImageJ) for roots in the pre- and post-treatment photos. For each seedling, the post-stress root growth was quantified as the length difference between the pre- and post-treatment images.

Hsp101 quantification

All seedlings used in the Hsp101 assay were collected immediately after heat treatment and were used for western blot Hsp101 quantification. Seedlings were collected into microcentrifuge tubes and immediately put into liquid nitrogen. A bulk Hsp101 expression standard was prepared by combining multiple leaves from a variety of genotypes that had been subjected to a 42 °C heat treatment. This bulk Hsp101 expression standard was used for comparison across individual gels to minimize gel-to-gel variation. Hsp101 accumulation level was quantified through image analysis of western blots following the procedure of Tonsor *et al.* (2008). We used N-terminal Hsp101 primary antibody from rabbit (Agrisera, AS07253) as the primary antibody and anti-rabbit antibody as the secondary antibody to capture Hsp101.

Climate quantification

Wolfe and Tonsor combined temperature and precipitation data at the 16 collection sites from the BIOCLIM data set (Hijmans *et al.* 2005, data available at http://www.worldclim.org) and used principal component (PC) analysis to reduce the dimensionality of this data set (Wolfe and Tonsor 2014). Two PCs were significant based on permutation tests. Their ClimatePC1 explained 75 % of the multivariate variance across the 19 BIOCLIM variables, while ClimatePC2 explained 17 % (see Fig. S3 in Wolfe and Tonsor 2014). ClimatePC1 was most strongly associated with temperature and precipitation (cool and moist vs. warm and dry), while ClimatePC2 was associated most strongly with seasonality. We used ClimatePC1 and ClimatePC2 to test hypotheses that climate of origin is associated with variation in both thermotolerance and Hsp101 expression.

Statistical analyses

One analysis was used for general characterization of responses to heat treatments and to test the hypothesis that populations differed in their responses to the heat treatments. The four heat treatments compose a factorial design in which there are two levels of acclimation (22 °C control or 38 °C acclimation) and two levels of heat challenge (42 or 45 °C). We classified all seedlings as to population of origin, survival or death and post-stress root growth. For all the tests of treatment and population effects on seedling survival described below, logistic regression analysis was applied using Proc GENMOD in SAS (SAS version 9.1, SAS Institute Copyright © 2005). Our logistic regression model, therefore, predicted seedling survival as a function of population, acclimation and heat challenge. Three population interaction effects were also included: population of origin × AT, population of origin × heat challenge and acclimation × heat challenges as well as the three-way interaction of population of origin × AT × heat challenge. Population, AT and heat challenge were all considered as fixed effects in this study. For testing heat treatment effects on post-stress root growth and Hsp101 expression, ANOVAs tested the effects of population, acclimation, heat challenge and their interactions in Proc GLM (SAS version 9.1, SAS Institute Copyright © 2005). In this case, we accounted for the effect of initial size on post-stress growth by using pre-stress root length as a covariate. Both pre- and post-stress root length were log transformed, achieving very close fits to normal distributions of residuals. When interaction tests were clearly non-significant, the analysis was rerun without them. The full-model ANOVA was followed by separate tests by 42 vs. 45 °C heat challenge. For all three measures, analyses were further dissected to examine within-treatment effect of population and its interactions **[see Supporting Information—Table S1]**; however, the power of the within-treatment tests is relatively low. All means comparisons conducted in Proc GENMOND or Proc GLM were followed by Bonferroni critical *P*-value adjustment including Holm's correction (Holland and Copenhaver 1987).

We tested for a relationship between Hsp101 expression and thermotolerance by conducting two regressions. Each regression treated Hsp101 expression (log transformed) as the putative causal variable. One regression tested for Hsp101 expression effects on population mean per cent survival, and one tested for Hsp101 expression effects on population mean post-stress root growth (log transformed). Data were pooled across treatments in these regressions. The regressions were conducted in Proc REG (SAS version 9.1, SAS Institute Copyright © 2005).

We further tested whether variation in thermotolerance and Hsp101 expression matched with variation in the climates of population origin. The climates of population origin, quantified as ClimatePC1 and ClimatePC2 values, were used to examine the evolved effect of past climate on both the population mean thermotolerance measures (per cent seedling survival and post-stress root growth) and on Hsp101 expression. Logistic regression was performed for testing the prediction of post-stress per cent seedling survival with ClimatePC1 and ClimatePC2 using Proc GENMOD (SAS version 9.1, SAS Institute Copyright © 2005). Generalized linear regression using Proc GLM (SAS version 9.1, SAS Institute Copyright © 2005) tested the prediction of past climate influences on post-stress root growth and Hsp101 expression separately. In all cases, the design effects (AT and heat challenge) were treated as fixed effects. The full-model regressions were followed by separate tests of 42 and 45 °C heat challenge.

Results

Thermotolerance phenotype depends on acclimation and heat challenge temperature

ANOVA provided the confidence behind the statements that follow in this paragraph; relevant here are the Heat Challenge and Acclimation × Heat Challenge rows in Table 2a, in conjunction with the associated means comparisons shown in Fig. 1. These analyses indicate that *A. thaliana* accessions from northeastern Spain suffered reduced survival and root growth when exposed directly to 45 °C (CT45 treatments), in keeping with prior experiments using standard laboratory lines of *A. thaliana* (e.g. Queitsch *et al.* 2000). Seedling survival and post-stress root growth differed between the 42 and 45 °C heat challenges (Fig. 1). This difference in heat challenge effect is further separable by the presence or absence of the 38 °C AT. Without acclimation, a 45 °C heat challenge (CT45) significantly decreased per cent seedling survival and post-stress root growth compared with a 42 °C heat challenge (BT42). However, with a 38 °C acclimation, a 45 °C heat challenge (AT45) had no discernable effect on seedling survival, but did show decreased root growth, both compared with the effects of a 42 °C heat challenge (AT42). The effect of a 38 °C AT varies according to response trait and heat challenge. At 42 °C, no effect of acclimation was detected for either response traits (CT42 vs. AT42). However, at 45 °C, acclimation improved thermotolerance for both traits (CT45 vs. AT45). Prior acclimation at 45 °C (AT45) significantly increased root growth by ~3-fold compared with CT45. However, this amount of growth is still only ~50 % of the total root length achieved in the CT42 treatment. Thus, at 45 °C, the induction of AT mechanisms is sufficient to insure high survival, but not sufficient for rapid post-stress growth recovery.

Hsp101 expression varies with acclimation and heat challenge temperature

Hsp101 expression was significantly up-regulated in acclimated plants that were later exposed to both 42 and 45 °C (Figs 2 and 3; Table 2; $P < 0.0001$). The content of Hsp101 as a result of acclimation at AT42 and AT45 was indistinguishable (Figs 2 and 3). The heat challenges also led to an increase in Hsp101 expression at both CT42 and CT45 (i.e. without prior acclimation). However, the amount of Hsp101 expressed at CT42 and CT45 was significantly less than that observed at AT42 and AT45: 34 and 85 % less Hsp101 at CT42 (Table 2b; Fig. 2; $P = 0.0004$) and CT45, respectively (Table 2c; Fig. 2; $P < 0.0001$). This difference in Hsp101 expression between the two CT treatments may be caused by strongly compromised cellular function at 45 °C and only partially compromised cellular function at 42 °C.

Table 2. ANOVA table for seedling survival, post-stress root growth and Hsp101 expression in full model and separate analysis by heat challenge temperature. Post-stress root growth was adjusted using pre-stress root growth as a covariate ($P < 0.0001$). —, not applicable. All models exhibit overall significance. When interaction terms were non-significant, the model was run again without the non-significant interaction terms (NS). All reported *P*-values are from the reanalysis.

Source	Seedling survival	Post-stress root growth	Hsp101 expression
(a) Full model			
Model r^2	—	0.89	0.50
Population	0.0007	<0.0001	0.02
Acclimation	<0.0001	0.002	<0.0001
Heat challenge	0.08	<0.0001	<0.0001
Acclimation × Heat challenge	0.0006	0.03	<0.0001
Population × Acclimation	NS	0.09	NS
Population × Heat challenge	0.009	<0.0001	NS
Population × Acclimation × Heat challenge	NS	0.0002	NS
(b) 42 °C heat stress			
Model r^2	—	0.89	0.36
Population	0.01	<0.0001	0.03
Acclimation	NS	NS	0.0004
Population × Acclimation	NS	0.005	NS
(c) 45 °C heat stress			
Model r^2	—	0.94	0.60
Population	0.0005	<0.0001	NS
Acclimation	<0.0001	<0.0001	<0.0001
Population × Acclimation	<0.0001	0.0002	NS

Yet even this reduced expression level at 42 °C was sufficient to enhance survival and post-stress root growth to an extent that was indistinguishable from that in seedlings in the AT42 treatment. However, the same was not true for CT45 plants; those plants had greatly reduced Hsp101 expression (85 % less) compared with AT45 plants, average seedling survival rates of only 40 % and post-stress root growth of only ~20 % of that observed at AT45. Our interpretation is that 42 °C heat challenge is mild enough that many plants can maintain sufficient cellular function to rapidly induce acquired thermotolerance mechanisms, including Hsps. In contrast, at 45 °C, cellular function appears to be compromised so quickly that induction of thermotolerance mechanisms is substantially impaired and the seedlings cannot protect cellular function from further damage. The differences among populations that we see in Hsp101 accumulation in acquired thermotolerance may be due in part to evolved differences in the maximum level accumulation of Hsp101. It may also be due to the differential response to 38 °C among these populations, that is, some populations may produce substantially more Hsp101, while some may produce less at 38 °C. Although we do not have a way to address this, the difference in Hsp101 response to the 38 °C acclimation temperature is very unlikely to contribute to the difference we see. Additionally, 42 °C might not represent a significant stress, while 45 °C represent a more stressful temperature.

Populations of *A. thaliana* vary in their responses to high temperature

Overall, we observed genetically based variation among populations in thermotolerance (seedling survival: $P = 0.0007$; post-stress root growth: $P < 0.0001$; Table 2a) and Hsp101 expression ($P = 0.02$; Table 2a). Within heat treatments, populations differ significantly except in seedling survival at CT42 **[see Supporting Information—Table S1]**. Populations significantly vary in post-stress root growth within all four treatments **[see Supporting Information—Table S1]**. Moreover, populations also differ significantly in their responses to the two heat challenges, as reflected in the Population × Heat challenge interaction (Table 2a), indicating complex among-population variation in stress-dependent responses.

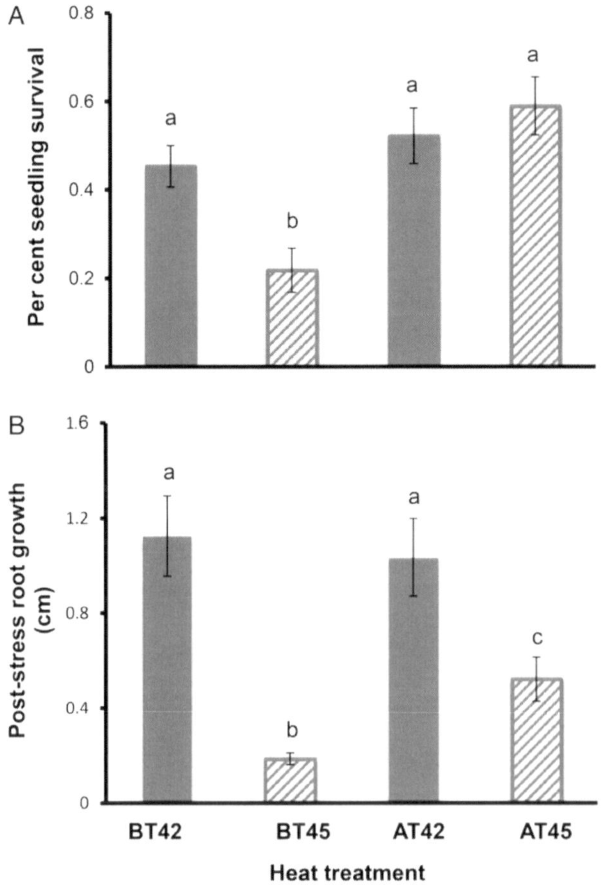

Figure 1. (A) Per cent seedling survival and (B) post-stress root growth 1 week after heat stress, by heat treatment. For per cent seedling survival (A), data displayed are treatment means based on population survival percentage within treatment. For post-stress root growth (B), data displayed are means of individuals within treatments. Vertical lines represent standard errors. Treatments with different letters above the bar are significantly different with an initial rejection criterion of $P < 0.05$ adjusted for multiple comparisons using the sequential Bonferroni method.

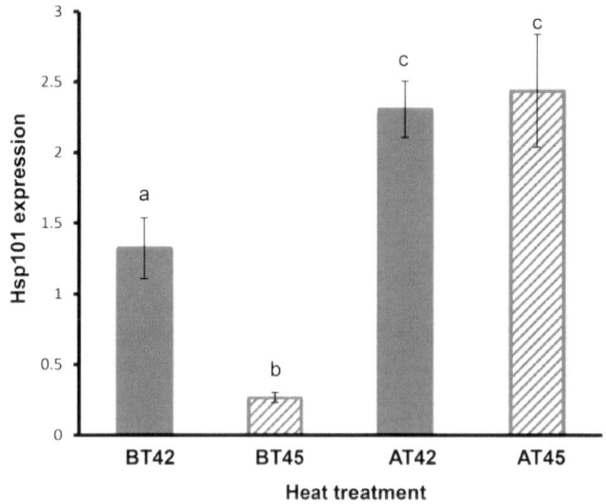

Figure 2. Hsp101 expression by heat treatment. Data displayed are means of individuals within treatments (vertical lines represent standard errors). Hsp101 expression is adjusted to a bulk Hsp101 expression standard to provide a standardized relative comparison among samples. Treatments with different letters above the bar are significantly different with an initial rejection criterion of $P < 0.05$ adjusted for multiple comparisons using the sequential Bonferroni method.

Significant variation in Hsp101 expression was observed among populations when all the data were pooled (Table 2a; Fig. 3). However, the inherent variability in Hsp101 expression and in quantification from western blots, together with limited sample number, resulted in low power to detect differences among populations when treatments were analysed separately **[see Supporting Information—Table S1]**. Thus, in further analyses and discussion, we focus on the general relationship with the pooled data between Hsp101 expression and the thermotolerance measures at the population level.

Variation in Hsp101 accumulation is positively associated with variation in thermotolerance

At 45 °C, higher Hsp101 accumulation was associated with higher per cent seedling survival, explaining 37 % of the variation (Fig. 4A; $P = 0.0003$). Similarly, higher Hsp101 accumulation was associated with increased post-stress root growth, accounting for 15 % of the variation (Fig. 4B; $P = 0.04$). Plants with low Hsp101 accumulation exhibited very low rates of survival and below average rates of post-stress root growth. In contrast, Fig. 4 also shows that plants with high Hsp101 accumulation range from very low to very high survival and post-stress root growth. This pattern demonstrates that Hsp101 up-regulation is necessary but not sufficient to ensure high levels of thermotolerance.

It is important to note that many other Hsps are co-regulated with Hsp101. Thus, we cannot ascribe a direct causal relationship to the significant regression of survival or post-stress growth on Hsp101 content. In fact, it is most likely that thermotolerance is conferred by a whole suite of genes, with Hsp101 playing a direct and essential role (Hong *et al.* 2003).

The observed variation in thermotolerance and Hsp101 expression appears to be adaptive

Both thermotolerance measures and Hsp101 accumulation co-varied with ClimatePC1 of the populations' sites of origin under some but not all treatments (Table 3). Overall, ClimatePC1 significantly predicted seedling survival (Table 3a; $P = 0.04$). That is, populations from the cooler, moister end of the climate gradient exhibited increased per cent seedling survival compared with populations from hotter and drier climates. This relationship was also significant at the 45 °C heat challenge (Table 3c;

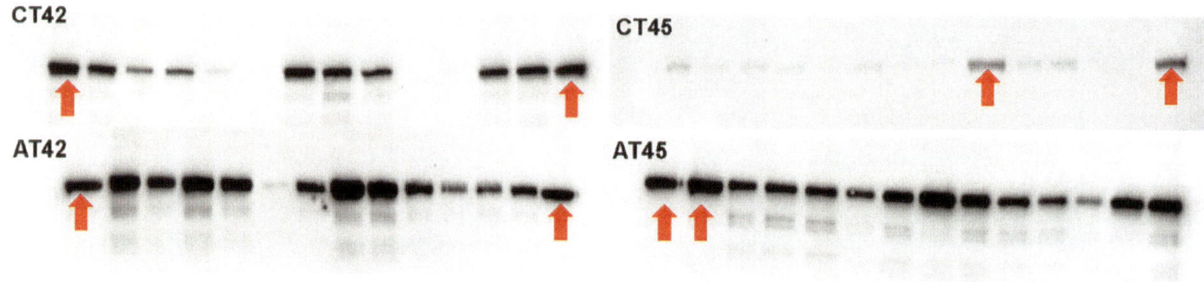

Figure 3. Western blots showing differences in Hsp101 expression among treatments. Variation within treatments is due to significant population variation in expression as well as sample variation in total protein content. A molecular ladder was used in each gel to identify Hsp101 position (ladder not visible in images). Within treatment, genotypes within populations were randomly assigned to the lanes of western blots. For analysis, variation in the amount of protein loaded was used to adjust Hsp101 expression per unit protein. A bulk sample was created and loaded in two lanes on each gel as a control for gel-level variation. Arrows indicate the bulk controls in each western blot.

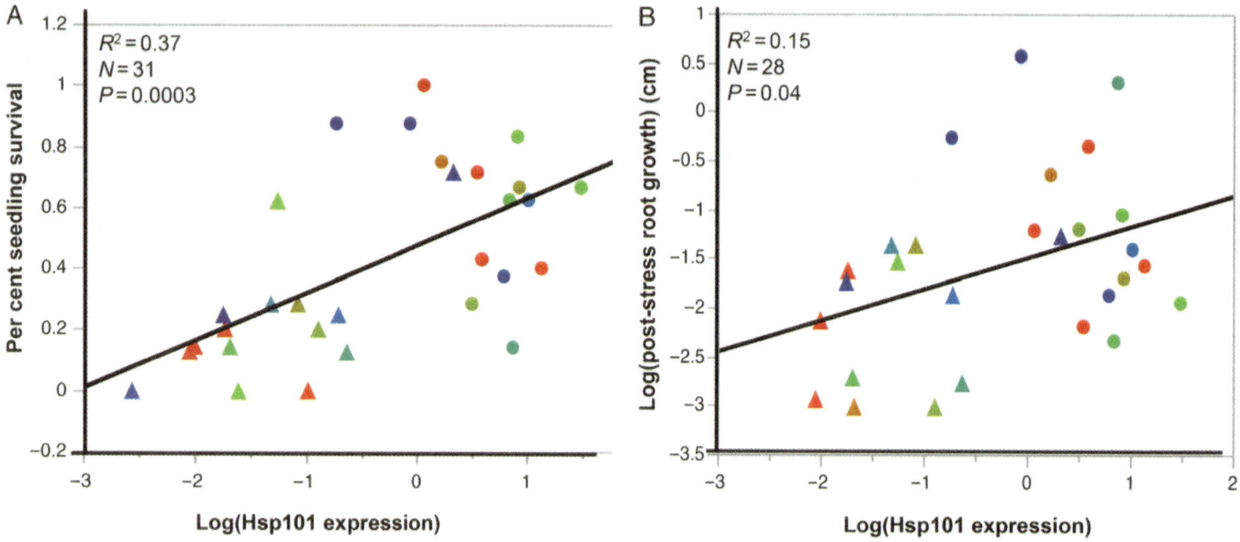

Figure 4. Regression of (A) per cent seedling survival and (B) post-stress root growth (cm, log transformed) on Hsp101 expression (log transformed) at 45 °C heat challenge (BT45 and AT45). Data displayed are population means of each treatment for all four treatments. (A) Per cent seedling survival = 0.47 + 0.16 × log (Hsp101 expression); (B) log (post-stress root growth) = −1.50 + 0.32 × log (Hsp101 expression). Triangle: BT45; circle: AT45. The colours of the populations were arranged along a red–green–blue gradient, from low to high elevation, e.g. red represents low-elevation populations, while blue represents high-elevation populations.

$P = 0.008$), and this effect at 45 °C appears to have driven the significance of the analysis overall. Regression of post-stress root growth on ClimatePC1 was not significant overall (Table 3a). However, for plants exposed to 45 °C, post-stress root growth increased significantly with ClimatePC1 (i.e. increased towards the cooler and moister end of the climate gradient compared with the hotter and drier end; Table 3c; $P = 0.0003$). Hsp101 expression was not significantly associated with ClimatePC1 overall (Table 3a) but was positively associated with ClimatePC1 at 42 °C (Table 3b; $P = 0.01$). ClimatePC2 was not a significant predictor of thermotolerance measures or Hsp101 accumulation in either the full model or in the separate treatment analyses (Table 3).

Discussion

This study demonstrated a general pattern among plant populations of natural variation in thermotolerance, here measured by seeding survival and post-stress root growth, and accompanied by increases in accumulation of Hsp. Furthermore, the pattern of variation in thermotolerance is associated with variation in the climates of the populations' sites in northeastern Spain. We found that pre-exposure to mild heat stress resulted in acquired thermotolerance and an associated up-regulation of Hsp101. The AT increased thermotolerance substantially. Hsp101 expression also varies among populations and shows the first evidence of genetically based variation in expression associated with climate.

Table 3. ANOVA table for seedling survival, post-stress root growth and Hsp101 expression with ClimatePCs in full model and in separate analyses by heat challenge temperature. Post-stress root growth was adjusted using pre-stress root growth as a covariate ($P < 0.0001$). —, not applicable. All models are significant. All reported P-values are from a reanalysed model after removing the non-significant interaction terms (NS).

Source	Seedling survival	Post-stress root growth	Hsp101 expression
(a) Full model			
Model r^2	—	0.86	0.41
ClimatePC1	0.04	NS	NS
ClimatePC2	NS	NS	NS
Acclimation	<0.0001	0.001	<0.0001
Heat challenge	0.09	<0.0001	<0.0001
Acclimation × Heat challenge	0.001	0.04	<0.0001
(b) 42 °C heat stress			
Model r^2	—	0.83	0.17
ClimatePC1	NS	NS	0.01
ClimatePC2	NS	NS	NS
Acclimation	NS	NS	0.001
(c) 45 °C heat stress			
Model r^2	—	0.92	0.49
ClimatePC1	0.008	0.0003	NS
ClimatePC2	NS	NS	NS
Acclimation	<0.0001	<0.0001	<0.0001
ClimatePC1 × Acclimation	0.02	NS	NS

It may seem counter-intuitive that the genetic lines from cooler, moister end of the climate gradient exhibited higher seedling survival and post-stress root growth than lines from the hotter, dryer end of the gradient. The lines from the cooler end of the gradient also showed greater accumulation of Hsp101. Importantly, these results are concordant with results of a previous study of geographic variation in Hsp101 expression (Tonsor *et al.* 2008). Genetic lines of *A. thaliana* from cooler, moister, northern latitudes exhibit greater Hsp101 expression than lines from the southern limits of the species' range (Tonsor *et al.* 2008). Populations of *Arabidopsis* appear to be differentiated in their mechanisms of response to heat stress with more southern populations relying less on acquired thermotolerance. Previous studies in the same populations used here suggest that low-elevation populations may instead avoid heat by maturing earlier (Montesinos-Navarro *et al.* 2012; Wolfe and Tonsor 2014). A similar pattern, which populations from cooler environments had higher induced/acquired thermotolerance, was also found by Barua *et al.* (2008). A number of other factors may also contribute to the seemly counter-intuitive results. Firstly, populations from higher altitudes and cooler climate are exposed to more variable environments—higher range of temperatures, such as high variability in annual or diurnal temperature. Barua *et al.* (2008) further found that variation in induced tolerance and Hsps was related to temperature variability, in that populations that are more likely to experience multiple heat stresses had higher tolerance and Hsp accumulation. Further analysis on the microclimate variability will provide more information on these populations' local climate conditions. Secondly, plants from lower altitudes and warmer climate may rely on basal mechanisms of thermotolerance that are thought to be lower cost compared with induced mechanisms. Thirdly, plants from lower altitudes and warmer climate may simply avoid heat stress by maturing earlier. Finally, plants from high altitudes are faced with other stresses like UV-B radiation, which may be correlated with thermotolerance and Hsps. However, our populations are not sufficiently high in altitude that annual mean UV is meaningfully greater along the elevation gradient (data extracted from DIVA-GIS 7.5.0).

The seeds used in this experiment are descended from field-collected genotypes through at least two, most of them three or four, generations of single seed descent in growth chambers. Within the growth chambers, individual seed plants were randomized in their locations. We can, therefore, be confident that the differences observed among populations reflect differences in population genetic composition rather than uncontrolled environmental differences during the experiment or environment maternal effects carried over from the field. Pico *et al.* (2008) demonstrated that the populations from this region of Spain share a common ancestry, no doubt resulting from their ancestor's emergence from their ice age refugium and their subsequent spread across the landscape of northeastern Spain (Pico *et al.* 2008). The significant association of seedling survival rates with the first PC of climate variation (ClimatePC1) suggests the signature of adaptation to the specific habitats from which the populations were collected. That we only detect an association of variation in seedling survival, post-heat root growth or Hsp101 expression with climate of origin under some circumstances suggests three additional explanations for the observed population variation. First, the power of this experiment may not have been sufficient, given the inherent variability in seedling root growth and Hsp101 quantification in our experimental design and execution. Second, macro-scale climate may not be the only factor involved in adaptation. In the highly dissected topography of northeastern Spain, many unmeasured site characteristics can influence microclimate, including slope, aspect, height of surrounding vegetation and thermal properties of the substrate. All of these could either directly influence the temperatures experienced or influence the trade-offs involved in implementing various potential thermotolerance mechanisms. Finally, founder effect and random genetic drift may account for some of the observed variations. However, the adaptive clines observed in a great many traits previously examined in these populations make it unlikely that random genetic drift is the predominant cause of the observed variation (Montesinos-Navarro *et al.* 2011, 2012; Wolfe and Tonsor 2014).

Nearly all of the existing knowledge of thermotolerance and heat shock response in *A. thaliana* has been gained from studies of seedlings. It is important to note that these studies cannot fully put cellular mechanisms of thermotolerance into the context of mechanisms of thermotolerance relevant in field settings. First, the seedling stage is the least likely stage to experience lethally high temperatures. *Arabidopsis thaliana* seeds germinate under the most benign conditions experienced by the plant—the cool, moist periods of late fall and early spring. Second, like most plants, *A. thaliana* possesses a number of mechanisms for avoiding high temperatures that cannot be used at the seedling stage because the necessary structures and functions simply are not present. These include avoidance of periods of high temperature through the evolution of altered life-history timing (Montesinos-Navarro *et al.* 2012), variation in constitutive high leaf angle, leaf hyponasty and petiole elongation in response to elevated temperature (Gray *et al.* 1998; Polko *et al.* 2011) and transpirational cooling. Full understanding of high-temperature responses and their consequences will require integrated study of cellular, physiological and developmental responses throughout the plant life cycle.

Conclusions

This study focuses on identifying the adaptive variation in response to high-temperature stress across an elevation gradient in natural *A. thaliana* populations. We show that the accumulation of Hsp101, an important Hsp known to be essential for acquired thermotolerance, was positively associated with seedling survival and post-stress root growth. Pre-acclimation significantly increased thermotolerance at 45 °C but not at 42 °C. Furthermore, both Hsp101 and thermotolerance were correlated with the climate variation of home sites. Our study contributes to growing knowledge on abiotic stress responses in natural plant populations.

Sources of Funding

Funding was provided by US National Science Foundation Grant IOS-1120383 to S.J.T. and a University of Pittsburgh Dietrich College of Arts and Sciences First Year Fellowship to N.Z.

Contributions by the Authors

N.Z. and S.J.T. designed the experiment. N.Z., B.B. and J.R. performed the experimental work. N.Z. and S.J.T. performed analyses. All co-authors discussed and interpreted results prior to writing. N.Z. and S.J.T. wrote the manuscript.

Acknowledgements

We are grateful to Tonsor Lab managers Adam Sinder and Kali Theis for excellent technical assistance and to S. Kalisz and the Tonsor Lab research discussion group for crucial discussion and critique. Elizabeth Vierling, University of Massachusetts, Amherst, generously provided Hsp101 antibody and advice. Jeff Brodsky, University of Pittsburgh, was generous in logistical support and advice for western blot assays of Hsp101. S.J.T. is extremely grateful to F. Xavier Picó of Estación Biológica de Doñana,

Seville, Spain, for introduction to the Spanish Arabidopsis system and for many hours of friendship during field characterization and collection of the populations.

Supporting Information

The following additional information is available in the online version of this article –

Table S1. Significance values for tests of within-population variation in seedling survival, post-stress root growth and Hsp101 expression. Each treatment was tested in a separate analysis.

Figure S1. Geographic location of the 16 populations used in this study (from Fig. 1, Montesinos-Navarro *et al.* 2011).

Literature Cited

Abbott RJ, Gomes MF. 1989. Population genetic structure and outcrossing rate of *Arabidopsis thaliana* (L.) Heynh. *Heredity* **62**: 411–418.

Amano M, Iida S, Kosuge K. 2012. Comparative studies of thermotolerance: different modes of heat acclimation between tolerant and intolerant aquatic plants of the genus *Potamogeton*. *Annals of Botany* **109**:443–452.

Barua D, Downs CA, Heckathorn SA. 2003. Variation in chloroplast small heat-shock protein function is a major determinant of variation in thermotolerance of photosynthetic electron transport among ecotypes of *Chenopodium album*. *Functional Plant Biology* **30**:1071–1079.

Barua D, Heckathorn SA, Coleman JS. 2008. Variation in heat-shock proteins and photosynthetic thermotolerance among natural populations of *Chenopodium album* L. from contrasting thermal environments: implications for plant responses to global warming. *Journal of Integrative Plant Biology* **50**:1440–1451.

Doyle SM, Wickner S. 2009. Hsp104 and ClpB: protein disaggregating machines. *Trends in Biochemical Sciences* **34**:40–48.

Duminil J, Fineschi S, Hampe A, Jordano P, Salvini D, Vendramin GG, Petit RJ. 2007. Can population genetic structure be predicted from life-history traits? *The American Naturalist* **169**:662–672.

Feder ME. 1999. Organismal, ecological, and evolutionary aspects of heat-shock proteins and the stress response: established conclusions and unresolved issues. *American Zoologist* **39**: 857–864.

Field C, Barros V, Mach K, Mastrandrea M. 2014. *Climate change 2014: impacts, adaptation, and vulnerability*. Fifth Assessment Report of the Intergovernmental Panel on Climate Change.

Garbuz D, Evgenev MB, Feder ME, Zatsepina OG. 2003. Evolution of thermotolerance and the heat-shock response: evidence from inter/intraspecific comparison and interspecific hybridization in the virilis species group of *Drosophila*. I. Thermal phenotype. *Journal of Experimental Biology* **206**:2399–2408.

Gaston KJ. 2009. Geographic range limits of species. *Proceedings of the Royal Society B Biological Sciences* **276**:1391–1393.

Gomaa NH, Montesinos-Navarro A, Alonso-Blanco C, Picó FX. 2011. Temporal variation in genetic diversity and effective population size of Mediterranean and subalpine *Arabidopsis thaliana* populations. *Molecular Ecology* **20**:3540–3554.

Gray WM, Östin A, Sandberg G, Romano CP, Estelle M. 1998. High temperature promotes auxin-mediated hypocotyl elongation in *Arabidopsis*. *Proceedings of the National Academy of Sciences of the USA* **95**:7197–7202.

Hamrick JL, Godt MJW. 1996. Effects of life history traits on genetic diversity in plant species. *Philosophical Transactions of the Royal Society B Biological Sciences* **351**:1291–1298.

Hannah MA, Wiese D, Freund S, Fiehn O, Heyer AG, Hincha DK. 2006. Natural genetic variation of freezing tolerance in *Arabidopsis*. *Plant Physiology* **142**:98–112.

Hijmans RJ, Cameron SE, Parra JL, Jones PG, Jarvis A. 2005. Very high resolution interpolated climate surfaces for global land areas. *International Journal of Climatology* **25**:1965–1978.

Holland BS, Copenhaver MD. 1987. An improved sequentially rejective Bonferroni test procedure. *Biometrics* **43**:417–423.

Hong SW, Vierling E. 2001. Hsp101 is necessary for heat tolerance but dispensable for development and germination in the absence of stress. *The Plant Journal* **27**:25–35.

Hong SW, Lee U, Vierling E. 2003. *Arabidopsis* hot mutants define multiple functions required for acclimation to high temperatures. *Plant Physiology* **132**:757–767.

Jensen LT, Cockerell FE, Kristensen TN, Rako L, Loeschcke V, McKechnie SW, Hoffmann AA. 2010. Adult heat tolerance variation in *Drosophila melanogaster* is not related to Hsp70 expression. *Journal of Experimental Zoology Part A: Ecological Genetics and Physiology* **313**:35–44.

Körner C. 2007. The use of 'altitude' in ecological research. *Trends in Ecology and Evolution* **22**:569–574.

Larkindale J, Knight MR. 2002. Protection against heat stress-induced oxidative damage in *Arabidopsis* involves calcium, abscisic acid, ethylene, and salicylic acid. *Plant Physiology* **128**:682–695.

Larkindale J, Vierling E. 2008. Core genome responses involved in acclimation to high temperature. *Plant Physiology* **146**: 748–761.

Larkindale J, Hall JD, Knight MR, Vierling E. 2005. Heat stress phenotypes of *Arabidopsis* mutants implicate multiple signaling pathways in the acquisition of thermotolerance. *Plant Physiology* **138**:882–897.

Lindquist S. 1986. The heat-shock response. *Annual Review of Biochemistry* **55**:1151–1191.

Lockwood BL, Sanders JG, Somero GN. 2010. Transcriptomic responses to heat stress in invasive and native blue mussels (genus *Mytilus*): molecular correlates of invasive success. *Journal of Experimental Biology* **213**:3548–3558.

Loveless MD, Hamrick JL. 1984. Ecological determinants of genetic structure in plant populations. *Annual Review of Ecology and Systematics* **15**:65–95.

Montesinos-Navarro A, Tonsor SJ, Alonso-Blanco CP, Picó FX. 2009. Demographic and genetic patterns of variation among populations of *Arabidopsis thaliana* from contrasting native environments. *PLoS ONE* **4**:e7213.

Montesinos-Navarro A, Wig J, Pico FX, Tonsor SJ. 2011. *Arabidopsis thaliana* populations show clinal variation in a climatic gradient associated with altitude. *New Phytologist* **189**:282–294.

Montesinos-Navarro A, Picó FX, Tonsor SJ. 2012. Clinal variation in seed traits influencing life cycle timing in *Arabidopsis thaliana*. *Evolution* **66**:3417–3431.

Nieto-Sotelo J, Martínez LM, Ponce G, Cassab GI, Alagón A, Meeley RB, Ribaut J-M, Yang R. 2002. Maize HSP101 plays important roles in both induced and basal thermotolerance and primary root growth. *The Plant Cell* **14**:1621–1633.

Nordborg M, Borevitz JO, Bergelson J, Berry CC, Chory J, Hagenblad J, Kreitman M, Maloof JN, Noyes T, Oefner PJ, Stahl EA, Weigel D. 2002. The extent of linkage disequilibrium in *Arabidopsis thaliana*. *Nature Genetics* **30**:190–193.

Paul JR, Sheth SN, Angert AL. 2011. Quantifying the impact of gene flow on phenotype-environment mismatch: a demonstration with the scarlet monkeyflower *Mimulus cardinalis*. *The American Naturalist* **178**:S62–S79.

Pico FX, Mendez-Vigo B, Martinez-Zapater JM, Alonso-Blanco C. 2008. Natural genetic variation of *Arabidopsis thaliana* is geographically structured in the Iberian peninsula. *Genetics* **180**:1009–1021.

Platt A, Horton M, Huang YS, Li Y, Anastasio AE, Mulyati NW, Ågren J, Bossdorf O, Byers D, Donohue K, Dunning M, Holub EB, Hudson A, Le Corre V, Loudet O, Roux F, Warthmann N, Weigel D, Rivero L, Scholl R, Nordborg M, Bergelson J, Borevitz JO. 2010. The scale of population structure in *Arabidopsis thaliana*. *PLoS Genetics* **6**: e1000843.

Polko JK, Voesenek LACJ, Peeters AJM, Pierik R. 2011. Petiole hyponasty: an ethylene-driven, adaptive response to changes in the environment. *AoB PLANTS* **2011**: plr031; doi:10.1093/aobpla/plr031.

Queitsch C, Hong SW, Vierling E, Lindquist S. 2000. Heat shock protein 101 plays a crucial role in thermotolerance in *Arabidopsis*. *The Plant Cell* **12**:479–492.

Shabala S. 2012. *Plant stress physiology*. Wallingford: CABI.

Sørensen JG, Dahlgaard J, Loeschcke V. 2001. Genetic variation in thermal tolerance among natural populations of *Drosophila buzzatii*: down regulation of Hsp70 expression and variation in heat stress resistance traits. *Functional Ecology* **15**: 289–296.

Sørensen JG, Nielsen MM, Kruhøffer M, Justesen J, Loeschcke V. 2005. Full genome gene expression analysis of the heat stress response in *Drosophila melanogaster*. *Cell Stress & Chaperones* **10**:312–328.

Tonsor SJ, Scott C, Boumaza I, Liss TR, Brodsky JL, Vierling E. 2008. Heat shock protein 101 effects in *A. thaliana*: genetic variation, fitness and pleiotropy in controlled temperature conditions. *Molecular Ecology* **17**:1614–1626.

Wang W, Vinocur B, Shoseyov O, Altman A. 2004. Role of plant heat-shock proteins and molecular chaperones in the abiotic stress response. *Trends in Plant Science* **9**:244–252.

Wehner DJ, Minner DD, Dernoeden PH, McIntosh MS. 1985. Heat tolerance of Kentucky bluegrass as influenced by pre- and post-stress environment. *Agronomy Journal* **77**: 376–378.

Wolfe MD, Tonsor SJ. 2014. Adaptation to spring heat and drought in northeastern Spanish *Arabidopsis thaliana*. *New Phytologist* **201**: 323–334.

Yeh CH, Kaplinsky NJ, Hu C, Charng Y. 2012. Some like it hot, some like it warm: phenotyping to explore thermotolerance diversity. *Plant Science* **195**:10–23.

16

Giving drought the cold shoulder: a relationship between drought tolerance and fall dormancy in an agriculturally important crop

Keith G. Pembleton[1*] and Puthigae Sathish[2]

[1] Tasmanian Institute of Agriculture, University of Tasmania, PO Box 3523, Burnie, TAS 7320, Australia

[2] Pastoral Genomics, c/o ViaLactia (NZ) Ltd, PO Box 109185, Newmarket 1149, New Zealand

Abstract. The growth of fall dormant/freezing tolerant plants often surpasses the growth of non-fall dormant/non-freezing tolerant types of the same species under water-limited conditions, while under irrigated conditions non-fall dormant types exhibit superior yield performance. To investigate the mechanism behind this phenomenon, we exposed seven diverse alfalfa (*Medicago sativa*) cultivars to water-limited and fully watered conditions and measured their shoot growth, shoot water potential and gas exchange parameters and the relative abundance of taproot RNA transcripts associated with chilling stress/freezing tolerance. Fall dormant cultivars had greater shoot growth relative to the fully watered controls under a mild water deficit (a cumulative water deficit of 625 mL pot^{-1}) and did not close their stomata until lower shoot water potentials compared with the more non-fall dormant cultivars. Several gene transcripts previously associated with freezing tolerance increased in abundance when plants were exposed to a mild water deficit. Two transcripts, *corF* (encodes galactinol synthase) and *cas18* (encodes a dehydrin-like protein), increased in abundance in fall dormant cultivars only. Once water deficit stress became severe (a cumulative water deficit of 2530 mL pot^{-1}), the difference between fall dormancy groups disappeared with the exception of the expression of a type 1 sucrose synthase gene, which decreased in fall dormant cultivars. The specific adaptation of fall dormant cultivars to mild water deficit conditions and the increase in abundance of specific genes typically associated with freezing tolerance in these cultivars is further evidence of a link between freezing tolerance/fall dormancy and adaption to drought conditions in this species.

Keywords: Alfalfa; forage legumes; gene expression; lucerne; moisture stress.

Introduction

Drought is a major limitation to agricultural production. Drought causes a cessation of plant growth and in extreme cases leads to the senescence of part or the whole of plants. There is a diverse range of adaptations to water-limited conditions within the plant kingdom, and environments that may be extreme for one species may be relatively benign for another. Adaptation to and tolerance of water deficit stress is often associated with adaptation to and tolerance of freezing temperatures as both environmental conditions lead to osmotic and oxidative stresses at a cellular level (Thomashow 1999; Zhu 2002) and disrupt plant water status (Chinnusamy *et al.* 2004). Cross-talk between the stress signalling pathways involved in both environmental stresses is a common phenomenon (Chinnusamy *et al.* 2004) and

* Corresponding author's e-mail address: Keith.Pembleton@utas.edu.au

often the same or similar metabolites accumulate in response to both stresses (Kaplan *et al.* 2004; Rizhsky *et al.* 2004; Usadel *et al.* 2008; Urano *et al.* 2009).

Alfalfa (syn. lucerne; *Medicago sativa*) is a globally important forage legume crop with a total area of production of ~32.5–35 million hectares worldwide (Russelle 2001; Radovic *et al.* 2009). Its importance is further increased by its ability to fix atmospheric nitrogen (through a symbiotic relationship with the rhizobia *Sinorhizobium (Ensifer) meliloti*), making it an important source of nitrogen within agricultural production systems (Vance *et al.* 1988). Based on their level of fall/winter growth, alfalfa cultivars can be separated into four fall dormancy groups: fall dormant, semi-fall dormant, non-fall dormant and highly non-fall dormant (Barnes *et al.* 1988). This in turn can be related back to the genetic pedigree of a cultivar (Barnes *et al.* 1988; He *et al.* 2009), with non-fall dormant types entirely comprised of ssp. *sativa* and fall dormant types containing some spp. *falcata* in their background. As part of the characterization process of new alfalfa cultivars, they are scored on a scale of 1–11, with 1 assigned to the most fall dormant cultivars and 11 assigned to the most non-fall dormant cultivars, based on their shoot length 1 month after defoliation in late autumn (Teuber *et al.* 1998). This score strongly correlates to the level of plant injury when the cultivar is exposed to freezing temperatures (Sheaffer *et al.* 1992; Schwab *et al.* 1996) and consequently is an important consideration when selecting cultivars to be grown in environments that expose plants to sub-zero temperatures over the winter.

A consistent trend in alfalfa field experiments undertaken in the cool temperate regions of Australia under water-limited conditions is the superior yield performance of fall dormant cultivars of alfalfa compared with non-fall dormant cultivars (Pembleton *et al.* 2010*a*, *b*). This is despite there being no difference in the soil water use pattern between cultivars under such conditions (Pembleton *et al.* 2011). Interestingly, in these environments there is no need for adaptation to freezing temperatures over the winter and when water deficit is alleviated non-fall dormant cultivars outperform or at least equal the yield of fall dormant cultivars (Pembleton *et al.* 2010*a*, *b*). Adaptation of fall dormant cultivars of alfalfa to water-limited conditions has been noted in the literature for some time (e.g. Grandfield 1943). When exposed to a water deficit, fall dormant and semi-fall dormant cultivars have been identified as being able to maintain a higher shoot water potential (Grimes *et al.* 1992; Pembleton *et al.* 2009), greater rates of stomatal conductance (Hattendorf *et al.* 1990) and taproot carbohydrate reserves (Boschma and Williams 2008) compared with non-fall dormant and highly non-fall dormant cultivars.

Many reports have highlighted the similarities between gene families (including dehydrins, protein kinases and transcription factors) that are upregulated in plants exposed to cold temperatures and water deficits (see Chinnusamy *et al.* 2004; Beck *et al.* 2007; Krasensky and Jonak 2012 for reviews). Transgenic modification of alfalfa with a superoxide dismutase from *Nicotiana plumbaginifolia* Viv. (McKersie *et al.* 1993, 1996, 1999) or a trehalose from yeast (Suarez *et al.* 2009) has improved tolerance to both freezing and water deficit stress. Furthermore, the alfalfa gene transcripts *cas17* (Wolfraim and Dhindsa 1993), *cas18* (Wolfraim *et al.* 1993) and CAR1/*cor*15 (Laberge *et al.* 1993) have been noted to increase in abundance when plants are exposed to either water deficit or freezing stress. In the field, it has been identified that the expression of freezing tolerance/fall dormancy, nitrogen storage and carbon partitioning genes during late autumn is enhanced by exposure to water deficit conditions over summer (Pembleton *et al.* 2010*c*). Clearly, several candidate genes may contribute to the superior performance of fall dormant cultivars when exposed to water deficits. To date, no attempt has been made to screen these genes to determine which are responsible for this specific adaptation. Given that alfalfa cultivars can be easily classified into a quantitative ranking of freezing tolerance based on their fall dormancy rating and that fall dormant cultivars exhibit superior adaptation to water-limited conditions compared with non-fall dormant and highly non-fall dormant cultivars, it is an ideal species to assess and explore the linkage between freezing tolerance and drought tolerance in perennial forages. This paper reports on a study undertaken to investigate the adaptation of seven alfalfa cultivars/lines representing a fall dormancy score from 3 (dormant) to 10 (highly non-fall dormant) to a water deficit and the expression profile of select genes. The aim of these experiments was to determine what level of fall dormancy is required to achieve superior tolerance to water deficit stress and to establish the relationship between the expression of genes known to be cold inducible or associated with freezing tolerance and the response of cultivars with differing levels of fall dormancy to a water deficit.

Methods

Plant material

Bare (uncoated/untreated) seeds of alfalfa cultivars were sourced from commercial seed companies. These cultivars represented a broad range of fall dormancy (FD) ratings and were Q31 (FD rating of 3; Seed Distributors, Wingfield, SA, Australia), Grasslands Kaituna (FD rating of 4.5; Wrightson Seeds Australia, Truganina, VIC,

Australia), A5225 (FD rating of 5; Cal/West Seeds, Woodland, CA, USA), SARDI 7 (FD rating of 7; Heritage Seeds, Mulgrave, VIC, Australia), Sequel HR (FD rating of 9; Wrightson Seeds Australia) and SARDI 10 (FD rating of 10; Heritage Seeds). In addition, an experimental line FFI DT1 (putative FD rating of 2; SARDI, Urrbrae, SA, Australia) was also included.

Screening test of fall dormancy

As plant material was sourced from a range of herbage development programmes, there was a need to standardize their fall dormancy classifications. Consequently, plants were grown outdoors, during the southern hemisphere autumn/fall and winter months, in 6-L (337 mm high, 75 mm radius) pots containing a proprietary potting mix (Earthcore all-purpose potting mix; Van Schaik's Bio-Gro Pty Ltd, Mt Gambia, SA, Australia) at the Cradle Coast Campus of the University of Tasmania in Burnie, TAS, Australia (41.06°S, 145.88°E, 100 m above sea level). Pots were placed on wooden frames elevated 150 mm above the ground. Five seeds of each cultivar were sown into four replicates of pots on 29 August 2011. Pots were arranged in a randomized complete block design. At the emergence of the first unifoliate leaf at 14 days after sowing (DAS), plants were spray inoculated with rhizobia bacteria (*S. (E.) meliloti*, Group AL; Becker Underwood Pty Ltd, Somersby, NSW, Australia) followed by a hand watering to wash the inoculant into the soil. At the emergence of the third trifoliate leaf (30 DAS), each pot was thinned by hand to one plant per pot. Plants were then grown until 24 March 2012 and were defoliated to a height of 50 mm each time crown bud elongation was observed to occur, giving a total of four defoliations (Fig. 1). Immediately after defoliation each pot received 2.7 g of slow-release fertilizer (Osmocote®; Scotts Australia Pty Ltd, Baulkham Hills, NSW, Australia; 19.4 % N, 1.6 % P, 5 % K, 9 % S, 1.8 % Fe, 0.5 % Ca, 0.5 % Mg, 0.3 % Mn, 0.1 % Cu, 0.1 % Zn, 142 mg kg^{-1} B and 90 mg kg^{-1} Mo). Similar to the standard test to characterize fall dormancy in alfalfa (Teuber *et al.* 1998), plants were defoliated on 24 March 2012 and were grown for another 42 days (5 May 2012). On this day the length of two shoots per plant was measured to assess the level of fall dormancy. Daily maximum and minimum temperatures and photoperiod (Fig. 1) were monitored and recorded using a weather station (HOBO weather station; Onset Computer Corp., Bourne, MA, USA). Pots were watered twice daily with a drip irrigation system to ensure that water supply did not limit plant growth.

Water deficit experiment

Growing conditions. The water deficit experiment was undertaken within the glasshouse research facility of the Tasmanian Institute of Agriculture at the Cradle Coast Campus of the University of Tasmania in Burnie, TAS, Australia. Within the glasshouse, air temperature was maintained at an average of 22 °C during the day (ranging between 25 and 18 °C) and at an average of 10 °C during the night (ranging between 14 and 8 °C). Photoperiod was extended beyond the natural day-length to 14 h with the aid of three 400-W halogen lamps suspended 2 m above the plant canopy, covering an area of 7 m × 7 m. These growing conditions were maintained with the aid of a computer-based glasshouse control system (Priva Maximizer; Priva Computers Inc., Vineland Station, ON, Canada). Temperatures and day-length were monitored using sensors and a data logger (HOBO microstation data logger and sensors;

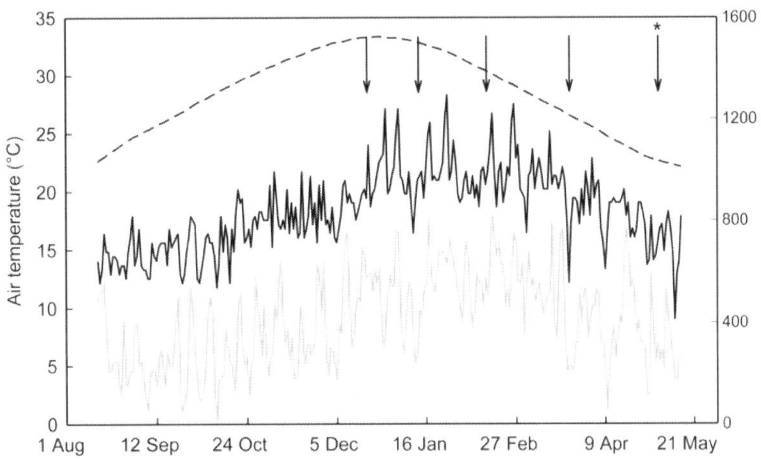

Figure 1. Daily maximum (black line) and minimum (grey line) air temperatures and photoperiod (broken line) at the Cradle Coast Campus of the University of Tasmania in Burnie, TAS, Australia from August 2011 to May 2012. Arrows indicate when the pots of alfalfa used in the fall dormancy screening test were defoliated. The arrow marked with the asterisk indicates when shoot length was assessed.

Onset Computer Corp.) separate to the glasshouse environmental control system.

Approximately 16 viable seeds were sown into 10-L (340 mm high, 96 mm radius) polyethylene bag pots containing 13.2 kg of dry sandy soil sourced from a local landscaping supplier. This soil had a baseline chemical fertility of 14 mg kg^{-1} of P (Olsen extraction; Olsen *et al.* 1954), 140 mg kg^{-1} of K (Colwell extraction; Colwell 1963), 0.15 mg kg^{-1} of Cu, 100 mg kg^{-1} of Fe, 5.6 mg kg^{-1} of Mn, 3.1 mg kg^{-1} of Zn (DTPA extraction; Lindsey and Norvell 1978) and 6.8 mg kg^{-1} of S (KCl40 extraction; Barrow 1967). The soil had a volumetric soil water content of 0.11 mm mm^{-1} at a soil water potential of −10 kPa. To address soil fertility limitations, each kilogram of soil received 0.19 g of Osmocote® (Scotts Australia Pty Ltd), 0.13 g of triple super phosphate (21 % P, 1 % S) and 0.08 g of muriate of potash (50 % K). Immediately following sowing, pots were watered by hand until water was observed to drain through the pot. Emergence was observed at 4 DAS. At 8, 14 and 21 DAS, plants were spray inoculated with rhizobia (*S.* (*E.*) *meliloti*, Group AL; Becker Underwood Pty Ltd) followed by a hand watering to wash the inoculant into the soil. At 28 DAS, plants were thinned by hand to eight plants per pot. From then on water was applied through a purpose-built drip emitter irrigation system designed and calibrated to deliver water to each pot at a rate of 2 ± 0.02 L h^{-1}. Plants were grown until 65 DAS and then defoliated to 50 mm. Plants were allowed to grow for a further 28 days (93 DAS) before they were once again defoliated to 50 mm. At this point the watering treatments commenced. Plants were grown for a further 35 days to the conclusion of the experiment (128 DAS). Following both defoliations, each pot was fertilized with 2.5 g of Osmocote®, 1.7 g of triple super phosphate and 1.0 g of muriate of potash.

Experimental design. The experiment was arranged as a completely randomized block design, with four replications. Blocking was based on a glasshouse bench and a buffer of pots of fully watered alfalfa surrounded each block. The buffer pots were maintained at a similar height to their adjacent experimental pots. To address within-block variation, pots were re-randomized within each block three times (65, 93 and 107 DAS) during the experiment. Each cultivar was exposed to two levels of water application, either 100 % of the plants' water requirement or 25 % of the plants' water requirement. Water requirement was determined as the amount of water required to return the pots receiving 100 % of their replacement water requirement to a soil water content of 0.11 mm mm^{-1}. Soil water content of the pots was determined at 1- to 2-day intervals by measurement with a theta probe (Delta-T Devices Ltd, Cambridge, UK) connected to a hand-held logger (Infield7; UMS GmbH, Munich, Germany). Soil water content was also determined following each water application. Over the 35 days of water deficit treatment, the pots that received 100 % of their water requirement received 3.37 L of water while the pots that received 25 % of their water requirement received 0.84 L of water (Fig. 2). There were three destructive harvests during the experiment: immediately prior to the experiment commencing, and at 14 and 35 days after the commencement of the water deficit treatments. At each harvest, one of each cultivar by water deficit treatment combination from each replicate was harvested. Excluding the buffer pots, there were a total of 168 pots in the experiment (seven cultivars, two water deficit treatments, three destructive harvests and four replications).

Measurements and sample collection. Shoot water potential and gas exchange measurements of plants commenced 3 days prior to the application of water deficit treatments and then occurred at 7-day intervals until the conclusion of the experiment. Measurements occurred between 1000 and 1400 h and were undertaken on plants that were scheduled to be destructively harvested at 35 days after the commencement of watering treatments. Shoot water potential was determined with a pressure chamber (Model 615; PMS Instrument Company, Albany, OR, USA) on the upper 50 mm of two shoots randomly selected per pot. Gas exchange parameters were determined on the middle leaflet on the third most recently emerged leaf on two randomly selected shoots using a portable infrared gas analysis system (LI6400XT; LI-COR Biosciences, Lincoln,

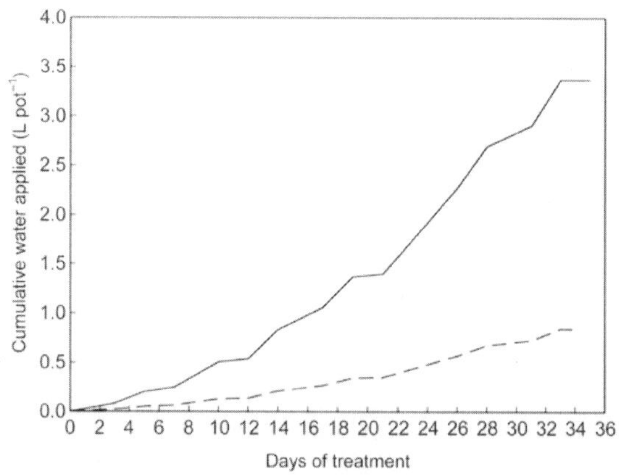

Figure 2. Cumulative water applied to the pots that received 100 % of their water requirement (solid line), and the pots that received 25 % of their water requirement (broken line) over the 35 days of water-deficient treatment.

NE, USA). Following measurement, the middle leaflet was severed at the petiole and scanned on a flatbed scanner (HP Scanjet 3670; Hewlett-Packard Australia Pty Ltd, Blackburn, VIC, Australia). The resulting image was then analysed using image analysis software (IMAGE J version 1.45h; National Institute of Health, Bethesda, MD, USA) to determine the area of the leaflet. Gas exchange measurements were adjusted based on leaf area, and the net carbon exchange rate and stomatal conductance were then calculated as per the equations outlined in the LI6400XT manual (LI-COR Biosciences).

At each destructive harvest, the pots selected for harvest were defoliated to 50 mm above the soil surface and the number of shoots per pot was counted. The leaves were hand separated from the stems. Total leaf area was determined using a flatbed scanner (HP Scanjet 3670; Hewlett-Packard Australia Pty Ltd) coupled with image analysis software (IMAGE J version 1.45h). Soil was washed free from the roots and crowns. Crowns were cut from the taproot at the lowest crown branch or crown bud. Roots were dried with a paper towel to remove excess water and were then weighed. The uppermost 50 mm of the roots were then diced and snap frozen by immersion in liquid nitrogen before being transferred to a −80 °C freezer. The upper 50 mm of roots were chosen as this was a consistent section of the taproot free of nodules and fibrous roots. The remainder of the root system was then weighed and immediately placed in a −20 °C freezer. Leaves, stems and crowns were dried at 60 °C in a fan-forced oven until a consistent weight was achieved while the roots frozen in the −20 °C freezer were freeze dried for 7 days. Once dry, all material was weighed. The original root dry mass was calculated using the root dry matter content and the original root weight.

RNA purification and cDNA synthesis. The taproot material that was snap frozen and stored at −80 °C was ground to a fine powder in liquid nitrogen with a mortar and pestle. Total RNA was extracted from 100 mg of ground material using the Invitrogen TriZol® Plus RNA extraction kit (Life Technologies Australia, Mulgrave, VIC, Australia). The manufacturer's instructions were modified by vortexing of the extraction tube containing the TriZol® reagent, chloroform and sample mixture when the procedure called for shaking by hand and by the addition of a second chloroform phase separation. Following extraction, RNA quantity was determined fluorometrically using a Qubit® 2.0 fluorometer with the broad-range total RNA assay kit (Life Technologies) and RNA integrity was visually checked with ultraviolet light after gel electrophoresis of 1 μg of RNA in 1.5 % agarose gels stained with SYBR® Green (SYBR® safe gel stain; Life Technologies). Following extraction, 20.5 μL of extracted RNA solution was DNase digested using the stringent protocol of the Ambion DNA-free™ kit (Life Technologies Australia). Following the digestion, RNA was re-quantified using the Qubit® 2.0 fluorometer. The absence of genomic DNA contamination was assessed by performing PCRs on 0.35 ng of RNA with forward (5′-GATCAGTGAACTTCG CAAAGTAC-3′) and reverse (5′-AGGGATGCTGCTACTTTG ATG-3′) primers designed to amplify a 154 base pair (bp) fragment of the alfalfa acetyl carboxylase gene (Alexander *et al.* 2007; GenBank accession L25042). Products from these PCR reactions were then visually compared with products from PCRs undertaken with alfalfa RNA samples known to contain genomic DNA contamination by electrophoresis in 2.0 % agarose gels stained with SYBR® Green. Any RNA samples that amplified a fragment were re-digested with DNase and re-quantified, and the second PCR products were visualized via gel electrophoresis. Synthesis of cDNA was undertaken using the iScript™ cDNA synthesis kit (Bio-Rad Laboratories Pty Ltd, Gladesville, NSW, Australia) using 0.35 μg of DNase-digested RNA. Following cDNA synthesis, 6 μL from each of the 192 samples was combined to create a 'calibrator' sample. This sample was then serially diluted to create 1 : 5, 1 : 50, 1 : 500 and 1 : 5000 dilutions to aid in generating a standard curve required for determining the amplification efficiency of the primer pairs. In addition to this, 2 μL of cDNA synthesized from each sample collected prior to the implementation of water treatments was combined and then diluted 1 : 50.

Primer design. Primers were designed for eight genes that were previously identified as having a role in the adaptation of alfalfa to freezing temperatures and eight other genes that were identified as potential reference genes for data normalization (Table 1). Potential reference genes for normalization were chosen from a list of 79 constitutively expressed probe sets identified in Affymetrix chip experiments as part of the *Medicago truncatula* Gaertn gene expression (MtGEA) atlas project (V. A. Benedito, pers. comm.). These probe sets were compared with known alfalfa RNA sequences in GenBank and eight homologues were identified. Primers were designed using the online primer design software 'Primer 3' (Rozen and Skaletsky 2000). Criteria for primer design were a PCR product size between 50 and 200 bp, a primer length between 20 and 24 bp, a predicted primer melting temperature between 57 and 59 °C, a GC content between 30 and 80 % and, wherever possible, a large proportion of the target sequence in

Table 1. Primer pairs (F: forward; R: reverse of the eight reference and eight genes of interest) used in the qPCR analysis of the relative abundance of reference genes (Ref) and genes of interest (GOI) in the taproot tissue of alfalfa cultivars exposed to a water deficit.

Ref or GOI	Gene name	GenBank accession number/ reference	Function/annotation	Primer sequence	Predicted fragment size (bp)	Tm (°C)	Region of fragment
Ref	ADP-ribosylation factor	AY466444	Punitive ADP-ribosylation factor	F: TTTCCTGAGTCTGGTGGTTC R: ACCCCAAGTAACACTGACGA	109	F: 57.68 R: 58.06	UTR: 100 % Coding: 0 %
Ref	Phosphoprotein phosphatase type 2A	X70399/Pirck *et al.* (1993)	Phosphoprotein phosphatase	F: TGCTGGATGTCATACTGAGG R: CTCCAGAAAGGGTGTCCAG	98	F: 57.17 R: 58.20	UTR: 100 % Coding: 0 %
Ref	Histone H3	X13674/Wu *et al.* (1989)	Encodes for histone H3 protein	F: CATTCATGCTAAGCGTGTCA R: CCCTAACAAAGCGAATCAAC	100	F: 58.45 R: 57.37	UTR: 25 % Coding: 75 %
Ref	GTP-binding protein	X79278/Jonak *et al.* (1995)	Protein product homologous to small GTP-binding proteins	F: TCCGAAAGTGAGAGAGATACAAA R: CCAGCAGCTCCTATTGAAAC	119	F: 58.14 R: 57.58	UTR: 100 % Coding: 0 %
Ref	Translationally controlled tumour protein	X63872/Pay *et al.* (1992)	Protein product homologous to a translationally controlled human tumour protein	F: ACTACAAGGATGGTGCTGCT R: TGATAAATTAGGGGCAGAACA	110	F: 57.41 R: 57.27	UTR: 36 % Coding: 64 %
Ref	Calmodulin	X52398/Barnett and Long (1990)	Calcium-binding protein	F: ACAGGGCAAATGAGTTTTGA R: AACAAACCGACCAACAAAAA	99	F: 58.20 R: 58.03	UTR: 100 % Coding: 0 %
Ref	Elongation initiation factor	X59441/Pay *et al.* (1991)	Homologue of eukaryotic translation initiation factor 4D	F: TCATTTCTCTAAGCTTTCACATTG R: CATAAACACCACCAACACCA	86	F: 57.79 R: 57.71	UTR: 100 % Coding: 0 %
Ref	Glyceraldehyde-3-phosphate dehydrogenase	GQ398120	Phosphate dehydrogenase	F: TGGAATCGTTGAGGGTCTTA R: AGCTCTTCCACCTCTCCAGT	100	F: 58.15 R: 58.06	UTR: 0 % Coding: 100 %
GOI	Cold acclimation response protein (CAR1/*cas15*)	AF072932/Haagenson *et al.* (2003)	Punitive nuclear regulatory protein	F: GCTTCATCATGTTGAGAGGTG R: TTTCTTTCCACACACACACG	98	F: 58.30 R: 58.14	UTR: 100 % Coding: 0 %
GOI	Bi-modular protein (corC)	L22305/Castonguay *et al.* (1994)	Regulation of plant development	F: GCACGATTGACTTTCACGA R: ACAGCGATACACCGTGATTT	87	F: 58.34 R: 58.10	UTR: 100 % Coding: 0 %
GOI	Cold- and drought-regulated protein (corA)	L03708/Laberge *et al.* (1993)	Cold- and drought-induced protein of unknown function	F: CACCATGCACTCTTTCTCAGT R: AACTGAAACTGCTGCACATCT	100	F: 57.94 R: 57.62	UTR: 100 % Coding: 0 %
GOI	Cold acclimation-specific protein (*cas18*)	L07516/Wolfraim *et al.* (1993)	Cold-induced punitive dehydrin protein	F: TCTGTTTTTGAGTAAGTTGGTTCA R: TGCCCCTACACTAAAATTCAA	123	F: 58.04 R: 57.33	UTR: 100 % Coding: 0 %
GOI	Protein kinase	X82270/Jonak *et al.* (1996)	Stress signalling protein	F: CACTGCTGGGAATTCAATCT R: CAACAGAAAGCAGGGTAAGC	99	F: 57.74 R: 57.63	UTR: 100 % Coding: 0 %
GOI	Galactinol synthase (*corF*)	AY126615/Cunningham *et al.* (2003)	Synthesis of raffinose family oligosaccharides (RFOs)	F: CAGCAATTTTGGAAGCTTATG R: AGACGATCATGCGGCTAATA	95	F: 57.62 R: 58.36	UTR: 42 % Coding: 58 %
GOI	Type 1 sucrose synthase mRNA	AF049487	Synthesis of sucrose	F: ATGCTCTCAAGTACCGCAAA R: CGGTTTCTCCATTTCTTCATT	96	F: 58.53 R: 58.17	UTR: 46 % Coding: 54 %
GOI	High-molecular-weight vegetative storage protein	AF530579/Volenec *et al.* (2002)	Vegetative storage protein	F: ACTTTGATTCCCTCCGTTTT R: AGCGCGCAATTCAATTTTA	106	F: 57.64 R: 59.44	UTR: 100 % Coding: 0 %

the 3′ untranslated region of the target transcript. Primers were checked for possible hairpins and self-complementation using the online tool 'Oligo Calc' (Kibbe 2007). Primers were custom ordered from a commercial supplier (GeneWorks Pty Ltd, Thebarton, SA, Australia). Primer pairs were tested by performing a PCR on the 1 : 50 dilution of the calibrator sample under the same conditions that would be used for the real-time qRT-PCR analysis of the samples. Following the reaction, PCR products were visualized after electrophoresis in 2 % agarose gels stained with SYBR® Green. Primer pairs that failed to produce a product, produced a product of the incorrect size or produced multiple products were discarded, and new primers were designed and re-ordered.

Determination of gene transcript abundance by real-time qRT-PCR. Real-time qRT-PCR reactions were undertaken in 384-well plates using a Roche 480 LightCycler® PCR system, using SYBR® Green I Master mix (Roche Diagnostics GmbH, Mannheim, Germany). Each reaction contained 5 μL of SYBR® Green I Master mix, 2 μL of a 1 : 50 dilution of template cDNA, 0.5 μL of each primer pair (10 $\mu mol\ L^{-1}$ concentrations) and 2 μL of water. Each plate also contained reactions for generating a standard curve for each primer pair, a plate calibrator sample, negative controls and the 1 : 50 dilution of the cDNA collected from each sample that was harvested before the commencement of water deficit treatments. Each reaction was undertaken in triplicate and all liquid handling was undertaken using an automated liquid handling system (epMotion 5075LH; Eppendorf, Hamburg, Germany). The conditions of real-time qRT-PCR reactions were as follows: 95 °C for 5 min, followed by 55 cycles of 95 °C for 10 s, 58 °C for 10 s and 72 °C for 8 s. At the end of the PCR reaction, a melting curve analysis (58–95 °C) was undertaken. LightCycler® 480 software (version 1.5; Roche Diagnostics) was used to determine the crossing point (Cp) values. Cp values and the standard curves were used to determine the concentration of each gene in each sample following adjustment using the plate calibrator sample. Relative concentration values for each of the genes of interest were normalized using the geometric mean of the concentration of the three most stably expressed reference genes identified using the GeNorm version 3.5 software package (Vandesompele *et al.* 2002). Normalized concentration values were divided by the normalized concentration values from the sample generated by taking an equal volume of cDNA from the samples collected prior to the implementation of the water deficit treatments to give a relative concentration.

Statistical analysis. Data from the fall dormancy screening test were subjected to an analysis of variance (ANOVA) of a randomized complete block design. Groups of cultivars with similar expression levels of fall dormancy were identified (from here on referred to as fall dormancy groups). Data from the water deficit experiment were subjected to an ANOVA of a randomized complete block design with cultivar, water treatment and harvest date as experimental factors. One-tail *t*-tests were used to identify significant decreases between the water deficit treatment and the fully watered controls of each cultivar and harvest. The relative concentration of each gene of interest from the real-time qRT-PCR analysis was log transformed prior to analysis. To explore the influence of fall dormancy group on the level of water deficit adaption and the abundance of freezing tolerance gene transcripts, growth parameters and gene transcript abundance of the water deficit stressed plants were expressed relative to their corresponding fully watered controls. These relative values were then subjected to an ANOVA with contrasts based on fall dormancy groups. Gene transcript abundance was log transformed prior to analysis. Net carbon exchange rate and stomatal conductance were fitted to the shoot water potential data using a logistic function (Equation 1). The shoot water potential that resulted in a 50 % reduction in carbon exchange rate and a 90 % reduction in stomatal conductance was then estimated from the fitted functions. All statistical analysis was undertaken with the software package GenStat 13th edition (VSN International Ltd, Hemel Hempstead, UK).

$$y = \frac{m}{1 + e^{b(x-c)}} \qquad (1)$$

Results

Fall dormancy screening test

Shoot length during autumn/fall when grown outdoors was influenced by cultivar ($P < 0.05$), with a general increase in shoot length with increasing fall dormancy scores (Fig. 3). The cultivars with the longest shoots were Sequel and SARDI 10. The cultivar SARDI 7 also had longer shoots compared with the more fall dormant cultivars of FFI DT1, Q31, Kaituna and A5225. Consequently, FFI DT1, Q31, Kaituna and A5225 were grouped together as the fall dormant group, SARDI 7 as the non-fall dormant group, and Sequel and SARDI 10 as the highly non-fall dormant group.

Soil water content under water deficit conditions

There was no influence ($P = 0.39$) from cultivar on the soil water content under each of the water treatments. The soil water content of each of the watering treatments differentiated from each other ($P < 0.001$) by 5 days of treatment (Fig. 4). The consistent refilling of the pots to a soil water content of 11 % (equivalent to a soil water potential of −10 kPa for the growing medium) maintained the soil water content in the 100 % treatment throughout the experiment. In contrast, for the 25 % watering treatment after 14 days of treatment the soil water content fluctuated between 6 and 3.5 %.

DM accumulation under water deficit conditions

Cultivar, water deficit and time of harvest interacted to affect shoot yield ($P < 0.05$). At Day 14 after the beginning of the water deficit treatment, the effect on shoot mass was dependent upon cultivar, with significant differences between plants receiving 100 and 25 % of their water requirement observed in the SARDI 7 and Sequel HR cultivars (Table 2). When cultivars were pooled based on fall dormancy group, the non-fall dormant and highly non-fall dormant groups showed a significant decrease in shoot mass per plant when exposed to a water deficit compared with the fully watered plants. At 35 days of treatment, the plants exposed to a water deficit had a lower shoot yield compared with the fully watered controls for all cultivars, and the same effect was observed when cultivars were grouped based on fall dormancy group. There were interaction effects between water deficit treatment and harvest ($P < 0.001$) and cultivar and water deficit treatment ($P < 0.01$) on total plant mass. Consequently, there was no difference in total plant mass between the plants of each cultivar receiving either 100 or 25 % of their water requirement, at 14 days of water deficit. However, after 35 days of exposure to water deficit treatment, total mass per plant was reduced in all the cultivars compared with their respective controls ($P < 0.05$; Table 2) and the same was noted when cultivars were grouped by fall dormancy groups.

There was no influence from cultivar or fall dormancy group on the shoots per plant, mass per shoot and leaf area per plant nor did cultivar interact with either harvest or water deficit treatment. Harvest and water deficit treatment did interact ($P < 0.05$) to affect these response variables. Shoots per plant, mass per shoot and leaf area per plant did not increase between 14 and 35 days of treatment for the plants that only received 25 % of their water requirement (Table 3). In contrast, shoots per plant, mass per shoot and leaf area per plant increased between Days 14 and 35 of treatment for the fully watered plants.

After 14 days of treatment, the cultivars classified into the fall dormant group had a greater ($P < 0.05$) shoot yield, mass per shoot, total plant mass and leaf area per plant relative to the fully watered controls compared with the non-fall dormant or highly non-fall dormant group (Fig. 5). However, this effect was transient and after Day 35 of exposure there was no difference between the fall dormancy groups for these variables.

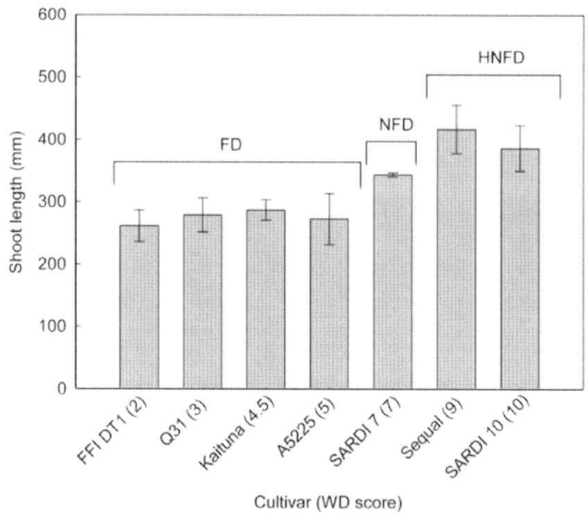

Figure 3. Shoot length (mm) of seven alfalfa cultivars (cultivar fall dormancy score given in parentheses) in late autumn (5 May 2012; 42 days after defoliation). Plants were grown outdoors in pots at Burnie, TAS, Australia. Error bars represent the standard errors of the means ($n = 4$). Cultivars grouped into fall dormant (FD), non-fall dormant (NFD) and highly non-fall dormant (HNFD) groups exhibited no significant difference in shoot length.

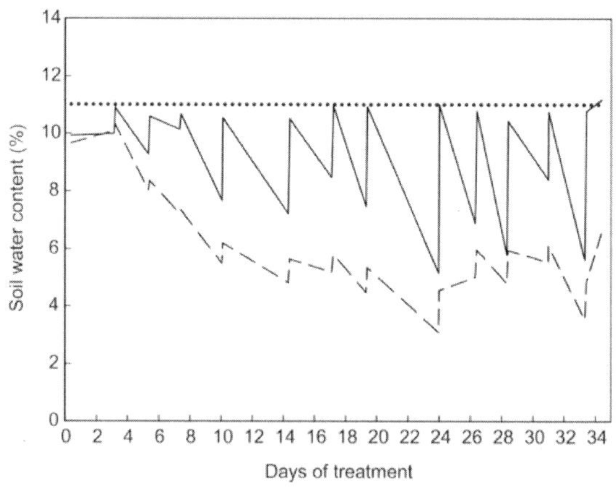

Figure 4. Soil water content in the pots that received 100 % of their water requirement (solid line), and the pots that received 25 % of their water requirement (broken line) over the 35 days of water-deficient treatment. The horizontal dotted line indicates a soil water content of 11 % (equivalent to a soil water potential of −10 kPa for the growing medium).

Shoot water potential and gas exchange under water deficit conditions

While differences in the shoot water potential occurred between cultivars when exposed to a water deficit,

Table 2. Shoot mass per plant and total plant mass (g DM plant^{-1}) of seven alfalfa cultivars (presented individually and grouped based on fall dormancy grouping) after 14 or 35 days of receiving either 100 or 25 % of their water requirement. One-tail *t*-tests were used to identify decreases between plants receiving 100 or 25 % of their water requirement within a cultivar. Differences are indicated by NS: $P > 0.05$, *$P < 0.05$, **$P < 0.01$, ***$P < 0.001$.

	Shoot mass per plant (g DM plant^{-1})						Total mass per plant (g DM plant^{-1})					
	14 days of drought			35 days of drought			14 days of drought			35 days of drought		
	100 %	25 %	t-Test	100 %	25 %	t-Test	100 %	25 %	t-Test	100 %	25 %	t-Test
Cultivar												
FFI DT1	0.27	0.18	NS	0.72	0.24	***	0.99	0.84	NS	2.07	1.21	***
Q31	0.20	0.19	NS	0.85	0.11	***	1.01	0.91	NS	2.28	1.09	***
Kaituna	0.26	0.22	NS	0.74	0.15	***	1.01	0.86	NS	2.28	1.06	***
A5225	0.29	0.19	NS	0.57	0.30	***	0.80	0.84	NS	1.57	1.25	**
SARDI 7	0.31	0.13	*	0.76	0.19	***	1.02	0.81	NS	2.03	1.10	***
Sequel HR	0.33	0.19	*	0.81	0.18	***	1.05	0.81	NS	2.21	1.07	***
SARDI 10	0.30	0.18	NS	0.67	0.19	***	1.01	0.86	NS	1.90	1.06	***
SED			0.0848						0.1576			
Fall dormancy group												
Fall dormant	0.26	0.20	NS	0.72	0.20	***	0.93	0.86	NS	2.05	1.15	***
Non-fall dormant	0.31	0.13	*	0.76	0.19	***	1.02	0.81	NS	2.03	1.10	***
Highly non-fall dormant	0.32	0.19	*	0.74	0.19	***	1.03	0.84	NS	2.06	1.07	***
SED			0.0695						0.1383			

these differences were inconsistent throughout the experiment (Fig. 6). For example, SARDI 10 had the lowest shoot water potential after 12 days of treatment but had the highest shoot water potential after 19 days of treatment. When grouped by fall dormancy, there were no differences ($P = 0.396$) in the shoot water potential within either watering treatments. There was a sharp decline in shoot water potential 19 days after the beginning of the water deficit treatments in all cultivars.

As shoot water potential decreased, so did net carbon exchange rate and stomatal conductance (Fig. 7). The point of stomatal closure (defined as a 90 % reduction in stomatal conductance) occurred at lower shoot water potentials (−2.54 MPa) in the fall dormant group compared with the highly non-fall dormant group (−1.05 MPa). There was no difference in the shoot water potential that resulted in a 50 % decline in photosynthesis among the fall dormancy groups.

Real-time qRT-PCR analysis of gene transcript abundance under water deficit conditions

The pairwise variation analysis undertaken with GeNorm identified that three reference genes would be required for the adequate normalization of transcript numbers for the genes of interest within this sample set (Fig. 8, upper panel). The candidate reference genes identified as being the most suitable for normalization under the current experimental conditions were GTP-binding protein, translationally controlled tumour protein and glyceraldehyde-3-phosphate dehydrogenase (Fig. 8, lower panel).

Of the eight gene transcripts investigated, five (galactinol synthase: *corF*; cold acclimation responsive protein: CAR1/*cas15*; cold acclimation-specific protein: *cas18*; cold- and drought-regulated protein: corA; and the vegetative storage protein: VSP) increased in abundance relative to the fully watered controls when plants were exposed to a water deficit for both 14 and 35 days (Fig. 9). The increase in abundance of gene transcripts for galactinol synthase (*corF*) and cold acclimation-specific protein (*cas18*) differed between fall dormancy groupings ($P < 0.05$). There was a similar effect with VSP gene transcript abundance; however, this effect had a *P* value of 0.061. Both these gene transcripts had a greater abundance relative to the fully watered controls in the fall dormant groups compared with the non-fall dormant or highly non-fall dormant groups after 14 days of treatment. However, this effect was transient with no influence from fall dormancy grouping present at 35 days of treatment.

Table 3. Shoots per plant, mass per shoot and leaf area per plant of alfalfa plants (the average of seven cultivars) receiving either 100 or 25 % of their water requirements after 14 and 35 days of treatment.

Days of treatment	Water treatment (%)	Shoots per plant	Mass per shoot (mg shoot^{-1})	Leaf area per plant (m^2 plant^{-1})
14	100	3.2	88.2	0.046
	25	2.3	80.9	0.024
35	100	3.8	197.5	0.096
	25	2.2	87.0	0.027
SED		0.2	8.8	0.005
P value		0.028	<0.001	<0.001

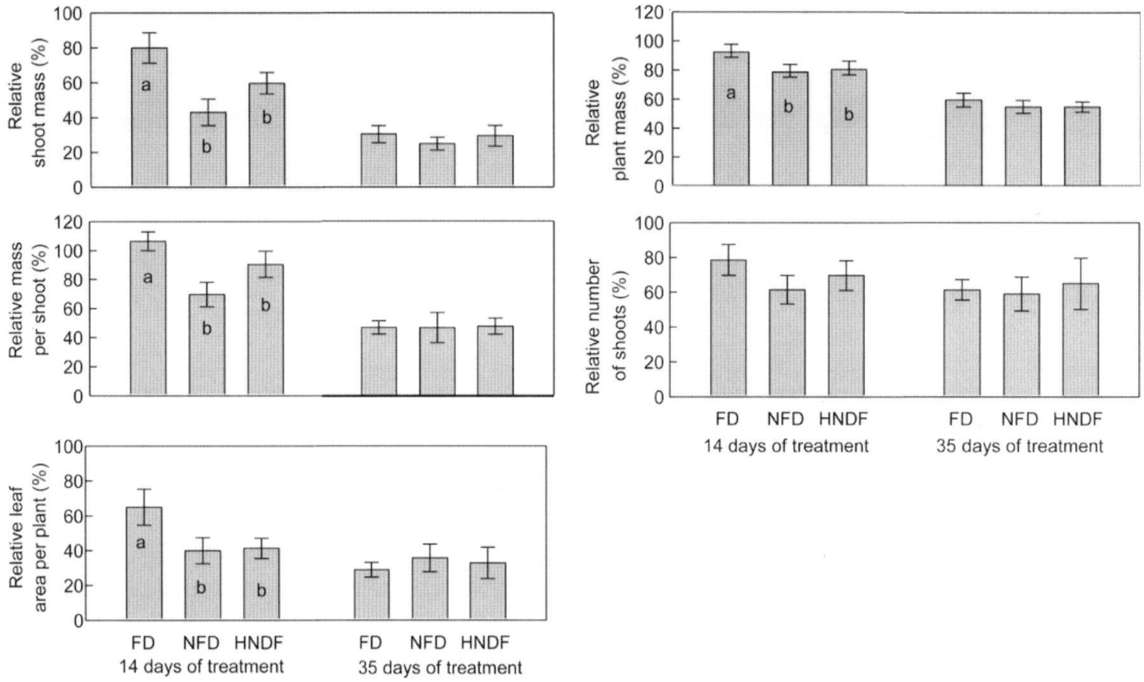

Figure 5. Shoot mass per plant, total plant mass, mass per shoot, number of shoots per plant and leaf area per plant expressed as a proportion of the value achieved by the fully watered controls (relative value) at 14 and 35 days after the initiation of treatments. Cultivars were grouped into fall dormant (FD), non-fall dormant (NFD) and highly non-fall dormant (HNFD) fall dormancy groups based on their level of fall dormancy identified in the fall dormancy screening test. Bars with different letters were identified as being different from each other using ANOVA with non-orthogonal contrasts ($P < 0.05$). Error bars represent the standard errors of the mean ($n = 4$ to 16).

The abundance of *sucrose synthase* gene transcripts (relative to the fully watered controls) was lower in the fall dormant cultivars compared with the non-fall dormant and highly non-fall dormant cultivars after 35 days of treatment. The expression of the sucrose synthase gene at 14 days of treatment was not statistically different among the fall dormancy groupings.

Discussion

The shoot growth of the water deficit stressed plants relative to fully watered controls at 14 days of treatment differed by fall dormancy grouping, with the fall dormant cultivars able to maintain shoot growth for longer periods under a water deficit (Fig. 5). Past research has noted that in the cool temperate environments of Australia that do not expose alfalfa crops to extreme winter temperatures, fall dormant cultivars outperform non-fall dormant cultivars in terms of forage yield in the face of mild to moderate drought (Pembleton *et al.* 2010*a*, *b*) despite having no need for the freezing tolerance that fall dormancy is associated with. Similarly, in certain temperate environments in China that receive nearly 1000 mm of annual precipitation, Wang *et al.* (2009) noted differences in annual forage yield among fall dormancy groupings during drought (600 mm annual rainfall) years but not during

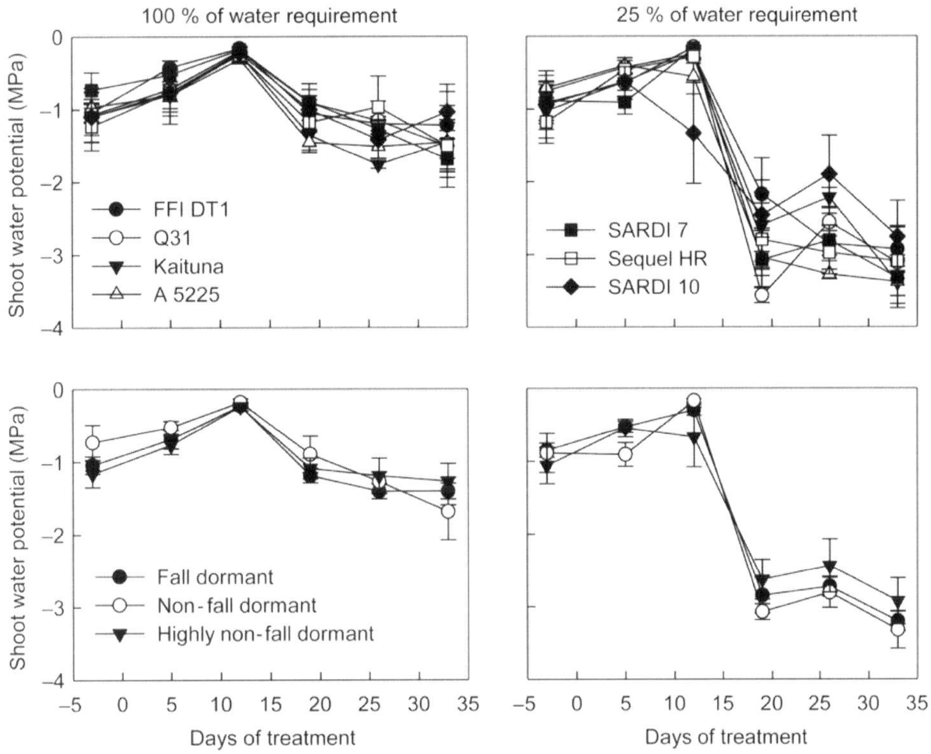

Figure 6. Shoot water potentials of seven alfalfa cultivars (upper panels) and the seven cultivars grouped by their fall dormancy groups (lower panels) through 35 days of receiving either 100 % of their water requirement or 25 % of their water requirement. Error bars represent the standard errors of the mean ($n = 4$).

years receiving rainfall close to the long-term average. Fall dormancy group by environmental interactions was observed in North America as well, with superior performance of fall dormant cultivars in water-limited environments (Orloff and Hanson 2008). However, the confounding effects of both summer droughts and freezing winter temperatures make it difficult to attribute this expression of fall dormancy group by environmental interactions to the superior freezing or to the drought tolerance of fall dormant cultivars.

The fall dormant group also did not fully close their stomata until lower shoot water potentials were reached when compared with the highly non-fall dormant types. By the 35th day of treatment, there was no difference in relative shoot growth among the fall dormancy groupings and the consequences of a low shoot water potential on stomatal conductance were similar among the fall dormancy groups. The cumulative water deficit (water applied to the 100 % replacement water requirement treatment minus water applied to the 25 % replacement water requirement treatment) by this time was four times as great as the water deficit at 14 days (a 2530-mL deficit compared with a 625-mL deficit). This was reflected in plant shoot water potential for which there was no difference between the fully watered and the water deficit treated plants up to the 12th day of treatment while a 2.5-fold decrease from the fully watered plants occurred between 19 and 33 days of treatment. The timing of when the acclimation to water deficit was observed to occur in relation to the severity of the water deficit stress suggests that the specific acclimation to water deficit by more fall dormant cultivars of alfalfa is expressed early after the onset of water deficits and that more extended stress restricts growth equally for all cultivars. This observation is further supported by the difference between the fall dormancy groupings in the relationship between shoot water potential and stomatal conductance. The superior vigour of the fall dormant cultivars over non-fall dormant cultivars under a mild water deficit as indicated in Fig. 5 helped this fall dormancy grouping maintain stomatal conductance and leaf area expansion under lower shoot water potentials (Fig. 7). This would enable increased net plant photosynthesis and a further continuation of shoot growth in these cultivars in the face of mild to moderate water deficits, as has been previously observed under field conditions (Orloff and Hanson 2008; Wang *et al.* 2009; Pembleton *et al.* 2010*a*, *b*). A similar stomatal regulation strategy has been observed in the water deficit tolerant forage grass tall fescue (*Festuca arundinacea*) compared with the water deficit

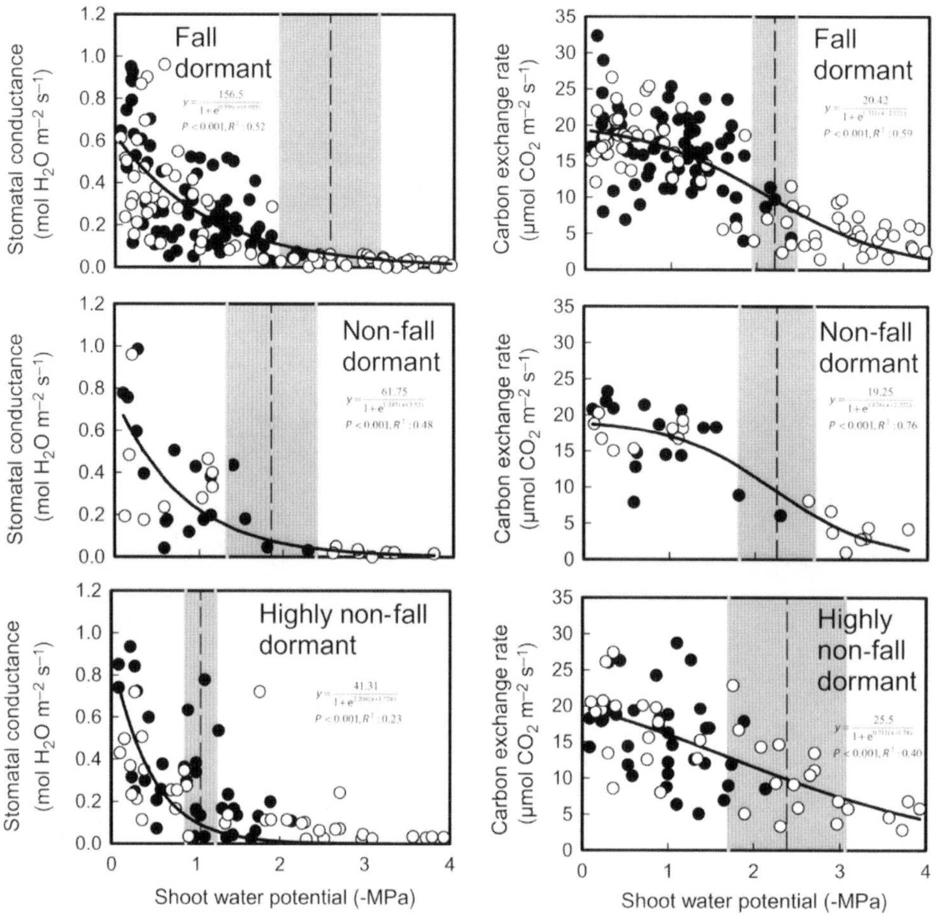

Figure 7. Relationship between shoot water potential and stomatal conductance/carbon exchange rate of alfalfa cultivars grouped into fall dormant, non-fall dormant or highly non-fall dormant dormancy groups. Plants received either 100 % (open circles) or 25 % (closed circles) of their water requirement over 35 days. A logistical function was fitted to the data (solid line). The vertical broken line indicates shoot water potentials at which a 50 % decline in carbon exchange rate and a 90 % decline in stomatal conductance (stomatal closure) occurred. The grey areas represent the standard errors around these estimates.

sensitive perennial ryegrass (*Lolium perenne*) (Holloway-Phillips and Brodribb 2011).

The maintenance of stomatal conductance, leaf area expansion and shoot growth at lower shoot water potentials would require osmotic adjustment within the plant. Osmotic adjustment is a key strategy that alfalfa plants use to minimize damage when exposed to freezing temperatures (Castonguay *et al.* 2006). The analysis of gene transcript abundance after 14 days of treatment identified that the transcripts for several genes that have been associated with the adaptation to freezing temperatures in alfalfa increased in abundance relative to the fully watered controls when exposed to a water deficit. The genetic basis of acclimation to freezing temperatures is common among all alfalfa fall dormancy groups (Castonguay *et al.* 1997). Consequently, the presence of freezing tolerance gene transcripts across all the fall dormancy groups as shown in Fig. 9 could be expected. After 14 days of water deficit, the *cas18* and *corF* gene transcripts had an increase in abundance that was related to fall dormancy grouping and their relative adaptation to water deficit indicated by the plant growth measurements. However, after 35 days of exposure to water deficit there was no pattern in gene transcript abundance that was related to fall dormancy and consequently water deficit tolerance. In Arabidopsis (*Arabidopsis thaliana*), wheat (*Triticum aestivum*), *Nicotiana benthamiana* and rice (*Oryza sativa*), homologues of the *cas18* and *corF* genes have been identified as early response genes following the exposure to a water deficit (Zhu 2002). The signalling pathways leading to the expression of these genes are constitutively maintained within the cell, leading to a rapid increase in their transcript abundance after exposure to stress (typical of dehydrins; Hanin *et al.* 2011). The other genes that exhibited an increase in transcript abundance are homologues of delayed response genes identified by Jaglo-Ottosen *et al.* (1998) and Kasuga *et al.* (1999). The signalling

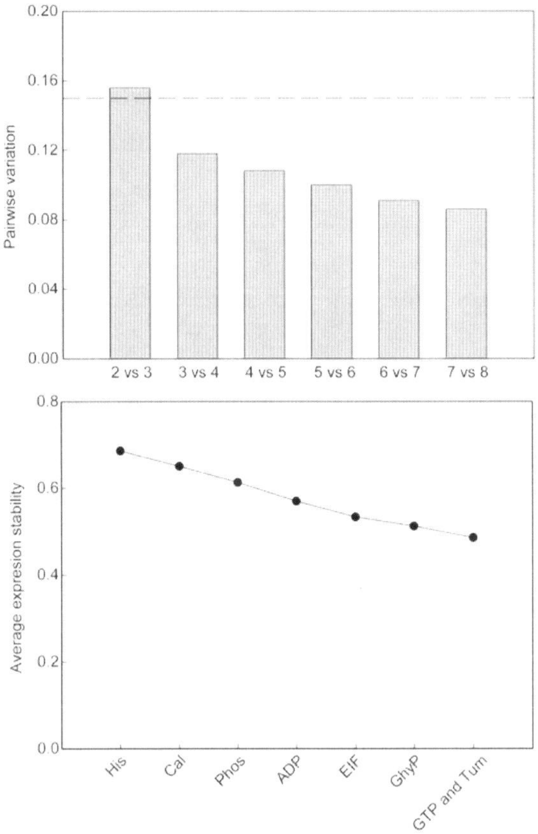

Figure 8. Results of pairwise variation analysis (upper panel) and average expression stability calculations (lower panel) undertaken using GeNorm from the quantitative PCR of alfalfa gene transcripts using primers designed for eight potential reference genes (Phos, phosphoprotein phosphatase type 2A; His, histone H3; Cal, calmodulin; ADP, ADP-ribosylation factor; EIF, elongation initiation factor; GhyP, glyceraldehyde-3-phosphate dehydrogenase; GTP, GTP-binding protein; and Tum, translationally controlled tumour protein). The broken horizontal line on the upper panel represents the 0.15 suggested cutoff for determining the optimum number of reference genes required for normalization.

pathways for these genes need to be constructed before they are expressed (Zhu 2002). The differential expression of the early response genes observed at Day 14 but not at Day 35 of treatment suggests that although the non-fall dormant cultivars possess a functional form of the freezing tolerance genes, they may not have an as efficient signalling pathway for these genes in place compared with the fall dormant cultivars. This may partly explain the difference in adaptation at a plant growth/physiology level between fall dormancy groups treated to mild water deficit stress (14 days of treatment) but not to severe water deficit stress (35 days of treatment) as observed in this experiment. The creation of plants that under/overexpress genes may be required to fully understand the impact these genes and their signalling pathways have on the adaptation of plants to a water deficit.

The *cas18* gene transcript encodes for a protein that has considerable homology to dehydrins isolated from other species (Wolfraim *et al.* 1993; Close 1997). Dehydrins are thought to maintain cellular membrane integrity during dehydration (Close 1997; Hanin *et al.* 2011) and this has been the hypothesized role for the cas18 protein during freezing adaptation. Given the upregulation of this gene transcript in the drought-tolerant fall dormant cultivars when exposed to a water deficit, it is conceivable that the role of the cas18 protein during adaption to a water deficit is similar to that of dehydrins in other species.

The *corF* transcript encodes for galactinol synthase, a key enzyme in the formation of raffinose family oligosaccharides (RFOs) (Peterbauer and Richter 2001). The concentration of RFOs in alfalfa taproots has previously been correlated with freezing tolerance and winter survival (Castonguay *et al.* 1995; Cunningham *et al.* 2003), and an increase in the tissue concentration of RFOs has been associated with dehydration tolerance in *Xerophyta viscosa* (Peters *et al.* 2007) and Arabidopsis (Taji *et al.* 2002). Differences in the levels of *corF* gene transcripts in the taproots of alfalfa cultivars grouped into fall dormancy groups during the onset of fall dormancy have been noted by Cunningham *et al.* (2003). The role of RFOs in both freezing and water deficit stress tolerance may be through the protection of cellular structures or the stabilization of cellular membranes (Leopold 1990; Crowe *et al.* 1992).

The reduced abundance of the type 1 sucrose synthase gene transcripts in the water deficit tolerant fall dormant alfalfa cultivars after 35 days of treatment is novel given the requirement for increased concentrations of osmolytes in the cytosol when the plant is exposed to a water deficit. Furthermore, simple sugars, including sucrose, are thought to help stabilize membranes during freezing desiccation (Crowe and Crowe 1992; Sun *et al.* 1994). Interestingly, Sus1 in Arabidopsis was also reported to be induced by changes in leaf osmotic potential via an ABA-independent mechanism during low-temperature stress and osmotic stress (Déjardin *et al.* 1999). Whereas transcript levels of *Sus1* could increase several fold after osmotic stress exposures, researchers failed to observe any changes to the sucrose synthase protein levels on a western blot even after long-term cold and drought treatments. Déjardin *et al.* (1999) proposed that *Sus1* is under tight post-transcriptional control during osmotic stress and that *Sus1* mRNA might perhaps be produced and stored for use during specific stages of cold hardening or during plant recovery from the stress. In alfalfa the maintenance of taproot starch reserves has been linked with the long-term persistence in drought-prone environments (Boschma and

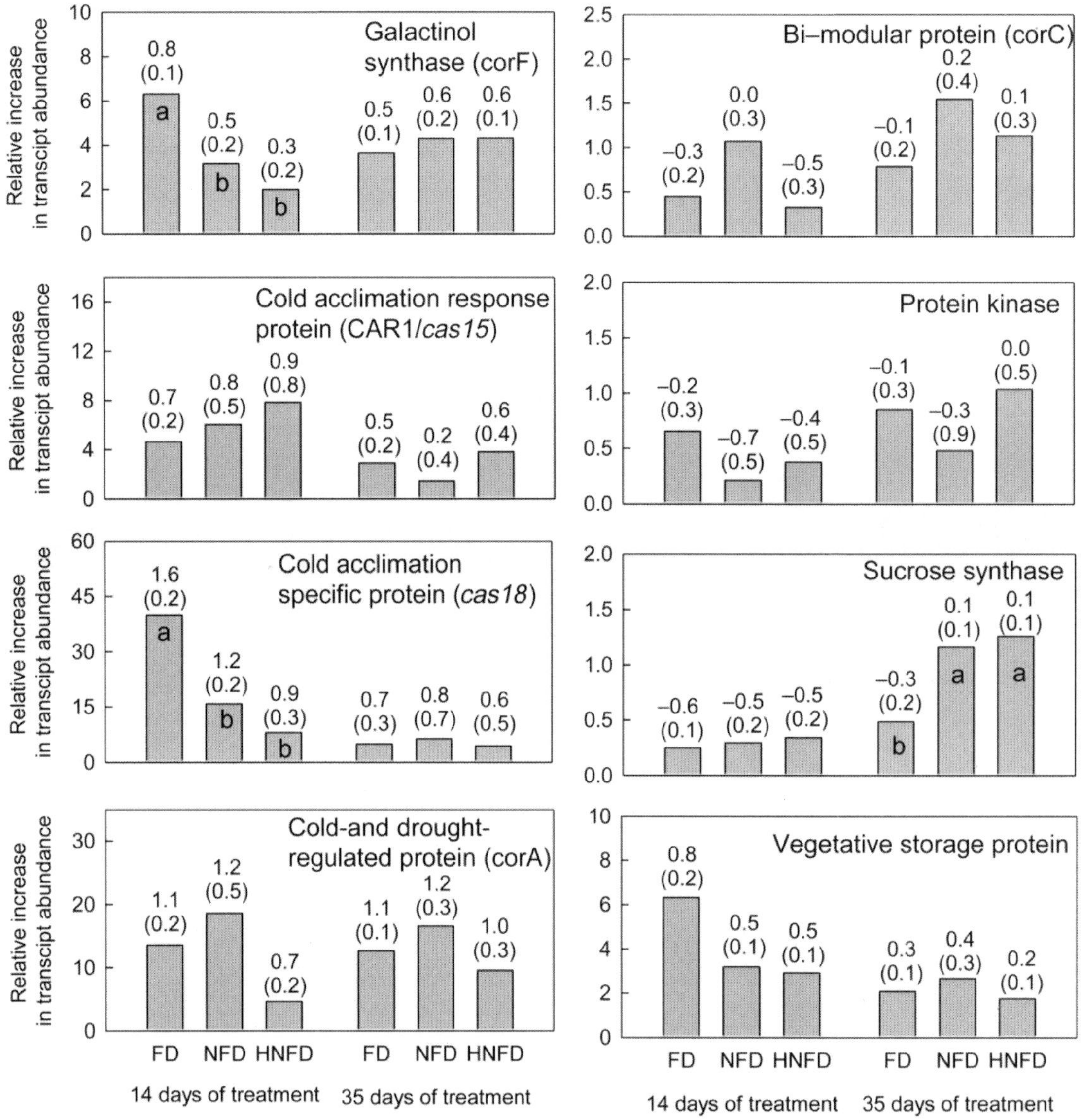

Figure 9. Relative increase in gene transcript abundance (relative to fully watered controls) of eight genes in the taproots of alfalfa cultivars (grouped into fall dormant (FD), non-fall dormant (NFD) or highly non-fall dormant (HNFD) groups based on their level of fall dormancy observed in the fall dormancy screening experiment) receiving 25 % of their water requirement. Values above the bars are the transformed means (log transformation) with the associated standard errors in parentheses. Bars with different letters within the same gene and number of days from treatment were identified as being different ($P < 0.05$) using ANOVA with non-orthogonal contrasts on transformed data.

Williams 2008). Given the role that sucrose synthase plays in utilizing the glucose released from starch degradation, the downregulation of this gene in the fall dormant cultivars could be a consequence of a long-term energy maintenance strategy under drought conditions in these cultivars. If true, this response indicates that fall dormant cultivars are not only more tolerant of mild water deficit conditions compared with more non-fall dormant cultivars but they may also be more persistent when exposed to severe long-term droughts. Further research is required to understand the role of sucrose synthase in alfalfa under different environmental conditions.

From the transcript abundance profiles and the punitive proteins these transcripts encode for, it can be concluded that the adaptation of the fall dormant cultivars is due to improved protection of cellular structures and membranes in the short term. The transcripts involved in stress signalling, protein storage and cellular regulation while also increased in abundance did not show any differential expression related to the adaptation of fall dormant cultivars to water deficit. The absence of differential expression of these transcripts is interesting as cross-talk between stress signalling pathways as well as the difference in cellular regulation has been linked to associations between abiotic stress tolerance including both freezing and drought (Zhu 2002; Seki *et al.* 2003; Chinnusamy *et al.* 2004). Consequently, if the drought tolerance of non-fall dormant alfalfa cultivars is to be improved without the use of gene modifications, focus should be placed on the improvement of the protective cellular structures and membranes by breeding together cultivars with desired gene expression patterns.

While this study is not the first to examine the responses of alfalfa cultivars to exposure to a water deficit (see Pembleton *et al.* 2009; Erice *et al.* 2011 for others), it is the first that has examined a broad enough range of fall dormancy to be able to conclude that decreasing levels of fall dormancy reduce adaptation to mild water deficit conditions. In support of this finding, transcripts of two genes that are associated with freezing tolerance increased in abundance in the more fall dormant types when exposed to a mild water deficit, while the transcript of a type 1 sucrose synthase gene associated with phloem loading and carbon channelling in roots decreased when the fall dormant types were exposed to a severe water deficit. The identification of a partial separation between fall dormancy and freezing tolerance by Brouwer *et al.* (2000) and Brummer *et al.* (2000) along with the linkage between freezing tolerance and water deficit tolerance at a molecular level identified in this study suggests that it will be possible to achieve greater drought tolerance in non-fall dormant cultivars. This study has identified that alfalfa cultivars must be of the fall dormant type to exhibit superior adaptation to a water deficit. The stress adaptation pathways associated with the upregulation of the freezing tolerance genes *cas18* and *corF* during mild water deficit stress and the downregulation of a type 1 sucrose synthase gene during severe water deficit stress appear to be the molecular mechanism by which fall dormant cultivars of alfalfa 'give drought the cold shoulder'.

Sources of Funding

This work was supported by Dairy Australia Ltd and the Minister for Agriculture, Fisheries and Forestry Awards of the 2011 Department of Agriculture, Fisheries and Forestry (DAFF) Science and Innovation Awards for Young People in Agriculture, Fisheries and Forestry.

Contributions by the Authors

K.G.P. conceived the study, designed the experiments and carried out the experiments. P.S. helped design and supervised the qRT-PCR analysis. Both authors drafted the paper and approved the final draft.

Acknowledgements

The authors appreciate the seed kindly provided by David Squib and James Sewell of PGG Wrightson Seeds Australia, David Barnett and Tom Damin of Seed Distributors Australia, Dr Jonathan Reich of Cal/West Seeds USA and Dr Allan Humphries of SARDI. The authors thank Prof. Derick Moot and the anonymous reviewers for their helpful comments on earlier drafts of this paper.

Literature Cited

Alexander TW, Reuter T, McAllister TA. 2007. Qualitative and quantitative polymerase chain reaction assays for an alfalfa (*Medicago sativa*)-specific reference gene to use in monitoring transgenic cultivars. *Journal of Agricultural and Food Chemistry* **55**:2918–2922.

Barnes DK, Goplen BP, Baylor JE. 1988. Highlights in the USA and Canada. In: Hanson AA, Barnes DK, Hill RR, eds. *Alfalfa and alfalfa improvement*. Madison: American Society Agronomy, 1–24.

Barnett MJ, Long SR. 1990. Nucleotide sequence of an alfalfa calmodulin cDNA. *Nucleic Acids Research* **18**:3395.

Barrow NJ. 1967. Studies on extraction and on availability to plants of adsorbed plus soluble sulfate. *Soil Science* **104**:242–249.

Beck EH, Fettig S, Knake C, Hartig K, Bhattarai T. 2007. Specific and unspecific responses of plants to cold and drought stress. *Journal of Biosciences* **32**:501–510.

Boschma SP, Williams RW. 2008. Using morphological traits to identify persistent alfalfas for dryland agriculture in NSW, Australia. *Australian Journal of Agricultural Research* **59**:69–79.

Brouwer DJ, Duke SH, Osborn TC. 2000. Mapping genetic factors associated with winter hardiness, fall growth, and freezing injury in autotetraploid alfalfa. *Crop Science* **40**:1387–1396.

Brummer EC, Shah MM, Luth D. 2000. Reexamining the relationship between fall dormancy and winter hardiness in alfalfa. *Crop Science* **40**:971–977.

Castonguay Y, Laberge S, Nadeau P, Vezina LP. 1994. A cold-induced gene from *Medicago sativa* encodes a bimodular protein similar to developmentally regulated proteins. *Plant Molecular Biology* **24**:799–804.

Castonguay Y, Nadeau P, Lechasseur P, Chouinard L. 1995. Differential accumulation of carbohydrates in alfalfa cultivars of contrasting winter hardiness. *Crop Science* **35**:509–516.

Castonguay Y, Laberge S, Nadeau P, Vezina LP. 1997. Temperature and drought stress. In: McKersie BD, Brown DCW, eds. *Biotechnology and the improvement of forage legumes*. New York: CAB International, 175–202.

Castonguay Y, Laberge S, Brummer EC, Volenec JJ. 2006. Alfalfa winter hardiness: a research retrospective and integrated perspective. *Advances in Agronomy* **90**:203–265.

Chinnusamy V, Schumaker K, Zhu JK. 2004. Molecular genetic perspectives on cross-talk and specificity in abiotic stress signalling in plants. *Journal of Experimental Botany* **55**:225–236.

Colwell J. 1963. The estimation of the phosphorus fertiliser requirements of wheat in southern New South Wales by soil analysis. *Australian Journal of Experimental Agriculture and Animal Husbandry* **6**:105–120.

Close TJ. 1997. Dehydrins: a commonality in the response of plants to dehydration and low temperature. *Physiologia Plantarum* **100**: 291–296.

Crowe JH, Crowe LM. 1992. Membrane integrity in anhydrobiotic organisms: toward a mechanism for stabilizing dry cells. In: Somero GN, Osmond CB, Bolis CL, eds. *Water and life*. New York: Springer, 87–103.

Crowe JH, Hoesktra FA, Crowe LM. 1992. Anhydrobiosis. *Annual Review of Physiology* **54**:579–599.

Cunningham SM, Nadeau P, Castonguay Y, Laberge S, Volenec JJ. 2003. Raffinose and stachyose accumulation, galactinol synthase expression, and winter injury of contrasting alfalfa germplasms. *Crop Science* **43**:562–570.

Déjardin A, Sokolov LN, Kleczkowski LA. 1999. Sugar/osmoticum levels modulate differential abscisic acid-independent expression of two stress-responsive sucrose synthase genes in Arabidopsis. *Biochemical Journal* **344**:503–509.

Erice G, Louahlia S, Irigoyen JJ, Sanchez-Diaz M, Alami IT, Avice JC. 2011. Water use efficiency, transpiration and net CO_2 exchange of four alfalfa genotypes submitted to progressive drought and subsequent recovery. *Environmental and Experimental Botany* **72**:123–130.

Grandfield CO. 1943. Food reserves and their translocation to the crown buds as related to cold and drought resistance in alfalfa. *Journal of Agricultural Research* **67**:33–47.

Grimes DW, Wiley PL, Sheesley WR. 1992. Alfalfa yield and plant water relations with variable irrigation. *Crop Science* **32**: 1381–1387.

Haagenson DM, Cunningham SM, Joern BC, Volenec JJ. 2003. Autumn defoliation effects on alfalfa winter survival, root physiology, and gene expression. *Crop Science* **43**:1340–1348.

Hanin M, Brini F, Ebel C, Toda Y, Takeda S, Masmoudi K. 2011. Plant dehydrins and stress tolerance: versatile proteins for complex mechanisms. *Plant Signalling & Behaviour* **6**: 1503–1509.

Hattendorf MJ, Evans DW, Peaden RN. 1990. Canopy temperature and stomatal conductance of water-stressed dormant and non-dormant alfalfa types. *Agronomy Journal* **82**:873–877.

He C, Xia ZL, Campbell TA, Bauchan GR. 2009. Development and characterization of SSR markers and their use to assess genetic relationships among alfalfa germplasms. *Crop Science* **49**: 2176–2186.

Holloway-Phillips M-M, Brodribb TJ. 2011. Contrasting hydraulic regulation in closely related forage grasses: implications for plant water use. *Functional Plant Biology* **38**:594–605.

Jaglo-Ottosen KR, Gilmour SJ, Zarka DG, Schabenberger O, Thomashow MF. 1998. Arabidopsis CBF1 overexpression induces COR genes and enhances freezing tolerance. *Science* **280**: 104–106.

Jonak C, Heberle-Bors E, Hirt H. 1995. A cDNA from *Medicago sativa* encodes a protein homologous to small GTP-binding proteins. *Plant Physiology* **107**:263–264.

Jonak C, Kiegerl S, Ligterink W, Barker PJ, Huskisson NS, Hirt H. 1996. Stress signaling in plants: a mitogen-activated protein kinase pathway is activated by cold and drought. *Proceedings of the National Academy of Sciences of the USA* **93**:11274–11279.

Kaplan F, Kopka J, Haskell DW, Zhao W, Schiller KC, Gatzke N, Sung DY, Guy CL. 2004. Exploring the temperature-stress metabolome of Arabidopsis. *Plant Physiology* **136**:4159–4168.

Kasuga M, Liu Q, Miura S, Yamaguchi-Shinozaki K, Shinozaki K. 1999. Improving plant drought, salt, and freezing tolerance by gene transfer of a single stress-inducible transcription factor. *Nature Biotechnology* **17**:287–291.

Kibbe WA. 2007. OligoCalc: an online oligonucleotide properties calculator. *Nucleic Acids Research* **35**:W43–W46.

Krasensky J, Jonak C. 2012. Drought, salt, and temperature stress-induced metabolic rearrangements and regulatory networks. *Journal of Experimental Botany* **63**:1593–1608.

Laberge S, Castonguay Y, Vezina LP. 1993. New cold-regulated and drought-regulated gene from *Medicago sativa*. *Plant Physiology* **101**:1411–1412.

Leopold AC. 1990. Coping with desiccation. In: Alscher JG, Cumming JR, eds. *Stress responses in plants: adaptation and acclimation mechanisms*. New York: Wiley-Liss Inc., 37–56.

Lindsey WL, Norvell WA. 1978. Development of a DTPA soil test for zinc, iron, manganese, and copper. *Soil Science Society of America Journal* **42**:421–428.

McKersie BD, Chen Y, de Beus M, Bowley SR, Bowler C, Inzé D, D'Halluin K, Botterman J. 1993. Superoxide dismutase enhances tolerance of freezing stress in transgenic alfalfa (*Medicago sativa* L.). *Plant Physiology* **103**:1155–1163.

McKersie BD, Bowley SR, Harjanto E, Leprince O. 1996. Water-deficit tolerance and field performance of transgenic alfalfa overexpressing superoxide dismutase. *Plant Physiology* **111**:1177–1181.

McKersie BD, Bowley SR, Jones KS. 1999. Winter survival of transgenic alfalfa overexpressing superoxide dismustase. *Plant Physiology* **119**:839–847.

Olsen SR, Cole CV, Watanbe FS, Dean LA. 1954. *Estimation of available phosphorus in soils by extraction with sodium bicarbonate*. Washington: U.S. Government Printing Office.

Orloff S, Hanson B. 2008. Conserving water through deficit irrigation of alfalfa in the intermountain area of California. *Forage and Grazinglands*. doi:10.1094/FG-2008-0421-2001-RS.

Pay A, Heberle-Bors E, Hirt H. 1991. Isolation and sequence determination of the plant homologue of the eukaryotic initiation factor 4D cDNA from alfalfa, *Medicago sativa*. *Plant Molecular Biology* **17**:927–929.

Pay A, Heberle-Bors E, Hirt H. 1992. An alfalfa cDNA encodes a protein with homology to translationally controlled human tumor protein. *Plant Molecular Biology* **19**:501–503.

Pembleton KG, Rawnsley RP, Donaghy DJ, Volenec JJ. 2009. Water deficit alters canopy structure but not photosynthesis in regrowing alfalfa (*Medicago sativa* L.). *Crop Science* **49**:722–731.

Pembleton KG, Donaghy DJ, Volenec JJ, Smith RS, Rawnsley RP. 2010*a*. Yield, yield components and shoot morphology of four contrasting alfalfa (*Medicago sativa*) cultivars grown in three cool temperate environments. *Crop & Pasture Science* **61**: 503–511.

Pembleton KG, Smith RS, Rawnsley RP, Donaghy DJ, Humphries AW. 2010*b*. Genotype by environment interactions of alfalfa (*Medicago sativa* L.) in a cool temperate climate. *Crop & Pasture Science* **61**:493–502.

Pembleton KG, Cunningham SM, Volenec JJ. 2010*c*. Effect of summer irrigation on seasonal changes in taproot reserves and the expression of winter dormancy/activity in four contrasting alfalfa cultivars. *Crop & Pasture Science* **61**:873–884.

Pembleton KG, Rawnsley RP, Donaghy DJ. 2011. Yield and water-use efficiency of contrasting lucerne genotypes grown in a cool temperate environment. *Crop and Pasture Science* **62**:610–623.

Peterbauer T, Richter A. 2001. Biochemistry and physiology of raffinose family oligosaccharides and galactosyl cyclitols in seeds. *Seed Science Research* **11**:185–197.

Peters S, Mundree SG, Thomson JA, Farrant JM, Keller F. 2007. Protection mechanisms in the resurrection plant *Xerophyta viscosa* (Baker): both sucrose and raffinose family oligosaccharides (RFOs) accumulate in leaves in response to water deficit. *Journal of Experimental Botany* **58**:1947–1956.

Pirck M, Pay A, Heberle-Bors E, Hirt H. 1993. Isolation and characterization of a phosphoprotein phosphatase type 2A gene from alfalfa. *Molecular and General Genetics* **240**:126–131.

Radovic J, Sokolovic D, Markovic J. 2009. Alfalfa—most important perennial forage legume in animal husbandry. *Biotechnology in Animal Husbandry* **25**:465–475.

Rizhsky L, Liang HJ, Shuman J, Shulaev V, Davletova S, Mittler R. 2004. When defense pathways collide. The response of Arabidopsis to a combination of drought and heat stress. *Plant Physiology* **134**: 1683–1696.

Rozen S, Skaletsky H. 2000. Primer3 on the WWW for general users and for biologist programmers. In: Krawetz S, Misener S, eds. *Bioinformatics methods and protocols: methods in molecular biology*. Totowa: Humana Press, 365–386.

Russelle MP. 2001. Alfalfa. *American Scientist* **89**:252–261.

Schwab PM, Barnes DK, Sheaffer CC. 1996. The relationship between field winter injury and fall growth score for 251 alfalfa cultivars. *Crop Science* **36**:418–426.

Seki M, Kamei A, Yamaguchi-Shinozaki K, Shinozaki K. 2003. Molecular responses to drought, salinity and frost: common and different paths for plant protection. *Current Opinion in Biotechnology* **14**:194–199.

Sheaffer CC, Barnes DK, Warnes DD, Lueschen WE, Ford HJ, Swanson DR. 1992. Seeding-year cutting affects winter survival and its association with fall growth score in alfalfa. *Crop Science* **32**:225–231.

Suarez R, Calderon C, Iturriaga G. 2009. Enhanced tolerance to multiple abiotic stresses in transgenic alfalfa accumulating trehalose. *Crop Science* **49**:1791–1799.

Sun WQ, Irving TC, Leopold AC. 1994. The role of sugar, vitrification and membrane phase transition in seed desiccation tolerance. *Physiologia Plantarum* **90**:621–628.

Taji T, Ohsumi C, Luchi S, Seki M, Kasuga M, Kobayashi M, Yamaguchi-Shinozaki K, Shinozaki K. 2002. Important roles of drought- and cold-inducible genes for galactinol synthase in stress tolerance in *Arabidopsis thaliana*. *The Plant Journal* **29**:417–426.

Teuber LR, Taggard KL, Gibbs LK, McCaslin MH, Peterson MA, Barnes DK. 1998. Fall dormancy. In: Fox CC, ed. *Standard test to characterize alfalfa cultivars*. North American Alfalfa Improvement Conference, Bozeman, MT, A-1-2.

Thomashow MF. 1999. Plant cold acclimation: freezing tolerance genes and regulatory mechanisms. *Annual Review of Plant Physiology and Plant Molecular Biology* **50**:571–599.

Urano K, Maruyama K, Ogata Y, Morishita Y, Takeda M, Sakurai N, Suzuki H, Saito K, Shibata D, Kobayashi M, Yamaguchi-Shinozaki K, Shinozaki K. 2009. Characterization of the ABA-regulated global responses to dehydration in Arabidopsis by metabolomics. *The Plant Journal* **57**:1065–1078.

Usadel B, Blasing OE, Gibon Y, Poree F, Hohne M, Gunter M, Trethewey R, Kamlage B, Poorter H, Stitt M. 2008. Multilevel genomic analysis of the response of transcripts, enzyme activities and metabolites in Arabidopsis rosettes to a progressive decrease of temperature in the non-freezing range. *Plant, Cell and Environment* **31**:518–547.

Vance CP, Heichel GH, Phillips DA. 1988. Nodulation and symbiotic dinitrogen fixation. In: Hanson AA, Barnes DK, Hill RRJ, eds. *Alfalfa and alfalfa improvement*. Madison: American Society of Agronomy, 229–257.

Vandesompele J, De Preter K, Pattyn F, Poppe B, Van Roy N, De Paepe A, Speleman F. 2002. Accurate normalization of real-time quantitative RT-PCR data by geometric averaging of multiple internal control genes. *Genome Biology* **3**:RESEARCH0034.

Volenec JJ, Cunningham SM, Haagenson DM, Berg WK, Joern BC, Wiersma DW. 2002. Physiological genetics of alfalfa improvement: past failures, future prospects. *Field Crops Research* **75**:97–110.

Wang C, Ma BL, Yan X, Han J, Guo Y, Wang Y, Li P. 2009. Yields of alfalfa varieties with different fall-dormancy levels in a temperate environment. *Agronomy Journal* **101**:1146–1152.

Wolfraim LA, Dhindsa RS. 1993. Cloning and sequencing of the cDNA for *cas17*, a cold acclimation-specific gene of alfalfa. *Plant Physiology* **103**:667–668.

Wolfraim LA, Langis R, Tyson H, Dhindsa RS. 1993. cDNA sequence, expression, and transcript stability of a cold acclimation-specific gene, *cas18*, of alfalfa (*Medicago falcata*) cells. *Plant Physiology* **101**:1275–1282.

Wu SC, Gyorgyey J, Dudits D. 1989. Polyadenylated H3 histone transcripts and H3 histone variants in alfalfa. *Nucleic Acids Research* **17**:3057–3063.

Zhu JK. 2002. Salt and drought stress signal transduction in plants. *Annual Review of Plant Biology* **53**:247–273.

PERMISSIONS

All chapters in this book were first published in AOBP, by Oxford University Press; hereby published with permission under the Creative Commons Attribution License or equivalent. Every chapter published in this book has been scrutinized by our experts. Their significance has been extensively debated. The topics covered herein carry significant findings which will fuel the growth of the discipline. They may even be implemented as practical applications or may be referred to as a beginning point for another development.

The contributors of this book come from diverse backgrounds, making this book a truly international effort. This book will bring forth new frontiers with its revolutionizing research information and detailed analysis of the nascent developments around the world.

We would like to thank all the contributing authors for lending their expertise to make the book truly unique. They have played a crucial role in the development of this book. Without their invaluable contributions this book wouldn't have been possible. They have made vital efforts to compile up to date information on the varied aspects of this subject to make this book a valuable addition to the collection of many professionals and students.

This book was conceptualized with the vision of imparting up-to-date information and advanced data in this field. To ensure the same, a matchless editorial board was set up. Every individual on the board went through rigorous rounds of assessment to prove their worth. After which they invested a large part of their time researching and compiling the most relevant data for our readers.

The editorial board has been involved in producing this book since its inception. They have spent rigorous hours researching and exploring the diverse topics which have resulted in the successful publishing of this book. They have passed on their knowledge of decades through this book. To expedite this challenging task, the publisher supported the team at every step. A small team of assistant editors was also appointed to further simplify the editing procedure and attain best results for the readers.

Apart from the editorial board, the designing team has also invested a significant amount of their time in understanding the subject and creating the most relevant covers. They scrutinized every image to scout for the most suitable representation of the subject and create an appropriate cover for the book.

The publishing team has been an ardent support to the editorial, designing and production team. Their endless efforts to recruit the best for this project, has resulted in the accomplishment of this book. They are a veteran in the field of academics and their pool of knowledge is as vast as their experience in printing. Their expertise and guidance has proved useful at every step. Their uncompromising quality standards have made this book an exceptional effort. Their encouragement from time to time has been an inspiration for everyone.

The publisher and the editorial board hope that this book will prove to be a valuable piece of knowledge for researchers, students, practitioners and scholars across the globe.

LIST OF CONTRIBUTORS

George A. Meindl and Tia-Lynn Ashman
Department of Biological Sciences, University of Pittsburgh, Pittsburgh, PA 15260, USA

Daniel J. Bain
Department of Geology and Planetary Science, University of Pittsburgh, Pittsburgh, PA 15260, USA

Lucy T. T. Tran-Nguyen and Mark N. Hearnden
Plant Industries, Northern Territory Department of Primary Industry and Fisheries, Darwin, NT 0801, Australia

Diana M. Hartley
CSIRO Ecosystem Sciences, Canberra, ACT 2601, Australia

Sean L. Bithell
Plant Industries, Northern Territory Department of Primary Industry and Fisheries, Darwin, NT 0801, Australia
New South Wales Department of Primary Industries, Tamworth Agricultural Institute, 4 Marsden Park Rd, Tamworth, NSW 2340, Australia

Marit Almvik, Nicholas Clarke, Anne Falk Øgaard and Jihong Liu Clarke
Norwegian Institute of Bioeconomy Research (NIBIO), N-1431 Ås, Norway

Susanne Eich-Greatorex and Tore Krogstad
Department of Environmental Sciences, Norwegian University of Life Sciences, N-1432 Ås, Norway

Hans Lambers
School of Plant Biology and Institute of Agriculture, University of Western Australia, Crawley (Perth), WA 6009, Australia

Yan-Liang Wang
Norwegian Institute of Bioeconomy Research (NIBIO), N-1431 Ås, Norway
Department of Environmental Sciences, Norwegian University of Life Sciences, N-1432 Ås, Norway

Jeffrey M. Warren, Anna M. Jensen and Richard J. Norby
Climate Change Science Institute and Environmental Sciences Division, Oak Ridge National Laboratory, Oak Ridge, TN 37831-6301, USA

Belinda E. Medlyn
School of Biological Sciences, Macquarie University, Sydney, NSW 2019, Australia

David T. Tissue
Hawkesbury Institute for the Environment, University of Western Sydney, Richmond, NSW 2753, Australia

Narayani Barve and A. Townsend Peterson
Biodiversity Institute, University of Kansas, 1345 Jayhawk Blvd, Lawrence, KS 66045, USA
Department of Ecology and Evolutionary Biology, University of Kansas, Lawrence, KS 66045, USA

Craig E. Martin
Department of Ecology and Evolutionary Biology, University of Kansas, Lawrence, KS 66045, USA

Johannes Isselstein
Department of Crop Sciences, University of Göttingen, Von-Siebold-Str. 8, D-37075 Göttingen, Germany

Ute Petersen
Department of Crop Sciences, University of Göttingen, Von-Siebold-Str. 8, D-37075 Göttingen, Germany
Johann Heinrich von Thünen Institute, Federal Research Institute for Rural Areas, Forestry and Fisheries, Institute of Climate-Smart Agriculture (AK), Bundesallee 50, D-38116 Braunschweig, Germany

Laura Sanderson and Pedro M. Antunes
Department of Biology, Algoma University, Sault Ste. Marie, ON, P6A 2G4 Canada

Vanessa Reininger
Department of Biology, Algoma University, Sault Ste. Marie, ON, P6A 2G4 Canada
Agroscope, Institute for Plant Production Sciences IPS, Schloss 1, 8820 Wädenswil, Switzerland
ETH Zurich, Institute of Integrative Biology, Forest Pathology and Dendrology, Universitätstrasse 16, 8092 Zurich, Switzerland

Laura B. Martinez-Garcia
Department of Biology, Algoma University, Sault Ste. Marie, ON, P6A 2G4 Canada
Department of Soil Quality, Wageningen University, 6700 AA, Wageningen, The Netherlands

Astrid Volder
Department of Plant Sciences, University of California - Davis, Davis, CA, USA

Roger M. Gifford
CSIRO Agriculture, Canberra, Australian Capital Territory 2601, Australia

John R. Evans
Division of Plant Sciences, Research School of Biology, The Australian National University, Linnaeus Building 134, Canberra, Australian Capital Territory 0200, Australia

Maria Teresa Ariza, Carmen Soria and Elsa Martínez-Ferri
IFAPA, Centro de Churriana, Cortijo de la Cruz s/n, 29140 Churriana, Málaga, Spain

Gaëtan Louarn, Ela Frak and Serge Zaka
INRA UR4 URP3F, BP6, F86600 Lusignan, France

Jorge Prieto
INTA EEA Mendoza, San Martín 3853, Luján de Cuyo (5507), Mendoza, Argentina

Eric Lebon
INRA, UMR 759, LEPSE, 2 place Viala, F34060 Montpellier, France

Victoria Vitali, Jorge Bellati and Gabriela Amodeo
Departamento de Biodiversidad y Biología Experimental, Facultad de Ciencias Exactas y Naturales, Instituto de Biodiversidad y Biología Experimental, Universidad de Buenos Aires and Consejo Nacional de Investigaciones Científicas y Técnicas, C1428EGA Buenos Aires, Argentina

Gabriela Soto and Nicolás D. Ayub
Instituto de Genética "Ewald A. Favret", CICVyA, INTA-Castelar and Consejo Nacional de Investigaciones Cientı´ficas y Técnicas, 1686
Buenos Aires, Argentina

James F. White, Qiang Chen, Mónica S. Torres, Robert Mattera, Ivelisse Irizarry, Mariusz Tadych and Marshall Bergen
Department of Plant Biology and Pathology, Rutgers University, New Brunswick, NJ, USA

Elvia Hernández-Gómez and Luis A. Valdez-Aguilar
Programa de Agricultura Protegida, Universidad Autónoma Agraria Antonio Narro, Calzada Antonio Narro 1923, Buenavista, Saltillo, Coahuila 25315, México

Donita L. Cartmill and Andrew D. Cartmill
School of Agriculture, University of Wisconsin-Platteville, 1 University Plaza, Platteville, WI 53818, USA

Irán Alia-Tajacal
Facultad de Ciencias Agropecuarias, Universidad Autónoma del Estado de Morelos, Avenida Universidad 1001, Cuernavaca, Morelos
62209, México

Krista Takkis, Thomas Tscheulin, Panagiotis Tsalkatis and Theodora Petanidou
Laboratory of Biogeography and Ecology, Department of Geography, University of the Aegean, University Hill, GR-81100 Mytilene, Greece

Nana Zhang, Brian Belsterling and Jesse Raszewski
Department of Biological Sciences, University of Pittsburgh, 4249 Fifth Ave., Pittsburgh, PA 15260, USA

Stephen J. Tonsor
Department of Biological Sciences, University of Pittsburgh, 4249 Fifth Ave., Pittsburgh, PA 15260, USA
Carnegie Museum of Natural History, 4400 Forbes Ave., Pittsburgh, PA 15213, USA

Keith G. Pembleton
Tasmanian Institute of Agriculture, University of Tasmania, Burnie, TAS 7320, Australia

Puthigae Sathish
Pastoral Genomics, c/o ViaLactia (NZ) Ltd, Newmarket 1149, New Zealand

Index

A

Acclimation, 34-35, 44-46, 100, 112, 118, 121-124, 128, 140, 174-175, 177-180, 183-185, 192, 195, 197-198, 202-203

Antioxidant Activity, 153

Aquaporins, 127-128, 130-131, 133-134, 136, 138-141, 158

Arbuscular Mycorrhizal Fungi, 24, 70, 74, 80-82

B

Ballota Acetabulosa, 161-162, 165, 167, 172

Beta Vulgaris, 127-129, 135, 137-140

Brassicaceae, 1-8, 55

C

C : N Ratio, 84, 86, 88-90, 92-93, 96

Canopy Structure, 57-59, 66, 68-69, 123-124, 202

Climate Change, 34, 45, 54, 56, 81, 84, 96-97, 161-163, 167-168, 170-174, 185

Community Ecology, 69-70, 81-82

Critical Periods, 99, 103, 106

Crops, 11, 22-24, 26-27, 29-33, 35, 99-100, 110, 112, 121-124, 140, 196, 203

D

Defoliation, 66, 84, 98, 113, 188-189, 194, 202

Dicots, 23-24, 57-59, 67-68

Diversity, 7-8, 21, 32, 55, 58, 60, 67-72, 74-82, 162, 176, 185-186

Dna Persistence, 10, 17, 19

Down-regulation, 34-35, 42, 45, 134

E

Electrolyte Leakage, 153, 155-157, 159-160

Elevated Temperatures, 161-162, 164-169, 171

Endemism, 1, 5-9

Era Data, 47, 49

F

Fine-roots, 10

Flower Development, 99-100, 106-108

Flower Differentiation, 99-100, 108

Flowers, 1, 3-6, 47-48, 50, 99-101, 106, 110, 114, 135, 137, 139, 161, 163-166, 168-169, 171-172

Fragaria × Ananassa, 99-100, 109, 158

Free-air Co2 Enrichment, 34-36, 98

G

Global Change, 44-46, 97-98, 161

Grasses, 11, 22, 35, 57, 59-61, 63, 66-67, 69-70, 72, 80, 85, 92, 112, 118, 124, 142-143, 146, 148-151, 202

Grassland, 57-59, 66-69, 76, 80-82, 84-85, 96, 98, 121, 123-124

H

Hydroponic Culture, 23-24, 30-31, 129

L

Light Interception, 57-58, 60, 67-69

Light; Medicago Sativa, 111

Lipid Peroxidation, 153, 155, 157, 159

Low Temperature; Pistil, 99

M

Metal Accumulation, 1-2, 6, 8

Microbiome, 142, 149, 151

Monocots, 57-59, 68

N

Nectar Production, 8, 161-162, 168-171, 173

Nectar Sugar Content, 161

Nickel, 1-2, 4-5, 7-9

Nitrogen Distribution, 111, 122-124

Nitrogen Limitation, 34, 44-45, 97-98

Nitrogen Nutrition Index, 111

Nitrogen Use Efficiency, 142, 151

Nutrient Imbalance, 153, 159

O

Oxidative Nitrogen Scavenging, 142, 149

P

Permanent Grassland, 57-59, 66-68

Phosphorus, 23-24, 29-30, 32-33, 82, 143, 153-155, 157, 201-202

Photosynthesis, 2, 34-39, 41-46, 69, 85, 98, 111-115, 118-120, 123-126, 153-156, 158, 160, 197, 202

Photosynthesis Rate, 153-154

Phrygana, 161, 171

Plant Growth-promoting, 142

Plant Species Composition, 57, 66

Plant-pollinator Interactions, 161-162, 167, 169, 171-173

Pollen, 1-4, 6, 8-9, 99-104, 106-110, 140

R

Rhizobacteria; Symbiosis, 142

Root Dna Concentration, 10, 12, 17
Root Dna Density (RDD), 10-11
Root Exudates, 23-25, 27, 29-33
Root Hydraulic Conductance, 140-141, 153-156, 158
Root Hydraulic Conductivity, 127, 131, 138-139, 160
Root Morphology, 23-24, 26, 29-32

S

Salt Stress, 127-129, 131, 134-137, 139-140, 160
Serpentine Soil, 1-2, 5-7
Soil-plant-atmosphere Continuum, 127
Spanish Moss, 47-55
Species Distribution, 8, 47
Stomatal Conductance, 36-37, 41, 43-44, 46, 113-114, 117, 121, 123, 125-127, 130, 133, 140, 156, 158, 193, 195, 197-198, 202
Stress Tolerance, 139-140, 153, 158, 199, 201-202
Suberization, 127, 133-134, 136-137
Sweetgum, 34-36, 38, 41-43, 45-46

T

Teucrium Divaricatum, 161, 165, 170, 172
Tillandsia Usneoides L., 47-48, 55
Tissue Quality, 84-86, 94
Transpiration, 111, 114, 118-121, 124, 127, 129, 131, 136, 138, 155-156, 158, 202

U

Upscaling, 111

W

Water Relations, 55, 98, 127, 138, 140-141, 153, 156, 159, 202
Within-canopy Variability, 111